Hydrogen as a Fuel

Hydrogen as a Fuel

Learning from Nature

Edited by
Richard Cammack
Michel Frey
Robert Robson

CRC Press
Taylor & Francis Group
Boca Raton London New York

CRC Press is an imprint of the
Taylor & Francis Group, an **informa** business

CRC Press
Taylor & Francis Group
6000 Broken Sound Parkway NW, Suite 300
Boca Raton, FL 33487-2742

First issued in paperback 2019

ISBN-13: 978-0-415-24242-4 (hbk)
ISBN-13: 978-0-367-39671-8 (pbk)

Library of Congress Cataloging-in-Publication Data

Catalog record is available from the Library of Congress

Visit the Taylor & Francis Web site at
http://www.taylorandfrancis.com

and the CRC Press Web site at
http://www.crcpress.com

This Book is Dedicated to David O. Hall (1935–99), Coordinator of the European Community Photobiology and Photochemical Research Programme, 1975–90.

Contents

List of figures

List of tables

List of contributors

Albracht, Simon P.J.
University of Amsterdam
Biochemistry
NL-1018 TV Amsterdam
The Netherlands

Afting, Christina
Max-Planck-Institüt für terrestrische
Mikrobiologie
D-35043 Marburg
Germany

Armstrong, Fraser A.
Inorganic Chemistry Laboratory
Oxford OX1 3QR
U.K.

Bachofen, Reinhard
Institute for Plant Biology
8008 Zurich
Switzerland

Bertrand, Patrick
CNRS
Lab. Bioénergétique et Ingéniérie des
Protéines
F-13402 Marseille
France

Bleijlevens, Boris
University of Amsterdam
Biochemistry
NL-1018 TV Amsterdam
The Netherlands

Böck, August
Institut für Genetik und Mikrobiologie
D-80638 München
Germany

Bothe, Hermann
Botanisches Institut
D-50923 Köln
Germany

Bouwman, Elisabeth
Gorlaeus Laboratories
NL-2300 RA Leiden
The Netherlands

Buurman, Gerrit
University of Amsterdam
Biochemistry
NL-1018 TV Amsterdam
The Netherlands

Cammack, Richard
King's College London
London SE1 9NN
U.K.

Colbeau, Annette
CEA/Grenoble
F-38054 Grenoble Cedex 9
France

De Lacey, L. Antonio
Instituto de Catalisis (C.S.I.C)
E-28049 Madrid
Spain

Dischert, Wanda
CEA/Grenoble
F-38054 Grenoble Cedex 9
France

Dole, François
Lab. Bioénergétique et Ingéniérie
des Protéines
F-13402 Marseille Cedex 20
France

Faber, Bart
University of Amsterdam
Biochemistry
NL-1018 TV Amsterdam
The Netherlands

Fernandez, Victor M.F.
Instituto de Catalisis (C.S.I.C)
E-28049 Madrid
Spain

Fiebig, Klaus
Freie Universität Berlin
Institut für Biologie – Mikrobiologie
D-14195 Berlin
Germany

Fontecilla-Camps, Juan Carlos
Institut de Biologie Structurale
F-38027 Grenoble Cedex
France

Frey, Michel
IBS/LCCP
Laboratoire de Cristallographie et de
Cristallogenese
38027 Grenoble Cedex 01
France

Friedrich, Bärbel
Humboldt-Universität zu Berlin
Institut für Biologie-Mikrobiologie
D-10115 Berlin
Germany

Garcin, Elsa
Institut de Biologie Structurale
F-38027 Grenoble Cedex
France

Gregoire Padró Catherine E.
National Renewable Energy Laboratory
Golden, CO 80401
USA

Guigliarelli, Bruno
CNRS Université de Provence
F-13402 Marseille Cedex 20
France

Hagen, Wilfred R.
Delft University of Technology
NL-2628 BC Delft
The Netherlands

Happe, Thomas
Botanisches Institut
D-53115 Bonn
Germany

Happe, Randolph P.
University of Amsterdam
Biochemistry
NL-1018 TV Amsterdam
The Netherlands

Hatchikian, E. Claude
Unité de Bioénergetiques et Ingéniérie
des Proteines
CNRS
13402 Marseille Cedex 20
France

Henderson, Richard K.
Leiden Institute of Chemistry
NL-2300 RA Leiden
The Netherlands

Higuchi, Yoshiki
Division of Chemistry
Sakyo-ku
Kyoto 606-8502
Japan

Holliger, Christof
Ecole Polytechnique Fédéral de
Lausanne
CH-1015 Lausanne
Switzerland

Hoppert, Michael
Institut für Mikrobiologie
D-37077 Gšttingen
Germany

Hüsing, Bärbel
Fraunhofer Institute for Systems and
Innovation Research
D-76139 Karlsruhe
Germany

Imperial, Juan
Universidad Politécnica de Madrid
E-28040 Madrid
Spain

Jones, Anne K.
Inorganic Chemistry Laboratory
Oxford OX1 3QR
U.K.

Klein, Albrecht
FB Biologie Molekuargenetik
D-35032 Marburg
Germany

Kovács, Kornél L.
Hungarian Academy of Sciences
Institute of Biophysics
H-6701 Szeged
Hungary

Krause, Birgit
Ecole Polytechnique
CH-1015 Lausanne
Switzerland

Lenz, Oliver
Institut für Biologie Mikrobiologie
D-10115 Berlin
Germany

Lindblad, Peter
Department of Physiological Botany
S-75236 Uppsala
Sweden

Lissolo, Thierry
Université de Savoie,
F-73376 Le Bourget du Lac
France

Lubitz, Wolfgang
Technische Universität Berlin
D-10623 Berlin
Germany

Macaskie, Lynne
School of Biosciences
Birmingham B15 2TT
U.K.

Maier, Robert J.
University of Georgia
Department of Microbiology
Athens, GA 30606
U.S.A

Mandrand-Berthelot, Marie-Andrée
CNRS UMR 5577, Bât. 406
F-69621 Villeurbanne Cedex
France

Maroney, Michael J.
University of Massachusetts
Department of Chemistry
Amherst, MA 01003-4510
U.S.A.

Massanz, Christian
Institut für Biologie, Mikrobiologie
Humboldt-Universität zu Berlin
D-10115 Berlin
Germany

Mayer, F.
Institut für Mikrobiologie
D-37077 Gšttingen
Germany

Meyer, Jacques
Institut de Biologie Structurale/CEA
F-38054 Grenoble Cedex 9
France

Mlejnek, K.
Institut für Mikrobiologie
Grisebachstr. 8
D-37077 Gšttingen
Germany

Mikheenko, Iryna
University of Birmingham
School of Biosciences
Birmingham B15 2TT
U.K.

Montet, Yaël
Laboratoire de Cristallographie et de
Cristallogenese des Proteines
F-38027 Grenoble Cedex
France

Moura, José J.G.
Universidade Nova de Lisboa
Faculdade de Ciências e Tecnologia
P-2825-144 Monte de Caparica
Portugal

Müller, Arnd
King's College London
London SE1 9NN
U.K.

Nicolet, Yvain
Institut de Biologie Structurale
F-38027 Grenoble Cedex 1
France

Pedroni, Paola M.
EniTechnologie S.p.A.
I-20097 S. Donato Milanese (MI)
Italy

Pereira, Alice S.
Universidade Nova de Lisboa
P-2825-144 Monte de Caparica
Portugal

Péringer, Paul
Institut du Génie de l'Environment
CH-1015 Lausanne
Switzerland

Pershad, Harsh R.
Inorganic Chemistry Laboratory
Oxford OX1 3QR
U.K.

Pierik, Antonio J.
University of Amsterdam
Biochemistry
NL-1018 TV Amsterdam
The Netherlands

Rákhely, Gábor
Hungarian Academy of Sciences
Institute of Biophysics
H-6726 Szeged
Hungary

Rao, K. Krishna
King's College London
Division of Life Sciences
London SE1 8WA
U.K.

Reedijk, Jan
Leiden Institute of Chemistry
NL-2300 RA Leiden
The Netherlands

Robson, Robert
School of Animal and Microbial Sciences
University of Reading
Reading RG6 6AJ
U.K.

Ruiz-Argüeso, Thomás
Universidad Politécnica de Madrid
E-28040 Madrid
Spain

Schulz, Rüdiger
Institut für Botanik
D-24098 Kiel
Germany

Schwitzguébel, Jean-Paul
Institut du Génie de l'Environment
CH-1015 Lausanne
Switzerland

Seibert, Michael
National Renewable Energy Laboratory
CO 80401 Golden
U.S.A.

Sellstedt, Anita
Department of Plant Physiology
S-90187 Umea
Sweden

Sorgenfrei, Oliver
Philipps-Universität-Marburg
D-35032 Marburg
Germany

Stein, Matthias
Technische Universität Berlin
D-10623 Berlin
Germany

Talabardon, Mylène
Institut du Génie de l'Environnement
CH-1015 Lausanne
Switzerland

Tamagnini, Paula
IBMC/Department of Botany
P-4150-180 Porto
Portugal

Tavares, Pedro
Universidade Nova de Lisboa
P-2825-144 Monte de Caparica
Portugal

Thauer, Rudolf K.
Max-Planck-Institüt für terrestrische
Mikrobiologie
D-35043 Marburg
Germany

Van Praag, Esther
Institute for Plant Biology
8008 Zurich
Switzerland

Van Soom, Carolien
Lab. Microbiologia, ETS Ing.
E-28040 Madrid
Spain

Vignais, Paulette M.
Laboratoire de Chimie Microbienne
F-38054 Grenoble Cedex 9
France

Volbeda, Anne
Institut de Biologie Structurale
F-38027 Grenoble Cedex 1
France

Wu, Long-Fei
Laboratoire de Chimie Bacterienne
F-13402 Marseille 20
France

Ping, Yong
School of Biosciences
Birmingham B15 2TT
U.K.

Preface

A fundamental and principal difficulty of the energy industry is that the rate of formation of fossil fuels is much slower than the rate of their exploitation. Therefore the reserves that can be recovered in an energetically feasible manner are shrinking, in parallel with an increasing worldwide energy demand. Among the alternative energy carriers, hydrogen appears to be most promising because it burns to environmentally friendly water when utilized and may be transported rather easily. Hydrogen can be produced in biological processes: in algae and cyanobacteria, solar energy captured by the photosynthetic apparatus is converted into chemical energy through water splitting, the reaction forms oxygen and can also produce hydrogen. Upon utilization, these components are combined to form water and energy is released in a cycle driven by the practically unlimited and safe energy source of the Sun.

In addition to offering an alternative for the global energy crisis, biologically produced hydrogen may also serve as reductant for numerous microbiological activities of environmental significance. Reductants are needed to convert CO_2, atmospheric nitrogen, nitrate or sulfate into useful products. For example, methane (biogas) is generated from industrial, agricultural or household waste, nitrogen is fixed by plants which decreases their need for fertilizers, and nitrate is reduced to nitrogen in drinking water thereby eliminating a public health hazard common in both developed and less developed countries.

The understanding of molecular fundamentals of hydrogen production and utilization in microbes is a goal of supreme importance both for basic and applied research applications. The key enzyme in biological hydrogen metabolism is hydrogenase, which catalyses the formation or decomposition of the simplest molecule occurring in biology: molecular hydrogen.

$$H_2 = 2H^+ + 2e^-$$

The simple-looking task is solved by a sophisticated molecular mechanism. Hydrogenase is a metalloenzyme, harbouring Ni and Fe atoms. Like most metalloenzymes, hydrogenases are extremely sensitive to inactivation by oxygen, high temperature and other environmental factors. These properties are not favourable for several potential biotechnological applications.

In metal-containing biological catalysts it is the protein matrix surrounding the metal centres that provides the unique environment for the Fe and Ni atoms which allows hydrogenases to function properly, selectively and effectively. Therefore, a major goal of hydrogenase basic research is to understand the protein–metal interaction.

The problem is not simple to address, as some of the methods for scientific investigation provide information on the metal atoms themselves without directly observing the protein matrix around them. Other modern techniques at our disposal reveal details of the protein core, but do not reveal the metal centres within. A combination of the various molecular approaches is expected to uncover the fine molecular details of the catalytic action of metalloenzymes.

Due to the complex nature of investigations the need for international collaboration was recognized at an early stage of research. The first international workshop on microbial gas metabolism was organized in Göttingen, Germany in 1978, which has been followed by a series of International Conferences on the Molecular Biology of Hydrogenase held in Szeged, Hungary (1985), Athens, GA, USA (1988), Troia, Portugal (1991), Noordwijkerhout, The Netherlands (1994), Albertville, France (1997) and Potsdam, Germany (2000).

In the European Union and Associated States, research networks are formed around topics of outstanding economic and/or scientific interest. One of these networks is COST Action 818 'Hydrogenases and their biotechnological applications'. COST Action 818 includes practically every laboratory of significant contribution to hydrogenase research, i.e. forty-five laboratories from fourteen European nations with invited laboratories from the USA and Russia. This book summarizes the state of the art at the time COST Action 818 ends as seen by the members of this research network. Many of the recent contributions come from the collaborations within this network, therefore the picture represents the leading edge in this field. It is our pleasure to announce that the network continues to function as COST Action 841, a recently approved project on 'Chemical and biological diversity of hydrogen metabolism'.

As a result of these cooperative activities, it was decided to assemble a book in which our state of knowledge was summarized. It includes contributions not only from Europe, but also from our collaborators around the world. We are at a significant point in hydrogenase studies, where we are reaching a consensus as to how the enzymes accomplish their functions, and are now facing the question as to how this knowledge can benefit mankind. A remarkable aspect of the studies is their multidisciplinary nature, including the activities of biochemists, biotechnologists, chemists, ecologists, microbiologists, molecular biologists and spectroscopists. One of the aims of this book was to allow specialists from the different disciplines, to understand the work of the others, and thereby to understand the significance of their work to the whole enterprise.

In addition to COST, the editors and authors of the book would like to acknowledge research support from national and international sources.

Kornél L. Kovács
Paulette M. Vignais

Chapter 1

Origins, evolution and the hydrogen biosphere

Richard Cammack

1.1. Life and oxidation–reduction processes

Hydrogen is the most abundant element in the Universe, and it is expected that the Earth's early atmosphere was a reducing one, in which hydrogen predominated. Now, over four billion years later, the atmosphere is an oxidizing one, and over most of the Earth's surface oxygen predominates. The gradual conversion from a hydrogen-rich environment to an oxygen-rich one was the result of a global redox cycle. Water was split into oxygen and hydrogen, and the hydrogen atoms escaped into Space. Biology played an important part in this process, which still continues. In photosynthesis, as is well known, carbohydrates are produced by reduction of carbon dioxide, while O_2 is released by the oxidation of water. Another, less well appreciated, part of this cycle is the fermentation of carbohydrates to produce H_2. This is carried out by species of bacteria, protozoa and fungi that live in anaerobic environments where the oxygen has been exhausted. These environments are not so far away from us. After a carbohydrate meal we exhale traces of H_2 in our breath, produced by the action of bacteria in our digestive tract. However, most of this hydrogen is undetectable, because it is immediately recycled by other bacteria, which reoxidize it to water.

The reaction between H_2 and O_2 is highly energetic. As is well known, the gases make explosive mixtures. Hydrogen will burn in oxygen, and oxygen will burn in hydrogen. For chemical rockets in the space programme, the combination of liquid H_2 and liquid O_2 was chosen as the fuel; only liquid H_2/liquid fluorine gives a higher specific thrust, and it is much more corrosive. Chemically speaking, H_2 is a strong reducing agent and O_2 a strong oxidizing agent. H_2 and O_2 in solution retain this potential to react, releasing energy. However, they will not react at ambient temperatures without a catalyst.

Reactions between molecules such as H_2 and O_2 are called oxidation–reduction or redox reactions. In this case, H_2 is oxidized, and O_2 is reduced. Almost all life processes derive their energy from redox reactions, either directly or indirectly. Organisms are classified by their type of energy source. Photosynthesis uses light-driven redox reactions of molecules such as chlorophyll, and organisms which exploit this energy source are called phototrophs. Away from the light, bacteria exploit the oxidation of H_2, sulfur and other compounds. These bacteria are called heterotrophs if the compounds oxidized are organic, or lithotrophs if the compounds are inorganic, including hydrogen. Dramatic examples are found among the prolific microbial and invertebrate ecosystems that thrive around the thermal vents in the depths of the oceans (Fig. 1.1).

Figure 1.1 Black smoker, at a depth of 2,500 m on the East Pacific Rise (by courtesy of John Baross, University of Washington). Hyperthermophilic organisms from such sources have been found to contain hydrogenases.

An astonishing recent discovery is that there are bacteria living deep in the Earth's crust. Colonies of anaerobic bacteria have been isolated from boreholes 1,500 m deep in basaltic rock formations. The bacteria use H_2 as electron donor, which may originate from fermentation of organic matter, or from a purely inorganic reaction of iron of the Earth's core with water (Stevens and McKinley 1995; Anderson *et al.* 1998).

1.1.1. Hydrogen as a microbial energy source

For living systems, as with space rockets, the highest yield of chemical energy is offered by the O_2-H_2 reaction. Many bacteria, including the molecular biologists' favourite organism *Escherichia coli*, can generate energy by oxidation of H_2. But the way in which they do this could hardly be more different from the heat engine. The principal mechanism is the generation of a transmembrane gradient of protons, which drives the formation of adenosine triphosphate (ATP) (Mitchell, 1961; Nicholls and Ferguson, 1992). ATP is the chemical energy intermediate in innumerable life processes, including the replication of the genetic code, synthesis of cellular materials, and movement of the cell. Bacteria exploit their energy sources by a sophisticated system of catalysts, the enzymes. Specific enzymes have evolved to oxidize hydrogen to protons (hydrons)s and electrons, and to use the electrons to reduce oxygen to water. The enzymes that react with dissolved gases contain metal ions: iron, nickel and copper.

It is possible to understand how the simple reaction of hydrogen and oxygen in a membrane could have created a transmembrane proton gradient in a primitive cell. The enzymes involved are embedded in the membrane surrounding the cell, and because they originally came from inside the cell, they each face in a particular direction. Hydrogenase, which faces outwards from the membrane, consumes hydrogen, and by its action ($H_2 = 2H^+ + 2e^-$) releases protons (or hydrons) on the *outside* of the cell. Meanwhile, an oxidase, facing inwards, reduces oxygen to water, and by its action ($O_2 + 4H^+ + 4e^- = 2H_2O$) takes up protons from the *inside* of the cell. The result, as pointed out by Lundegardh (1940), is the creation of a gradient of protons across the membrane. The gradient of protons across the cell membranes represents a form of stored energy. Mitchell (1961) was the first to describe how the gradient could be exploited in a simple way to synthesize ATP.

Fig. 1.2 illustrates how this works in practice. The action takes place across the inner membrane of a bacterium, between the cytoplasm and periplasm. Oxidation of H_2 by a hydrogenase facing outwards releases two H^+ ions into the periplasmic space, and ATP is produced by a complex rotating enzyme in the membrane, ATP synthase, driven by the proton gradient. Different types of cells generate the proton gradient in different ways. The hydrogen-cycling mechanism (Fig. 1.2A) for an anaerobic bacterium such as *Desulfovibrio* (Odom and Peck, 1984) relies on the fact that H_2 diffuses easily through the membrane, whereas the electrons are conducted through specific electron-transfer proteins such as hydrogenase. H^+ ions cannot diffuse through the membrane and have to pass through the ATP synthase, making ATP. In present-day organisms the transmembrane proton gradient is generated by highly sophisticated enzymes that actually pump protons across the membrane (Fig. 1.2 B). Fig. 1.2 B. shows a more complex sequence of enzymes known as a respiratory chain, in which the membrane-bound electron-transfer proteins are proton pumps. This type of mechanism is now used by most living organisms, and is used in the mitochondria of our own cells. Although we do not use hydrogen as our food, some of the proteins in our mitochondria show evidence of their hydrogenase ancestry.

Unlike animals, bacteria are not restricted to O_2 as an oxidant. Within a community of microorganisms, when the O_2 runs out, alternative oxidizing compounds are exploited. The best oxidants are used preferentially, and the ones that are preferred are those that generate the largest amount of energy for growth. This is illustrated by the hierarchy of bacteria that live in a pond (Fig. 1.3). At the top, conditions are aerobic and there is access to light. Here, Cyanobacteria and green plants oxidize water, producing O_2 and fixing CO_2 to produce organic matter (biomass). Aerobic bacteria use O_2 to oxidize H_2 to water, methane to CO_2 and so on. Near the bottom of the pond, where the O_2 is exhausted, anaerobic photosynthetic bacteria such as the purple bacterium *Allochromatium vinosum* use reductants such as H_2, or sulfide which they oxidize to sulfate. At the same time they synthesize organic matter from CO_2. In the mud at the bottom, the organic matter is fermented by anaerobic bacteria such as *Clostridium pasteurianum*, releasing H_2 and acetate. In the anaerobic environment, H_2 is a central source of reducing power.

If hydrogen plays such an important role in the anaerobic world, why do we not *see* H_2 produced, in the same way that bubbles of O_2 are produced by water plants on a

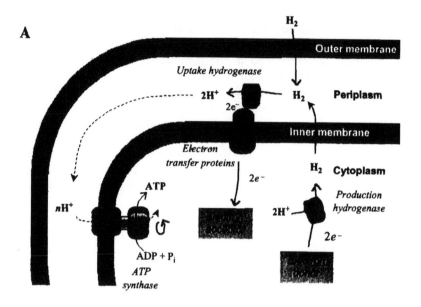

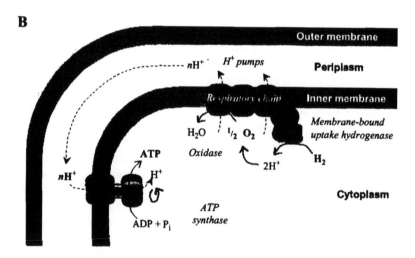

Figure 1.2 Mechanisms of energy conservation in cells, involving transmembrane potentials. It shows a section through a bacterial cell (only one corner of the cell is shown). ATP synthase links the discharge of the transmembrane proton gradient to the formation of ATP, A: simple mechanism (hydrogen cycling) B. a mechanism in which the membrane-bound electron-transfer proteins are proton pumps.

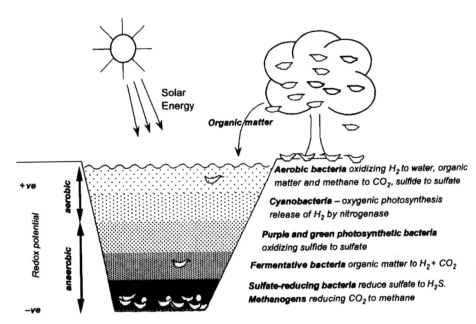

Figure 1.3 Aerobic and anaerobic metabolism in an aquatic bacterial ecosystem.

sunny day? It is because H_2 is such a valuable commodity that it does not accumulate. In microbial communities, there is a flow of H_2 from one type of cell to another, the interspecies hydrogen transfer. Fermentative bacteria excrete H_2 as a waste product, while chemolithotrophic bacteria use it as their fuel (Fig. 1.4). Denitrifying bacteria oxidize H_2 with nitrate as oxidant, reducing it to N_2 gas. Sulfate-reducing bacteria such as *Desulfovibrio desulfuricans* reduce sulfate to sulfide, and acetogenic bacteria such as *Moorella thermoautotrophica* reduce CO_2 to acetic acid. Finally, at the bottom of the pond and the end of the microbial food chain, methanogens reduce CO_2 and acetate to methane. Many of these organisms will play a part in the story of hydrogenase, which is a major subject of this book.

Symbiotic associations between nitrogen-fixing organisms and plants also involve hydrogenase (Fig. 1.5). Only prokaryotes contain the nitrogen-fixing enzyme nitrogenase, which releases H_2 as it reduces N_2 to ammonia. This has been exploited (see Chapter 10) in biological systems for splitting water to H_2 and O_2. Hydrogenases in these organisms often recover the H_2 and recover some energy by respiration.

The subcellular organelles, mitochondria and hydrogenosomes, illustrate the interplay between O_2 and H_2 metabolism. In the great majority of eukaryotic organisms (those having a cell nucleus), including ourselves, foodstuffs are converted to organic acids such as citrate, and oxidized by the citric acid cycle in the mitochondria. Within these energy-producing subcellular organelles, O_2 is consumed by the enzyme cytochrome oxidase, and energy is produced in the form of ATP (Fig. 1.2B). But some

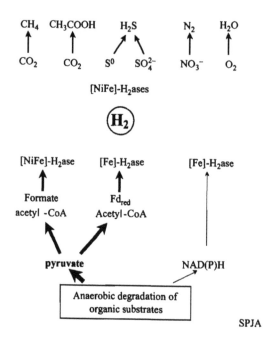

[NiFe]-H$_2$ases

(H$_2$)

[NiFe]-H$_2$ase   [Fe]-H$_2$ase   [Fe]-H$_2$ase

Formate   Fd$_{red}$
acetyl -CoA   Acetyl-CoA

pyruvate   NAD(P)H

Anaerobic degradation of
organic substrates

SPJA

Figure 1.4 Microbial production and consumption of hydrogen by chemolithotrophic bacteria.

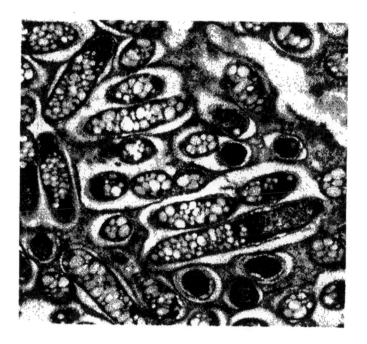

Figure 1.5 Bacteroids of *Rhizobium leguminosarum* in root nodules in the pea (Dr P. Poole, University of Reading).

Figure 1.6 Nyctotherus ovalis (an anaerobic heterotrichous ciliate from the hindgut of an American Cockroach): hydrogenosome with mitochondrial-type cristae (white arrows); r: putative ribosomes; m: membrane; me: methanogenic endosymbiont (methanogenic bacterium) (J.H.P. Hackstein, Catholic University of Nijmegen, The Netherlands).

protozoa and fungi that live anaerobically, for example in the intestinal tracts of insects, have subcellular organelles known as hydrogenosomes (Fig. 1.6). A fermentation process, similar to that used in some anaerobic bacteria, provides the ATP for growth. The oxidation is carried out by the release of H_2, using the enzyme hydrogenase. There has been recent speculation that the mitochondria and hydrogenosomes have a common evolutionary origin, and acquired their enzymes by genetic transfer from the aerobic and anaerobic bacteria, respectively (Embley and Martin 1998).

Bacterial H_2 metabolism impacts on human activities in other ways. The corrosion of iron, for example, is accelerated by bacteria, particularly sulfate-reducing bacteria. They do not appear to interact with the iron directly, but with the hydrogen that is produced on the surface of iron in contact with water. For this reason antibacterial agents are used in preservative solutions for heating systems.

1.2. Technological implications

Hydrogen is increasingly recognized as a potential fuel for industry and transport. It can be produced by electrolysis or photolysis of water, and its oxidation produces no greenhouse gases. Moreover, it is the best fuel for fuel cells, which generate electricity directly by the reversal of electrolysis. Fuel cells have been known for almost two centuries; they are a type of battery in which electricity is produced by the redox reaction between H_2 and O_2 in solution. They offer high thermodynamic efficiencies (theoretical 100 per cent, actual 45–60 per cent) compared to internal combustion engines (theoretical 35 per cent, actual 15 per cent). Fuel cells are a clean form of technology: small-scale, low-voltage, non-polluting (Cleghorn et al . 1998).

Fuel cells have found specialized uses, for example as an electricity source for manned spacecraft. An increasing number of applications are envisaged, replacing batteries in equipment, vehicles and domestic electricity supplies. However, there are two principal limitations, which have prevented the wider use of hydrogen in fuel cells. One of the main limitations to the use of hydrogen as a fuel is one of storage. It is a flammable, low-density gas, and storage as a liquid is much more difficult than for liquid hydrocarbons. However, progress has been made in the use of metals or carbon as adsorbants for H_2 and in future this problem may become tractable. The second is the expense of the catalysts used to convert the reducing equivalents from H_2 in solution to electrons within a conductor. Traditional fuel cells have used expensive metals such as platinum.

What lessons can technology learn from the use of H_2 by living systems?

The advantage of handling H_2 in solution. In thermodynamic terms, the most efficient processes are those that are close to equilibrium. Hydrogenases catalyse the oxidation of H_2 reversibly, in contrast to flames or explosions. Fuel cells can achieve higher efficiencies than heat engines.

In Nature, storage of hydrogen hardly ever occurs; it is present in small quantities and is consumed almost immediately. In other words, hydrogen is a good fuel but is stored in other forms.

Nature uses the transition-metal elements iron and nickel, rather than noble metals, and in their ionic form rather than the metals. As will be seen in this book, for the simplest chemical reaction, the metal–ion centres in hydrogenases are some of the most complex catalysts known. Their structures, which have just been elucidated, have proved to be an elegant and totally unexpected solution to the problem. The construction of these catalysts is in itself a molecular assembly line of extraordinary sophistication.

Chapter 2

Biodiversity of hydrogenases

Robert Robson

With contributions from *Hermann Bothe, Klaus Fiebig, Juan Imperial, Albrecht Klein, Kornél L. Kovács, Jacques Meyer, Thomás Ruiz-Argüeso, Anita Sellstedt, Rüdiger Schulz, Paulette M. Vignais*

2.1. Introduction

The study of biodiversity is a relatively recent subject concerned not only with recording and monitoring changes in the diversity of life but also with the levels on which biology is organised and the interrelationships between them. Biodiversity is measured on three fundamental levels: ecosystems, species and genes. Questions concerning evolutionary relationships underlie all of biodiversity. This chapter considers the biodiversity of H_2 metabolism by reviewing the species in which H_2 metabolism has been studied, the importance of H_2 metabolism to those organisms and to different ecosystems, and the occurrence, function and evolution of different hydrogenases and the genes which encode them. Questions concerning the analysis of the diversity of H_2 metabolising bacteria in different environments and the potential existence of novel systems are also considered.

2.2. Species diversity

2.2.1. Overview

It is remarkable that various aspects of H_2 metabolism have been studied in some detail in over sixty species (Table 2.1). The spread of interest can be judged in relation to the universal phylogenetic tree based on 16S/18S rRNA sequence comparisons (Fig. 2.1) which shows that life (excluding viruses) divides into three domains, two of which, the Eubacteria and the Archaea, are prokaryotic, i.e. contain organisms with no nuclear membrane and the third consists of all the Eukarya, those organisms which contain a membrane-bound nucleus. Most of the organisms in which H_2 metabolism has been studied are prokaryotes and the physiological range is wide and includes both aerobes and anaerobes, autotrophs and heterotrophs, prokaryotic and eukaryoytic photosynthetic organisms, knallgas bacteria, methanogens, sulfate reducers, N_2 fixers, fermentative organisms, hyperthermophiles, parasitic protozoans, and anaerobic fungi. Particular groups of organisms have been studied because H_2 consumption is essential to their survival, e.g. the methanogens such as *Methanobacterium*, *Methanococcus* and *Methanosarcina* sp. Other organisms (e.g. *Ralstonia eutropha*, *Bradyrhizobium japonicum*, *Rhodobacter capsulatus*) can adapt to use H_2 as a sole fuel source. Particular organisms have been studied because of close interrelationships between H_2 metabolism and other biochemical and physiological

Table 2.1 Organisms in which hydrogenase metabolism, biochemistry and/or genetics have been studied

Domain	Group	Order	Genus	Species	Abbreviation
Archaea	Crenarchaeota		*Pyrodictium*	*brockii*	*Pbr*
	Euryarchaeota	Archaeglobales	*Archaeoglobus*	*fulgidus*	*Afu*
		Methanobacteriales	*Methanobacterium*	*marburgensis*	*Mtm*
			Methanothermus	*fervidus*	*Mfe*
		Methanococcales	*Methanococcus*	*jannaschii*	*Mja*
			Methanococcus	*thermolithotrophicus*	*Mtl*
			Methanococcus	*vanielli*	*Mva*
			Methanococcus	*voltae*	*Mvo*
		Methanomicrobiales	*Methanosarcina*	*mazei*	*Mma*
		Methanopyrales	*Methanopyrus*	*kandleri*	*Mka*
		Methanosarcinacaea	*Methanosarcina*	*barkeri*	*Mba*
		Thermococcales	*Pyrococcus*	*furiosus*	*Pfu*
			Pyrococcus	*hirokoshii*	*Phi*
			Thermococcus	*litoralis*	*Tli*
			Thermococcus	*stetteri*	*Tst*
Bacteria	Aquificales		*Aquifex*	*aeolicus*	*Aae*
			Aquifex	*pyrophilus*	*Apy*
			Calderobacterium	*hydrogenophilim*	*Chy*
			Hydrogenobacter	*thermophilus*	*Hth*
	Cyanobacteria	Chroococcales	*Synechococcus*	PCC6301	*S6301*
			Synechocystis	PCC6803	*S6803*
		Nostocales	*Anabaena*	PCC29413	*A29413*
			Anabaena	PCC7120	*A7120*
			Anabaena	*variabilis*	*Ava*
			Nostoc	PCC73102	*N73102*
		Oscillatoriales	*Oscillatoria*	sp.	*Osp*
		Prochlophytes	*Prochlorothrix*	*hollandica*	*Pho*
	Firmicutes	Actinomycetales	*Frankia*	sp.	*Fsp*
			Streptomyces	*thermoautotrophicus*	*Sth*
		Bacillus/Clostridium	*Bacillus*	*schlegelii*	*Bsc*
			Bacillus	*tusciae*	*Btu*
			Clostridium	*acetobutylicum*	*Cac*
			Clostridium	*pasteurianum*	*Cpa*
		Corynibacterinae	*Rhodococcus*	*opacus*	*Rop*
	Proteobacteria	α	*Acetobacter*	*flavidum*	*Afl*
			Bradyrhizobium	*japonicum*	*Bja*
			Paracoccus	*denitrificans*	*Pde*
			Rhizobium	*leguminosarum*	*Rle*
			Rhodobacter	*capsulatus*	*Rca*
			Rhodobacter	*sphaeroides*	*Rsp*
		β	*Acidovorax*	*facilis*	*Afa*
			Alcaligenes	*eutrophus*	*Aeu*
			Alcaligenes	*hydrogenophilus*	*Ahy*
			Ralstonia	*eutrophus*	*Reu*
			Rhodocyclus	*gelatinosus*	*Rge*
			Thiobacillus	*plumbophilus*	*Tpl*
		δ	*Desulfomicrobium*	*baculatus*	*Dba*
			Desulfovibrio	*fructosovorans*	*Dfr*
			Desulfovibrio	*gigas*	*Dgi*
		ε	*Helicobacter*	*pylori*	*Hpy*
			Wolinella	*succinogenes*	*Wsu*
		γ	*Azotobacter*	*chroococcum*	*Ach*
			Azotobacter	*vinelandii*	*Avi*

Table 2.1 Continued

Domain	Group	Order	Genus	Species	Abbreviation
			Citrobacter	freundii	Cfr
			Chromatium	vinosum	Cvi
			Escherichia	coli	Eco
			Desulfovibrio	vulgaris	Dvu
			Pseudomonas	carboxydovorans	Pca
			Salmonella	typhimurium	Sty
			Thiocapsa	roseopersicina	Tro
	Thermotogales		Thermotoga	maritima	Tma
Eukarya	Ciliophora	Litostomatea	Dasytricha	ruminantium	Dru
		Spirotrichea	Nyctotherus	ovalis	Nov
	Parabasalidea	Trichomonidida	Trichomonas	vaginalis	Tva
	Chlorophyta	Volvocales	Chlamydomonas	reinhardtii	Cre
			Chlamydomonas	moewsii	Cmo
		Chlorococcoccales	Scenedesmus	obliquus	Sob
	Fungi	Chytridiomycota	Neocallimastix	sp. L2	Nsp

processes. N_2-fixing organisms (e.g. *Rhizobium, Anabaena, Azotobacter* and *Frankia* species) have been of special interest because H_2 is both an obligatory product and a potential inhibitor of N_2 reduction by the enzyme nitrogenase and hydrogenases may therefore enhance the efficiency of N_2 fixation (the H_2-recycling hypothesis). Some organisms have proved amenable because they are easily grown and their enzymes easily isolated. Some have good genetic systems and, once gene probes or DNA sequences become available for hydrogenases from one group of organisms, an explosion of interest in closely related species usually follows.

2.2.2. The species coverage

Whilst the number of organisms in which H_2 metabolism has been studied is already large, many gaps exist in our coverage. The exploration of new phylogenetic or physiological groups has often led to the discovery of new hydrogenases which in turn have advanced structural and mechanistic understanding of these enzymes and also provided potentially interesting biotechnological opportunities.

Amongst the Eubacteria (Table 2.1), the coverage is dominated particularly by studies in two major divisions, the Proteobacteria (purple photosynthetic bacteria) in which all five subdivisions (α, β, δ, ϵ, and γ) are represented, and the Cyanobacteria (of which at least three groups are represented). However, many widely divergent groups have been neglected. They include the Spirochaetes, the Cytophaga/Flexibacter/Bacteroides group, Fibrobacters, and both high and low G + C content Gram +ve bacteria in which studies of Frankia and the saccharolytic clostridia respectively almost stand alone. The green non-sulfur bacteria and the planctomyces are also neglected groups. Genome sequencing projects are filling some gaps but it is doubtful whether we will recognise truly novel systems by sequence similarity searches alone. Amongst the Archaea, studies on species in the sub-domain Euryarchaeota dominate, partly because some of these are the well-known mesophilic or moderately thermophilic methanogens. However, relatively few members of the Crenarchaeota have been studied probably because most were discovered only recently and many are extreme thermophiles which require specialised growth facilities.

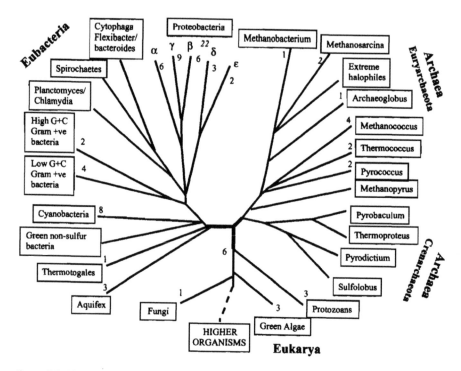

Figure 2.1 Universal phylogenetic tree. The figure shows the schematic evolutionary relationships between various organisms focusing especially on the prokaryotic eubacterial and archaeal domains. The boxes show either the various groups of Eubacteria or genera of Archaea. No details are provided for the evolutionary relationships amongst the members of the Eukarya. Numbers at the ends of the branches represent the number of species of that Group or Genus in which hydrogenases or hydrogen metabolism has been studied or in which putative hydrogenase gene sequences have been discovered through genome sequence projects. For more details see Table 2.1. It should be noted that evolutionary distances are not to scale.

Now, a growing number of complete and partial genome sequences are available for members of this sub-domain, e.g. *Sulfolobus solfataricus*, *Pyrobaculum aerophilum* and *Aeropyrum pernix*. Compared to the prokaryotes, there are relatively few studies of Eukaryotic hydrogenases despite the fact that H_2 evolution is important in some lower orders of the Eukarya including anaerobic protozoa, e.g. *Trichomonas vaginalis* and *Dasytricha ruminantium*, unicellular green algae, e.g. *Chlamydomonas* and *Scenedesmus*, and anaerobic rumen fungi, e.g. *Neocallimastix* sp.

2.3. Functional diversity of hydrogenases

2.3.1. Functions of hydrogenases: An overview

At least 13 families of hydrogenases are known. All but one are involved directly or indirectly in energy metabolism (see Table 2.2) and two main physiological functions can be discerned: either they catalyse H_2 oxidation (H_2 uptake/consumption) linked

Table 2.2 Families of hydrogenases, their occurrence and function

Number	Hydrogenase family	Occurrence	Function
1	Fe-only hydrogenases	Obligately anaerobic bacteria and Eukaryotes	Fermentation/ energy conservation?
2	NiFe(Se) membrane-bound or periplasmic/respiratory hydrogenases	Aerobes, facultative anaerobes, obligate anaerobes of the Proteobacteria	Energy conservation
3	NiFe-(thylakoid)uptake hydrogenases	Cyanobacteria	Energy conservation?
4	Bidirectional NAD(P)-reactive hydrogenases	Cyanobacteria	Energy conservation, Redox poising?
5	NAD(P)-reactive hydrogenases	Facultative and obligately anaerobic Eubacteria	Energy conservation
6	NAD(P)-reactive hydrogenases	Obligately anaerobic Archaea	Fermentation
7	F420-non-reactive hydrogenases	Methanogens	Energy conservation
8	F420-reactive hydrogenases	Methanogens	Energy conservation
9	NiFe-sensor hydrogenases	Chemolithotrophic/ phototrophic Proteobacteria	Hydrogen sensing components in genetic regulation of hydrogenase expression
10	NiFe-hydrogenases associated with the formate hydrogen lyases complex	Facultative and obligate anaerobes, Archaea	Fermentation
11	Ech hydrogenases	Methanogens	Methanogenesis pathway
12	Non-metal hydrogenases	Methanogens	Energy conservation
13	Soluble hydrogenase	*Anabaena cylindrica*	Unknown

to energy conserving reactions (e.g. respiration, NAD(P)H formation, methanogenesis) or they catalyse H^+ reduction (H_2 evolution) coupled to the disposal of excess reducing potential through the re-oxidation of reduced pyridine nucleotides and electron carriers. However, two other potentially related functions have emerged in recent years. One family of hydrogenases present in several autotrophic Proteobacteria (e.g. *R. eutropha*, *R. capsulatus* and *Bradyrhizobium leguminosarum*) appear to act as the H_2 sensing components of a complex genetic relay controlling the expression of other hydrogenases in these organisms (see Chapter 4). A fourth function for hydrogenases has been suggested for the so-called bidirectional hydrogenases in Cyanobacteria which may serve to poise the redox of photosynthetic and respiratory electron transport chains.

2.3.2. The multiplicity of hydrogenases

The importance of H_2 metabolism to some organisms is highlighted by their possession of more than one hydrogenase system. Four hydrogenase systems are known in *Escherichia coli* and *Methanococcus voltae*, three in *Desufovibrio vulgaris* and

R. eutropha and two in the Eukaryote, *T. vaginalis*. Several systems may be expressed simultaneously in some organisms but in others enzymes are expressed differentially depending on the growth conditions (Chapter 4). Moreover, the complement of hydrogenases may even differ between strains of a particular species as shown for sulfate reducing bacteria (Voordouw *et al.* 1990). However, some organisms contain a single system, e.g. *Azotobacter vinelandii*, and many more organisms apparently need no hydrogenase at all. No recognisable hydrogenase genes have been found in the genomes of important organisms, e.g. yeast, amongst the lower Eukaryotes, *Haemophilus influenzae* and *Mycobacterium tuberculosis* amongst the Eubacteria. All Archaea examined appear to contain hydrogenases.

2.3.3. The diversity of prosthetic groups in hydrogenases

All but two of the known families of hydrogenases are known to be metalloenzymes usually containing several metal centres including [Fe-S] clusters, NiFe centres with remarkable coordination properties, or haem groups (Chapter 6). Some also contain non-metal prosthetic groups, e.g. FAD, FMN. All the metallo hydrogenases contain Fe (Fe-only hydrogenases) but many also contain Ni (the NiFe(Se) hydrogenases) some of which, as their title suggests, contain Se in the form of selenocysteine which substitutes for one cysteinyl ligand to the Ni atom. One non-metal hydrogenase is the so-called metal-free hydrogenase involved in methanogenesis under Ni-starved conditions in *Methanobacter marburgensis* (Afting *et al.* 1998). A second potential non-metal hydrogenase has been isolated from *Anabaena cylindrica* (Ewart and Smith 1989; Ewart *et al.* 1990). It exhibits tritium exchange activity, the defining activity of members of the hydrogenase enzyme family, but its function is unknown. Generally the Fe-only enzymes are structurally related, have high turnover numbers and evolve H_2. They are found in obligate anaerobes such as saccharolytic clostridia and protozoan parasites such as *T. vaginalis* where they function in the reoxidation of reduced cofactors. However, other Fe-only enzymes function in H_2 oxidation as in the examples from the sulfate reducing bacteria such as *D. vulgaris*. The NiFe(Se) enzymes are the most diverse group and serve in energy conserving reactions in diverse organisms such as sulfate reducers, chemolithotrophs, methanogens, chemoheterotrophs and phototrophs. In addition the H_2-sensing and bidirectional hydrogenases are also NiFe enzymes. Whereas their subunit structures are quite diverse, the Ni-containing subunit appears to be conserved (Reeve and Beckler 1990). The NiFe(Se) enzymes do not appear to be especially structurally related to the Fe enzymes by primary amino acid sequence although it has emerged recently that the active site may be similar (Chapter 6). The five basic groups of hydrogenases described above can be further subdivided to give the twelve groups (Table 2.2) described in more detail below.

Group 1. The Fe-only hydrogenases

These are classically represented by the enzymes from obligate anaerobes including the sulfate reducer *D. vulgaris* (Vourdouw and Brenner 1985) and the saccharolytic *Clostridium pasteurianum* (Meyer and Gagnon 1991) in which the enzymes act in uptake or evolution modes respectively. Many examples are αβ heterodimers like the NiFe-membrane or periplasmic enzymes yet the primary sequences of the subunits of these groups of enzymes are not similar. Nevertheless it has recently been revealed that their active sites show a great deal in common (Chapter 6). They occur in the periplasm in the case of *D. vulgaris* where they may reduce cytochrome c_3 or in the cytoplasm in the case of *C. pasteurianum*, where low-potential ferredoxins (e.g. the 2[4Fe-4S] cluster containing species) may provide the electrons *in vivo*. A novel three-subunit NADH-oxidising Fe-only hydrogenase has recently been found in *Thermotoga maritima* (Verhagen *et al.* 1999).

Group 2. The periplasmic and membrane-bound NiFe(Se) hydrogenases in the Proteobacteria

These enzymes are αβ heterodimers which are widely distributed in a variety of different organisms. These enzymes share two subunits in common, the large (α subunit) which contains the NiFe centre and the smaller (β subunit) of which usually contains three [Fe-S] clusters forming an intraprotein electron conduit, leading electrons produced at the NiFe centre to an accessory and specific electron carrier protein usually a *c* or *b*-type cytochrome. They occur either free in the periplasm (Fig. 2.2A) as in the sulfate-reducing bacteria such as *Desulfovibrio gigas* (Li *et al.* 1987) where the role of the enzyme is unclear, or most probably attached to the outer surface of the cytoplasmic membrane as in the classic examples from *B. japonicum* (Sayavedra-Soto *et al.* 1988), *E. coli* (Ballantine and Boxer 1985; Menon *et al.* 1990a,b; Menon *et al.* 1994b), *R. eutropha* (Schink and Schlegel 1979; Kortlüke *et al.* 1992), *R. capsulatus* (Leclerc *et al.* 1988), *A. vinelandii* (Seefeldt and Arp 1986; Menon *et al.* 1990a) and *Wolinella succinogenes* (Dross *et al.* 1992) (Fig. 2.2B). Based on immunological and molecular biological evidence, similar enzymes may also occur in Frankia species, the N_2-fixing symbionts of woody dicotyledenous plants such as *Alnus* and *Casuarina* (Fig. 2.3) (Sellstedt and Lindblad 1990). Though direct localisation studies have been performed for only a few cases, these enzymes are believed to be localised on the outer face of the cytoplasmic membrane because they all have signal peptides at the N-terminal sequences of the β subunit which are absent in the as isolated, mature proteins. In addition the C-terminal sequence of the β subunit is a highly hydrophobic domain missing in the soluble periplasmic *D. gigas* enzyme and which may be in part responsible for membrane attachment. The genes for the β and α subunits (*hupSL* or *hoxKG*) are usually at the proximal end of an operon (see Chapter 4) which contains a third conserved gene potentially encoding a *b*-type cytochrome as established in *W. succinogenes* where the three protein complex, as isolated, is capable of reducing quinols (Dross *et al.* 1992). The activity of these enzymes which show relatively high affinities for H_2 can be measured in membrane preparations of aerobes by following H_2-dependent O_2 consumption. Therefore they presumably act as proximal electron donors to the respiratory chain.

Group 3. The NiFe-(thylakoid) uptake hydrogenases of Cyanobacteria

These enzymes are found in filamentous Cyanobacteria, e.g. *Anabaena* 7120 (Houchins and Burris 1981), *Nostoc* sp. strain PCC73102 (Oxelfelt *et al.* 1998) and *Anabaena variabilis* (Schmitz *et al.* 1995) where they occur in the heterocysts. They may also occur in vegetative *Anabaena* cells grown under microaerobic or anaerobic non-N_2-fixing

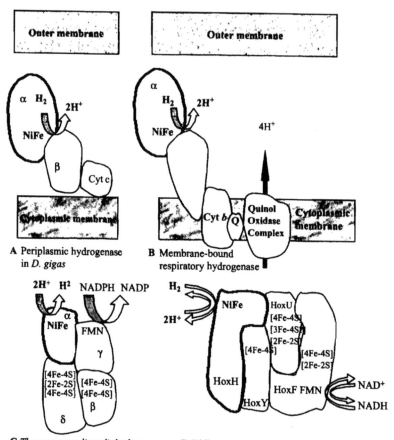

Figure 2.2 Hydrogenases: structure and function. The figure illustrates structure and functional relationships for various members of the hydrogenase enzyme family. Subunit and gene nomenclatures and composition of prosthetic groups are generally indicated. (A,B) Soluble periplasmic and membrane-bound examples of the Group 2 enzymes from Proteobacteria. (C) The Group 6 tetrameric hydrogenase from the archaean *T. litoralis*. (D) The Group 4 bidirectional hydrogenase from Cyanobacteria.

conditions. Physiological and biochemical evidence also suggests that a similar enzyme may occur in the unicellular non-N$_2$-fixing species *Anacystis nidulans* (*Synechococcus* sp. strain PCC6301) (Peschek 1979a,b). Although as yet these membrane-attached enzymes have not been purified to homogeneity, there is physiological evidence to support a requirement for Ni for their activity and the sequences of the *hupSL* genes from several organisms suggests that they are $\alpha\beta$ dimers, resembling the periplasmic and membrane-bound NiFe hydrogenases except that the small subunit has no transit peptide. This suggests that they are not exported to the periplasm but attached instead to the inner surface of the cytoplasmic membrane or, more likely, to the thylakoid membrane in the heterocysts. Their physiological role (Fig. 2.2B) may be to recycle H$_2$ produced by nitrogenase and a role in energy conservation is supported by observations which suggest that electrons first reduce cytochrome *b* or

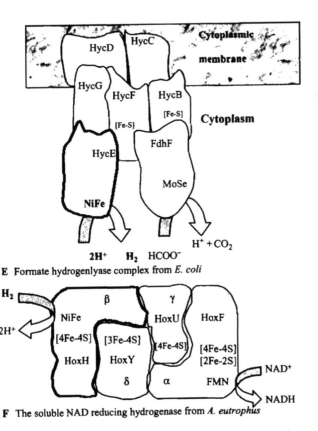

E Formate hydrogenlyase complex from *E. coli*

F The soluble NAD reducing hydrogenase from *A. eutrophus*

Figure 2.2 (continued) (E) The Group 10 hydrogenase associated with the formate hydrogen lyase complex from *E. coli*. (F) The Group 5 tetrameric soluble cytoplasmic hydrogenase from *R. eutropha*.

plastoquinone and are then allocated via the cytochrome *bc* complex to photosystem I or the respiratory terminal oxidase. In *Anabaena* 7120, induction of heterocyst development involves an extensive DNA excision event that affects the *hupL* gene for the NiFe subunit (Carrasco *et al.* 1995a,b), but this type of genetic rearrangement involving the *hupSL* genes is not found in all heterocyst-forming strains.

Group 4. The bidirectional-NAD(P)-reactive hydrogenases in Cyanobacteria

The so-called bidirectional or reversible hydrogenases found in some Cyanobacteria exhibit both $Na_2S_2O_4$ and methyl viologen (MV)-dependent H_2 evolution and also phenazine methosulfate or methylene-blue-dependent H_2 oxidation (uptake). They occur in filamentous N_2-fixing organisms such as *Anabaena* species (Serebrykova *et al.* 1996; Boison *et al.* 1998) and also in the unicellular non-N_2-fixing organism, e.g. *A. nidulans* (Boison *et al.* 1996) but not in *Nostoc* sp. PCC73120. These enzymes reduce NAD(P)$^+$, and sequence alignments show a distinct structural similarity between some subunits (e.g. HoxF and HoxU) of this enzyme and those of the NADH:ubiquinone oxidoreductase (Complex I) of the respiratory chain. This suggests

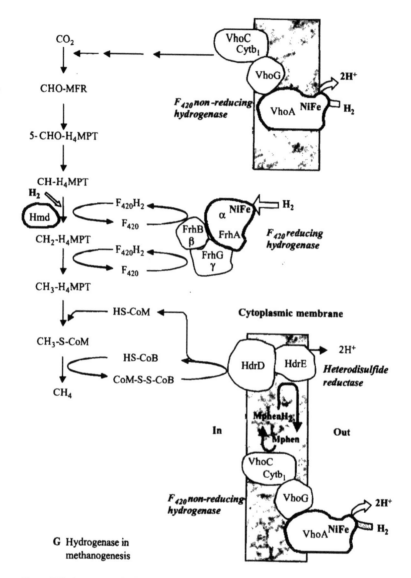

Figure 2.2 (continued) (G) The Group 7 (F_{420} non-reactive), Group 8 (F_{420} reactive), and Group II (non-metal) hydrogenases from methanogens and their roles in metahanogenesis. CHO-MFR, N-formylmethanofuran; 5-CHO-H_4MPT, N^5-formyltetra dromethanopterin; CH-H_4MPT, N^5,N^{10}-methenyltrahydromethanopterin CH2-H_4MPT, N^5,N^{10}-methylenetrahydro- methanopterin; CH$_3$-H_4MPT N^5-methyltrahydromethanopterin; CH$_3$-ScoM, methyl coenzyme M; HS-CoB, Co-enzyme B or N-7-mercaptoheptanoyl-)-phospho-L-threonine (HS-HTP); Hmd is the metal-free hydrogenase.

that these enzymes are involved in energy conservation. However, the physiological role of these enzymes is unclear since these particular Cyanobacteria are incapable of autotrophic growth on H_2 and a mutant of the *hoxH* gene encoding the NiFe subunit of this hydrogenase in *A. nidulans* is unaffected in growth. In the filamentous

Figure 2.3 Micrograph of *Frankia* strain R43 grown in nitrogen free medium. The thread-like structures are hyphae, the small swollen spheres are vesicles and the larger structures are sporangia (Dr A. Sellstedt, University of Umeå).

N_2-fixing organisms, the enzyme is found in both the vegetative cells and the heterocysts. In *A. variabilis* and *A. nidulans* the enzyme is thought to be membrane bound. Examination of the interrelationship between photosynthesis and respiration has led to speculation that the role of the enzyme may be to control electron flow in respiration and photosynthesis (Appel and Schulz 1998, 2000) (Fig. 2.2D).

Group 5. The NAD(P)-reactive hydrogenases from bacteria

These membrane-bound or soluble multi-subunit enzymes (usually heterotetramers) are capable of reducing or oxidising nicotinamide adenine dinucleotides. They play an important part in the metabolism of knallgas bacteria which are a group of bacteria which can use H_2 as a sole energy source with O_2 as terminal electron acceptor. Also in autotrophs, the NADH formed from H_2 oxidation probably serves as the reductant for CO_2 reduction. The group includes organisms such as *R. eutropha*, *Nocardia opaca* and related organisms (Schneider and Schlegel 1976; Schneider *et al.* 1984; Hornhardt *et al.* 1990; Tran-Beckte *et al.* 1990) where H_2 can serve as sole energy source. In this well-studied system, the enzyme is soluble, reduces NAD to NADH and consists of four subunits (Fig. 2.2F) of which HoxH (β subunit) is the NiFe-containing subunit, HoxF (α subunit) contains FMN and two [Fe-S] clusters and the likely active site for NADH reduction and two other subunits, HoxY (δ subunit) and HoxU (γ subunit) both of which contain Fe-S centres which probably are involved in electron transfer between the active sites. These enzymes are structurally related to the bidirectional hydrogenases found in Cyanobacteria.

Group 6. The NAD(P)-reactive hydrogenases from Archaea

In the hyperthermophilic Archaea, NAD(P)-reactive enzymes are involved in recycling the reduced cofactors to produce H_2 as a waste product as in the case of the NADPH oxidising hydrogenases from the hyperthermophilic Archaea, e.g. *Pyrococcus* species (Bryant and Adams 1989; Pedroni *et al.* 1995) and *Thermococcus litoralis* (Rakhely *et al.* 1999). These enzymes are also heterotetramers (Fig. 2.2C) with an apparently similar organisation of subunits and prosthetic groups to the Eubacterial examples of Group 5.

Group 7. The F_{420}-non-reducing hydrogenase of methanogens

Methanogens have an obligate requirement for H_2, which is oxidised by at least two types of hydrogenases which in turn reduce cofactors or substrates involved in the reduction of C_1 units (e.g. CO_2) to CH_4. Factor F_{420} (8-hydroxy 5-deazaflavin) is one of several unique cofactors required for methanogenesis (Fig. 2.2G). The two metal-containing hydrogenases found in methanogens can be distinguished biochemically by their ability to reduce F_{420}. The F_{420}-non-reducing hydrogenases cannot reduce F_{420} but reduce the artificial electron acceptor MV (Deppenmeier *et al.* 1992; Woo *et al.* 1993). They are thought to be important in two steps of methanogenesis: the reductive carboxylation of methanofuran to form N-formylmethanofuran, and the provision of electrons for the reductive cleavage by heterodisulfide reductase of the heterodisulfide CoM-S-S-CoB formed in the release of methane and the production of a membrane potential which is used in the synthesis of ATP (Setzke *et al.* 1994) (Fig. 2.2G). In *Methanococcus mazeii*, the F_{420} non-reducing hydrogenase and the heterodisulfide reductase may associate to form a complex known as the H_2:heterodisulfide oxidoreductase which may provide an energy-conserving proton pump (Ide *et al.* 1999). These hydrogenases are membrane bound and are of either the NiFe or NiFeSe types. Both types occur in *M. voltae* and are encoded by the *vhc* or *vhu* genes respectively (Halboth and Klein 1992). The *vhu*-encoded NiFeSe enzyme is constitutive but as the NiFe enzyme is made only when Se is limiting it appears to serve as a contingency for Se deficiency. More details of the regulation of this system appear in Chapter 4. Mostly these enzymes contain three subunits, one of which consists of the *vhc* (*vht*, *vho*)A gene which contains the NiFe(Se) centre, *vhcG* and *vhcC* which is deduced to encode a *b*-type cytochrome. However, interestingly, in *M. voltae*, VhuA is truncated at the C-terminal and lacks the cysteinyl and selenocysteinyl ligands to Ni. However, this domain is supplied by a twenty-five amino acid residue peptide encoded by a fourth gene, *vhuU. Methanosarcina mazei* Göl also appears to contain a pair of these enzymes encoded by the constitutively expressed *vho* genes and the *vht* genes which are not expressed when cells are grown on acetate. There is no evidence for either hydrogenase containing Se in this organism (Deppenmaier *et al.* 1992). It is interesting to note that, based on genome sequencing data, heterodisulfide reductase and hydrogenases which resemble the F_{420}-non-reactive hydrogenase also occur in the non-methanogenic hyperthermophilic, sulfate-reducing archaeon, *Archaeoglobus fulgidus*.

Group 8. The F$_{420}$-reducing hydrogenases of methanogens

Examples of these hydrogenases and the *frhA, B* and *G* genes which encode them were first characterised from *Methanothermobacter* sp. (Jacobson *et al.* 1982; Alex *et al.* 1990) and somewhat later from *Methanobacterium formicicum* (Baron and Ferry 1989). The role in methanogenesis of the F$_{420}$ reduced by these enzymes is shown in Fig 2.2G. These are all membrane-bound NiFe(Se) enzymes which contain FAD and comprise three subunits: FrhB (β subunit), FrhG (γ subunit) and FrhA (α subunit) which bears the NiFe centre. The stoichiometry of the enzyme appears to be $\alpha_2\beta_2\gamma_1$. Most methanogens appear to possess parallel systems, one of which may be constitutive and the other expressed only during particular growth conditions, e.g. selenium deficiency, or during growth with a particular C$_1$ compound, e.g. with methanol but not with acetate. As for the F$_{420}$-non-reducing hydrogenases some methanogens contain two forms: an NiFe only form (encoded by the *frc, fre,* or *frh* genes) or an NiFeSe form (encoded by the *fru* genes). Both enzymes exist in *M. voltae* but the presence of Se in the growth medium suppresses expression of the NiFe-only enzyme (Halboth and Klein 1992) (see Chaper 4). *Methanosarcina barkeri* also contains a second F$_{420}$-reducing counterpart encoded by the *fre* genes which has a high degree of identity to the *frh*-encoded system (Fiebig and Friedrich 1989; Michel *et al.* 1995; Vaupel and Thauer 1998).

Group 9. The NiFe H$_2$-sensing hydrogenases

These enzymes are known in *R. eutropha* (Lenz *et al.* 1997), *B. japonicum* (Black *et al.* 1994) and *R. capsulatus* (Elsen *et al.* 1996; Vignais *et al.* 1997) which can use H$_2$ as an alternative sole energy source in lithotrophic situations. They have many features in common with the Group 2 enzymes in that they are heterodimers except that they are not known to couple to a cytochrome and from the gene sequences, their small subunits lack both C-terminal anchor-like and N-terminal signal domains and so presumably they are cytoplasmic rather than membrane bound or periplasmic. More details of this recently discovered group of enzymes thought to be involved in sensing H$_2$ can be found in Chapter 4.

Group 10. NiFe hydrogenase in the formate hydrogen lyase complex

This H$_2$-evolving NiFe enzyme is part of a multi-component membrane-bound enzyme system studied extensively in Enterobacteria, especially *E. coli* where it is known as hydrogenase 3 and expressed under predominantly anaerobic conditions (Sawers *et al.* 1985, 1986). The entire complex couples the reduction of protons to the oxidation of formate to form CO$_2$ and serves to void excess reducing potential (Fig. 2.2E). It is believed to lie on the inner face of the cytoplasmic membrane. Purification and further characterisation has been difficult because it is intrinsically unstable. However, much is known about the contiguous cluster of 8 *hyc* genes required for this hydrogenase (Bohm *et al.* 1990) of which *hycE* encodes the Ni-containing subunit. The role of a number of accessory genes in the maturation of this system will be discussed in Chapter 3. Sequence analysis of the genome of *E. coli* revealed a fourth potential NiFe hydrogenase encoded by the *hyf* gene cluster which appears to be related to hydrogenase 3 (Andrews *et al.* 1997).

Group 11. The membrane-bound Ech hydrogenases in methanogens

Relatively recently a NiFe-containing membrane-bound hydrogenase which appears to be related to the formate hydrogenase lyases has been described from *M. barkeri*. This enzyme is composed of four hydrophilic and two membrane-spanning subunits all of which show sequence identity to Complex I of respiratory chains (Künkel *et al.* 1998; Meuer *et al.* 1999). Ech does not reduce quinones but catalyses the reduction of protons to H_2 with reduced 2[4Fe-4S] ferredoxin as electron donor. This enzyme plays an important function during growth of *M. barkeri* on acetate as sole energy source in which it appears to couple to the generation of a proton motive force (Bott and Thauer 1989). The enzyme also appears to be present in *Methanothermobacter* sp. Two transcriptional units each encoding Ech-like enzymes have been identified in the genome sequence (Tersteegen and Hedderich 1999).

Group 12. The metal-free hydrogenase in methanogens

A unique metal-free hydrogenase was found first in *M. marburgensis*. It catalyses the reversible reaction of N^5,N^{10}-methenyltetrahydromethanopterin with H_2 to give N^5,N^{10}-methylenetetrahydromethanopterin and a H^+ (Fig 2.2G) (Hartmann *et al.* 1996). Several lines of evidence support the conclusion that it lacks Fe or Ni. The enzyme does not catalyse an exchange between H_2 and the protons of water nor the conversion of para H_2 to the ortho-form unless methenyl H_4MPT is present. The specific activity of the enzyme increases in cells growing under nickel-limiting conditions which suggests that it is important in methanogenesis in Ni deficient environments (Afting *et al.* 1998). The protein is encoded by the *hmd* gene which also appears in the genomes of *Methanococcus jannaschii* and *Methanopyrus kandleri*.

Group 13. The soluble hydrogenase in A. cylindrica

A. cylindrica contains two soluble hydrogenases one of 100 kDa which exhibits MV-dependent H_2 evolution and the other of 42 kDa which exhibits tritium exchange activity only. The MV-reactive enzyme is a heterodimer composed of a 50 kDa and the 42 kDa subunit form. Both proteins are required for MV reactivity (Ewart and Smith 1989). The function of this enzyme is unknown at present and the gene sequence for the tritium exchanging subunit is unrelated to any of the hydrogenases described above (Ewart *et al.* 1990).

2.4. Genetic diversity

2.4.1. Describing the genetic diversity

It is not the intention of this chapter to review the genetics of hydrogenases. However, there has been an explosion of information in this area in recent years. Table 2.3 shows that more than 27 genotypes are used to describe genes involved in the metabolism of H_2. Some are synonyms, e.g. *hox, hup, hyn, hya* and *hyd* have all been applied to the structural genes for the periplasmic, heterodimeric NiFe(Se) hydrogenases which form a coherent phylogenetic cluster. Unfortunately this adds to the impression of diversity for the specialist and non-specialist alike. Some of the genotypes such as *hyd*

Table 2.3 Genotypes used to describe hydrogenases, their accessory genes and related functions

Genotype	Functions (known or presumed)	Organisms
anf	Alternative nitrogenase (Fe-only)	Avi, A7120, Rca, Rru
eha	Energy converting hydrogenase a	Mta
ehb	Energy converting hydrogenase b	Mta
fhl	Formate hydrogen lyase	Eco
frc	F_{420}-reducing hydrogenases	Mvo
fre	F_{420}-reducing hydrogenases	Mba
frh	F_{420}-reducing hydrogenases	Mba, Mtm, Mvo, Mja
fru	F_{420}-reducing hydrogenases	Mvo
hdr	Heterodisulfide reductase	Mka
hev	Hydrogen evolving hydrogenase	Eco
hox	Membrane bound/Periplasmic/ soluble NADH-reducing hydrogenases (structural and accessory genes)	Aeu, Ava, Avi, Pca, Pho, S63801, S6803, Rop
hmd	Metal-free hydrogenase	Mtl
hup	Membrane-bound/Periplasmic hydrogenases (structural and accessory genes)	A7120, Ach, Ahy, Ava, Avi, Bja, N73102, Phy, Rle, Rca, Rsp, Rge, Tro
hya	Membrane-bound hydrogenase 1	Eco, Cfr, Hpy
hyb	Membrane-bound hydrogenase 2	Eco
hyc	Membrane-bound hydrogenase 3 associated with formate-hydrogen lyase	Eco
hyd	Various hydrogenases	Cac, Dfr, Dgi, Dvu, Hpy, Pfu, Tma, Tva, Wsu
hyf	Postulated hydrogenase 4 related to formate hydrogen lyase	Eco
hyp	Pleiotropic hydrogenase maturation genes	
hyn	Periplasmic (NiFe)	Dvu
hys	Periplasmic (NiFeSe)	Dvu
mbhl	Methylene blue reducing	Aae
mvh	Methyl viologen reducing	Mta, Mfe
nif	Nitrogen fixation (Mo system)	A7120, Ach, Avi, Bja, Dvu, Rle, Rca, Mvo, Mta, N73102, Tro
vhc	F_{420} non-reducing	Mvo
vho	F_{420} non-reducing	Mma
vht	F_{420} non-reducing	Afu, Mma,
vhu	Methyl viologen reducing	Mja
vnf	Alternative nitrogenase (V system)	A7120, Ach, Avi

Organisms: A7120, Anabaena 7120, Aae, Aquifex aeolicus; Afl, Acetobacter flavidum; Aeu, Alcaligenes eutropha; Ahy, Alcaligenes hydrogenophilus; Ach, Azotobacter chroococcum; Afu, Archaeoglobus fulgidus; Ava, Anabaena variabilis; Avi, Azotobacter vinelandii; Bja, Bradyrhizobium japonicum; Cfr, Citrobacter freundii; Cac, Clostridium ocetobutylicum; Cpa, Clostridium pasteurianum; Dba, Desulfomicrobium baculatus; Dfr, Desulfovibrio fructosovorans; Dgi, Desulfovibrio gigas; Dvu, Desulfovibrio vulgaris; Eco, Escherichia coli; Hpy, Helicobacter pylori; Mba, Methanosarcina barkeri; Mfe, Methanothermus fervidus; Mja, Methanococcus jannaschii; Mka, Methanopyrus kandleri; Mma, Methanosarcina mazei, Mta, Methanobacterium thermoautotrophicum; Mtl, Methanococcus thermolithotrophicus; Mvo, Methanococcus voltae; N73102, Nostoc 73102; Pho, Prochlorothrix hollandica; Pca, Pseudomonas carboxydovorans; Phy, Pseudomonas hydrogenovora; Pfu, Pyrococcus furiosus; Phl, Pyrococcus horikoshii; Reu, Ralstonia eutrophus; Rle, Rhizobium leguminosarum; Rca, Rhodobacter capsulatus; Rsp, Rhodobacter sphaeroides; Rop, Rhodococcus opacus; Rge, Rhodocyclus gelatinosus; S6301, Synechococcus 6301; S6803, Synechocystis 6803; Tma, Thermotoga maritima; Tro, Thiocapsa roseopersicina; Tva, Trichomonas vaginalis; Wsu, Wolinella succinogenes.

are confusing and are often employed as a generic term for various groups of hydrogenases. Rationalisation would appear to be required but the adoption of a simpler nomenclatures would add further confusion and tend to obscure earlier literature. A similar situation appears to be developing in the nomenclature for various hydrogenases in the Archaea and especially in the methanogens.

One of the fascinating insights to have emerged from the study of hydrogenase genetics is that many genes, in addition to regulatory and structural genes, are required for the functioning of the various hydrogenases. These accessory genes are required for the uptake of metals such as Ni, or the synthesis and insertion of metal clusters and their ligands into the nascent enzymes, the targeting of the assembled enzymes into cell compartments such as the periplasm in Eubacteria, the chloroplast of green algae, or the hydrogenosome of certain anaerobic fungi and protozoans. In addition, many enzymes need accessory proteins to become physiologically relevant, e.g. to couple to their cognate redox chains, energy conserving complexes or fermentation pathways (see Chapter 3). What we have learned from this of biotechnological importance is that the structural genes cannot be transferred from one organism to another without the appropriate accessory genes.

In the prokaryotes, structural, regulatory and accessory genes are usually organised into polycistronic, transcriptional units (operons) (Chapter 4). In Eukaryotes all genes appear to be monocistronic and there appears to be little grouping of related genes. The significance of this is that the location of accessory genes is much more difficult in the Eukaryotes and as yet though hydrogenase structural genes have been characterised, no accessory genes have been described though they must surely be present. Different operon structures occur in different organisms (see Section 5.5 and Chapter 4) and these reflect functional diversification or different regulatory strategies.

2.5. The evolution of hydrogenases

2.5.1. H_2 metabolism and the origins of life

The ability to catalyse the evolution or oxidation of H_2 may have been exploited by the earliest life forms as H_2 would have been present in the early prebiotic environments. The origins of the proton-dependent chemiosmotic mechanism for ATP synthesis may also reflect the formation of proton gradients created by hydrogenases on either side of the cytoplasmic membrane. In addition, it has been speculated that the coupling of H_2 and S metabolisms was also of fundamental importance in the origin of life. These two processes seem intimately coupled in the bifunctional sulfhydrogenase found in *Pyrococcus furiosus* (a combination of subunits for hydrogenase and sulfite reductase) which can dispose of excess reductant either by the reduction of protons to H_2 or S^0 to H_2S (Ma *et al.* 1993; Pedroni *et al.* 1995).

2.5.2. Revealing hydrogenase evolution by protein comparison

Insights into the evolution of H_2 metabolism can be gained by the comparison of different hydrogenases at the primary sequence level. The choice of which proteins to compare is influenced by the number of sequences available, their occurrence and the presumption of functional analogy. This approach was thoroughly explored by Wu and

Mandrand-Berthelot (1993). Since that time many more sequences and several new hydrogenases have emerged. Also, not only can the hydrogenase structural genes be compared but also the accessory genes, e.g. those involved in hydrogenase maturation (see Chapter 3). Other information (e.g. the comparative arrangement of hydrogenase gene clusters in different organisms) can be incorporated into the analysis.

2.5.3. How phylogenetic trees are built

The construction of phylogenetic trees relies on the computer-aided alignment of amino acid sequences and the calculation of the degree of similarity between the sequences. This results in a similarity matrix from which trees of relatedness can be constructed. Entire gene/protein sequences may be used but some degree of editing of the sequences is performed to maximise the alignment. Alternatively, it is possible to compare only those segments of molecules which are most easily aligned, i.e. they contain few or no gaps and show high sequence identity and may also represent important conserved domains within a protein. The statistical significance of such trees is usually assessed by determining how many times the tree or elements of the tree are likely to have been produced at random. The significance of such trees increases when a large number of truly homologous proteins or genes are available for comparison as for the 16S/18S ribosomal RNA subunit sequences mentioned earlier. Two basic types of trees can be constructed. Rooted trees compute or assume a common ancestor. However the rooting of trees becomes highly problematic where sequences are quite divergent, and where functionality is not analogous or unknown. This applies in most cases when considering both the hydrogenase structural and accessory gene products. Alternatively, the construction of unrooted trees avoids many such assumptions.

2.5.4. Relationships between the basic groups of hydrogenases

The first conclusion to emerge from attempts to align the sequences of the H_2-activating subunits of the metal-containing hydrogenases (Group 11 and 12 hydrogenases are not considered here) is that the Fe-only hydrogenases, the NiFe(Se) hydrogenases, and the hydrogenase component of the formate hydrogen lyases form three distinct families. It is tempting to speculate that all hydrogenases diverged from a single ancestral enzyme. However, there is little evidence to distinguish between convergent or divergent evolutionary pathways. This picture may change if novel intermediate enzymes are discovered. It is possible that certain domains, e.g. [Fe-S] cluster binding domains within the proteins, have been derived from common ancestral proteins which at one time may have been encoded by smaller genes. In this respect it is interesting to note that two of the metal-ligands forming residues of the NiFeSe centre in the F_{420}-reducing hydrogenase from *M. voltae* are borne on a separately encoded peptide. So it is possible that the Ni subunits of the NiFe(Se) enzyme family are the product of one or more gene fusion events.

Relationships in the NiFe(Se) enzyme family

Within the NiFe(Se) hydrogenase family, the unrooted tree (Fig. 2.4) clearly reveals several major lineages. As might be expected, the enzyme groups discussed above all emerge as distinct clades which reflect the major prokaryotic groups and the enzyme

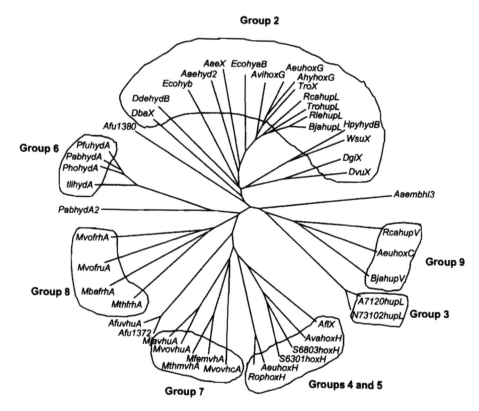

Figure 2.4 Evolutionary relationships between various members of the NiFe(Se) hydrogenase families based on Ni-binding subunits. Amino acid sequences for Ni-bearing (large) subunits of various members of the NiFe(Se) family were aligned by CLUSTAL W (Thompson *et al.* 1994). Trees were drawn using the program Treeview (Page 1996). Numbers refer to the various hydrogenase family groups shown in Table 2.2. Gene products are indicated as follows. The first three letters indicate the species (see Table 2.1) with the exception of the Cyanobacteria where the Genus and the species number is indicated. The following four letters indicate the gene product derived from the gene name where defined.

functionality within that group. It is difficult to root such trees with any degree of certainty and so comment on the very early origins of this enzyme family. For example, it is not possible to deduce that they emerged originally in an ancestor of the hyperthermophilic Archaea which are proposed to be the closest descendents of the universal common ancestor based on 16S rRNA data. However, the arrangement within each clade broadly accords with the phylogenetic relationships between the constituent organisms. This is clearly illustrated in the lineage of the Group 2 periplasmic and membrane-bound NiFe(Se) hydrogenases where the overall clustering follows the likely evolutionary origin of the α, β, δ, γ and ε groups of the Proteobacteria in which all these enzymes occur. Enzymes from the α, β and γ groups of the Proteobacteria (represented by *R. capsulatus*, *R. eutropha* and *E. coli* hydrogenase 1 respectively) form one cluster (Group 2a) whilst functionally comparable enzymes from the δ and ε groups of the Proteobacteria (Group 2b represented by *D. gigas* and *W. succinogenes* respectively) form two other divergent clusters. This large group is easily expanded to include hydrogenase 2 (encoded by the *hyb* genes) from *E. coli*. But it is interesting to note that hydrogenases 1 and 2 of *E. coli* (encoded by the *hya* and *hyb* genes) clearly did not arise from a gene duplication event in a recent ancestor of the Enterobacteraceae. Probably these enzymes are not functionally interchangeable. But if hydrogenase 2 from *E. coli* is to be included in this group, then two hydrogenase sequences from *Desulfitobacter dehalogens* and *Aquifex aeolicus* must also be included. Yet more distantly rooted is the important NiFeSe enzyme from *Desulfomi crobium baculatum*, which does not group with the other enzymes from the δ group of the Proteobacteria and which appears as one of the most ancient of this functionally related group.

Another especially interesting feature of the tree concerns the position of the Group 3 thylakoid uptake NiFe hydrogenases from *Anabaena* and *Nostoc*, which are presumed to be functionally analogous to the Group 2 enzymes in H_2 recycling from nitrogenase in the N_2-fixing Proteobacteria (e.g. members *Azotobacteraceae* and *Rhizobiaceaea*). However, they cluster most closely with the cytoplasmic H_2-sensing hydrogenases (Group 9) found in the Proteobacteria. One common feature of these two enzyme types is that neither has a transit peptides and so they must be assumed to be cytoplasmic, or, at least in the Cyanobacteria, to be bound to thylakoid membranes. These data suggest that the sensor enzymes and the thylakoid hydrogenases may have evolved from a common ancestor.

As might be expected, the heterotetrameric NAD(P)-reactive enzymes in the Eubacteria and the Archaea form two distinct clusters within a broad lineage, which more distantly includes the F_{420}-reducing and F_{420}-non-reducing hydrogenases from methanogens. Amongst the bacteria, the cyanobacterial bidirectional Group 4 enzymes are clearly distinct from the Group 5 enzymes of the knallgas bacteria. The *Acetobacter flavidum* hydrogenase is also related to this group.

The F_{420}-reducing hydrogenases found in methanogens form a lineage (Group 7) which is quite distinct from the F_{420}-non-reducing hydrogenases (Group 6) yet both enzymes appear to have emerged from a common ancestor. Included in this group are the deduced hydrogenase sequences from *M. jannaschii*, which lies close to the universal ancestor. Also, it appears that the NiFe (*frhA*) and NiFeSe (*fruA*) enzymes which both occur in *M. voltae*, though quite closely related evolved, after a gene duplication event which occurred after the divergence of the major groups of methanogens. It is interesting that the relatively ancient *M. jannaschii* appears from the genome sequence data to contain both NiFe and NiFeSe enzymes.

The H_2-evolving multi-subunit NAD(P)-reactive hydrogenases from the Thermococcales group of the hyperthermophilic Crenarchaeota, including *Pyrococcus* sp. and *T. litoralis*, also form a distinct lineage (Group 6). Total genome sequence data suggests that *Pyrococcus abyssi* contains an isoenzyme which may have arisen as a result of a very early gene duplication in an ancestor of this group.

Despite the multiplicity of hydrogenases which occur in many organisms, there are relatively few examples of very recent gene duplications within the NiFe(Se) enzyme group. The situation is different amongst the Fe hydrogenases (see later). Hydrogenases 1 and 2 of *E. coli* emerged before the Enterobacteriaceaea appeared as a distinct lineage. Also, uptake hydrogenases 1 and 2 of *Thiocapsa roseopersicina* clearly fall

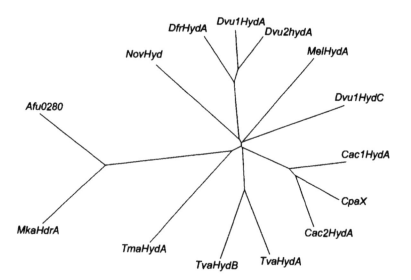

Figure 2.5 Evolutionary relationships amongst the Fe-only hydrogenases. Amino acid sequences of large subunits of Fe hydrogenases from Archaea, Eubacteria and Eukaryotes were used to construct the tree, as described for Fig. 2.3.

into two different lineages originating after the divergence of the δ, ε and α, β, γ clades but prior to the divergence of the latter three groups. The origin of the multiplicity of hydrogenases in the methanogens is also particularly interesting especially since it raises the question of the origin of the Se-containing enzymes. Although relatively few sequences are available at present, Se hydrogenase are found only in the Archaea within the methanococci which are the oldest of the methanogen lineages. In *M. voltae* the Se-containing enzyme appears to expressed unless Se is absent whereupon the NiFe only enzymes are expressed. *M. jannaschii* which is one of the most ancient of the methanococci also apparently contains pairs of NiFe and NiFeSe enzymes. Despite the sparse evidence, it appears as if the Se enzymes of methanogens are ancient.

Relationships amongst the Fe-only hydrogenases

Only the Fe-only hydrogenases are found in all three domains (Eubacteria, Archaea, and Eukaryotes). Therefore the evolutionary history of these enzymes is particularly interesting. They may have been present in the universal ancestor and radiated throughout all three domains or alternatively they may have evolved in a prokaryotic lineage and transferred to the Eukaryotes as a result of prokaryotic endosymbiotic events such as led to the origin of mitochondria and chloroplasts. In this respect it is particularly interesting to note that the hydrogenosome, an organelle-like structure in the protozoan *Nyctotherus ovalis*, contains its own genome much as the chloroplast and the mitochondria do. The tree for the Fe hydrogenases (Fig 2.5) shows that though the *N. ovalis* hydrogenase is deeply rooted it falls in the same lineage as the Fe hydrogenases from the sulfate-reducing γ Proteobacteria bacteria but does not group with the hydrogenases from the other protozoan *T. vaginalis*. Fe-hydrogenase isozymes are

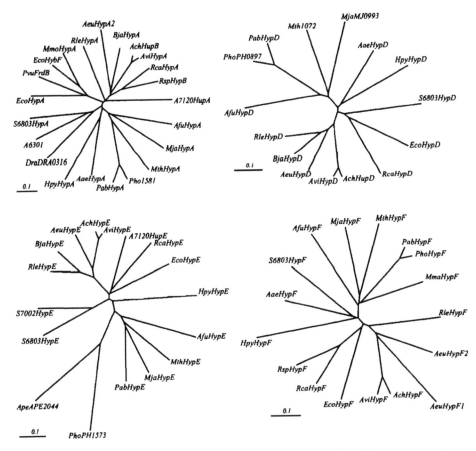

Figure 2.6 Evolutionary relationships between various *hyp* gene products. Trees were constructed for each of the *hypA, hypD, hypE,* and *hypF* encoded proteins as described in the legend for Fig. 2.3. The HypA tree includes the FrdB protein from the fumarate reductase operon of *Proteus vulgaris*.

found in the Eukaryote *T. vaginalis* and the Eubacterium *Clostridium acetobutylicum*. These isozymes appear to have arisen from gene duplications but not particularly recently in evolutionary terms. Studies of hydrogenase(s) in protozoans such as *Giardia* would be interesting especially from an evolutionary point of view since it possesses a prokaryotic 16S-type rather than eukaryotic 18S-type rRNA and may be one of the least evolved Eukaryotes.

2.5.5. Relationships amongst accessory (hyp) genes

Hydrogenase accessory genes can also be used to study the evolutionary relationships between various hydrogenase systems. This is particularly the case with the *hypA, B, C, D, E* and *F* genes most of which were originally identified as being required for the maturation of the several NiFe hydrogenases in *E. coli* (Bohm *et al.* 1990; Lutz *et al.* 1991).

Now, they are known to occur in Eubacteria and Archaea though at present none have been found in any Eukaryote. However the *hyp* genes are apparently absent in the genome of *T. maritima* which contains only an Fe-only hydrogenase. Hence the *hyp* genes may only provide information about the history of the NiFe(Se) hydrogenases. Comparison of the trees for each of the related genes should prove informative. If all the genes have followed the same evolutionary histories, then the trees should be consistent. However, at present relatively few *hyp* gene sequences are available. Those derived from genome sequencing projects can be included, but with caution, since we can only assume analogous functions. For example, from genomic sequence data we can recognise putative *hyp* genes in methanogens and hyperthermophiles but as yet there remains no direct demonstration that these genes are required for any of the hydrogenases in these organisms.

Hyp genes in the Eubacteria

The trees derived for the HypA, D, E and F proteins (Fig. 2.6) clearly divide the Archaea and eubacterial examples. Close inspection of the trees reveals interesting features. For example, in the Proteobacteria, for which the greatest number of sequences are available, the trees are remarkably consistent for each of the Hyp proteins examined. This is evident in the groupings of all four proteins in *Ralstonia*, *Azotobacter*, *R. leguminosarum*, *R. capsulatus* and *Rhodobacter sphaeroides*. But these trees do not mirror the evolutionary pathway deduced from 16S rRNA sequences (see Fig. 2.1). It is interesting to note that in all these organisms, the structural, regulatory and accessory genes are all tightly clustered in the genomes, and moreover in *R. eutropha*, and *B. leguminosarum*, the hydrogenase genes are even plasmid borne. Therefore it is likely that such hydrogenase gene clusters have undergone lateral gene transfer events, especially between the Proteobacteria. The positions of the *E. coli* sequences are interesting because these might be expected to group consistently with other members of the γ subdivision (e.g. the Azotobacters). However, the *E. coli* Hyp proteins are consistently more deeply rooted and more closely match the relative positions of the hydrogenase 2 (*hyb*) genes rather than the hydrogenase 1 (*hya*) genes which group more closely with other γ group members. It is also interesting to note that the *hyp* genes in *E. coli* unlike those in other Proteobacteria cluster with the *hyc* (formate hydrogenase lyase) genes rather than the *hya*, *hyb*, or *hyf* genes. In *E. coli*, *hypF* is separate from the main *hyp* gene cluster (Maier *et al.* 1996) but nevertheless HypF has an evolutionary history which also closely matches those of hydrogenase 2.

Another interesting observation that arises from the Hyp protein trees is the inconsistency in the apparent evolutionary history of the cyanobacterial proteins despite the fact that the Nostocales (*Anabaena* 7120) and the Chroococcales (*Synechococcus* and *Synechocystis*) are not greatly divergent according to 16S rRNA sequences. However, although there are only a few sequences available as yet for this group, a clear difference has already emerged between the positions of the HypAs and HypEs. In *Anabaena* they group with the Proteobacteria which more closely accords with the locations of the Group 2 uptake hydrogenase genes but the Hyp proteins from the Chroococcales are more deeply rooted. It appears as if the structural genes and accessory genes in the Cyanobacteria do not have parallel evolutionary histories.

Relationships between hyp genes in the Archaea

Unlike the situation with the Eubacteria, the archaeal Hyp protein trees are surprisingly inconsistent with one another. This may be because the relatively few sequences available derive from widely divergent organisms.

Diversity in the organisation of the Hyp gene clusters

There is a remarkable variation in the organisation of *hyp* genes in different organisms. In the relatively ancient organism M. *jannaschii*, the *hypA, B, C, D, E* and *F* genes are all present but they are scattered around the genome in a remarkable way and moreover none is linked to any of the hydrogenase structural gene clusters. In M. *marburgensis* the situation is similar with the exception that *hypA* and *B* are contiguous and no *hypF* has been found. It is argued that polycistronic operons allow functionally related genes to be controlled coordinately with economic use of promoter regions and regulatory proteins. However in the methanogens, expression of the *hyp* genes is probably constitutive and there seems to have been no selective advantage to be gained from the clustering of the genes. Alternatively, no particular selective disadvantage has arisen from their scattering throughout the genome. This is in striking contrast to the situation in A. *fulgidus*, their obligately sulfate reducing non-methanogenic relative, where the *hyp* genes are contiguous and sandwiched between sets of putative hydrogenase and H_2:heterodisulfide oxido reductase genes.

One can envisage how the evolutionary histories of structural and accessory genes could be clearly distinct and how the ancestry of the accessory genes may better reflect the early evolutionary origins of the NiF(Se) hydrogenase systems. In the cases of gene clusters introduced into any organism by lateral transfer events, the pre-existence of a functional set of *hyp* genes in that organism capable of supporting the newly arrived structural genes would most likely lead to the loss of the incoming set whereas the structural genes would be retained if of selective value. It is therefore interesting to note that in the remarkable assembly of megaplasmid-borne hydrogenase genes in R. *eutropha*, duplications of the *hypA, B* and *F* genes occur (Wolf et al. 1998). These genes are even partially interchangeable. Inspection of the HypF tree shows that the gene duplication was not recent but possibly occurred prior to the separation of the γ and β groups of the Proteobacteria. Therefore it appears most likely that R. *eutropha* is in the process of losing one set of *hyp* genes rather than gaining a second set by gene duplication.

2.6. Extending the horizons

One goal of biodiversity is the development of tools with which to survey and catalogue hitherto unknown genes and species. Biologists have accumulated knowledge about an astonishing 1.5 million species but an even more surprising statistic is that only 5,000 of these (~3 per cent) are prokaryotes despite the relative antiquity of these organisms. Though different species concepts are used by different groups of biologists, nevertheless, it appears that microbiologists may only have isolated and cultivated 1 per cent or less of all the existing prokaryotes. Clearly this poses an enormous challenge but offers a potential wealth of novel systems. Modern technologies have provided new tools with which to explore this 'great unknown'. The use of the polymerase chain reaction (PCR) has enabled amplification and sequencing of genes from DNA extracted directly from environmental samples. The existence of many potential novel organisms in different environments has been predicted on the basis of the amplification and sequencing of their 16S rRNA genes which are so highly conserved that universal primers can be designed. Although hydrogenase genes are not so well conserved, it should still be possible to design sets of primers for the amplification of different hydrogenase genes from environmental samples. Such an approach has already been used with environmental samples and bioreactors to survey the diversity and expression of a 440 bp fragment of genes potentially encoding large (α) subunits of [NiFe] hydrogenases most similar to those in *Desulfovibrio* (Wawer *et al.* 1997). A considerable diversity of genes and therefore species was observed in the environmental samples but less in bioreactor samples. Also, cDNA amplified by RT-PCR of RNA obtained from the samples was used to show differential expression of the NiFe-hydrogenase genes under denitrifying and methanogenic growth conditions.

The increasing number of genome sequencing projects being undertaken means that more examples of hydrogenase genes will be uncovered. In some cases the roles of these genes may be reasonably deduced but in other cases matches to the existing databases may be ambiguous. Robust phylogenetic trees (e.g. Fig. 2.4) can help to resolve and identify potentially interesting systems. The example of *A. fulgidus*, one of the first members of the Archaea to be entirely sequenced, is illustrative. The sequence reveals a large contiguous cluster of seventeen genes encoding two sets of putative F_{420}-non-reducing hydrogenases, a putative heterodisulfide reductase and a complete cluster of *hyp* genes. However, the phylogenetic tree (Fig. 2.4) shows that in *Archaeoglobus*, the putative Ni-binding subunits AF1372 (VhuA) and AF1380 (VhtA) have roots which are quite distinct from their methanogenic counterparts with which they share closest sequence identity and which they should be most closely related on the basis of 16S rRNA analysis. Whereas the roles of these genes in methanogens is linked to methanogenesis, their role in *Archaeoglobus* is unclear since it is an obligately sulfate-reducing bacterium. The contiguity of the genes in *Archaeoglobus* also suggests that they may only be expressed under certain growth conditions unlike the constitutive expression likely in their methanogen counterparts. In this case the database searches identify convincing matches but neither the physiology of the organism nor the detailed location of these genes in the overall phylogenetic tree could convince one of the actual functions of the hydrogenases in this organism which clearly remain to be established.

Chapter 3

Regulation of hydrogenase gene expression

Bärbel Friedrich, Paulette M. Vignais, Oliver Lenz and Annette Colbeau

With contributions *from August Böck, Wanda Dischert, Albrecht Klein, Thomás Ruiz-Argüeso and Carolien Van Soom*

3.1. Hydrogenase regulation is guided by environmental factors and physiological requirements

Prokaryotes have the capacity to rapidly adapt to changes in their chemical environment. Proteins needed for growth are normally synthesized at levels sufficient to support maximal growth rate on a given substrate. In Nature, however, the cells are usually exposed to a mixture of nutrients, and complex regulatory networks guarantee a hierarchical utilization of the substrates, guided by metabolic efficiency.

The ability of microbes to take up or to evolve H_2 is usually a facultative trait. Therefore it is economically well designed that hydrogenases are predominantly formed only when the substrate is available, and that there are mechanisms of regulated gene expression. Nevertheless, a few specialists organisms (e.g. methanogens) whose metabolism is strictly adapted to H_2 activation are probably synthesizing hydrogenase constitutively. Little is known about gene regulation in archaea, extremophilic organisms such as *Aquifex aeolicus* or in obligate anaerobes, such as sulfate reducers. Regulated hydrogenase gene expression in these organisms, however, cannot be excluded *per se* since many of these specialized microbes harbour multiple isofunctional hydrogenases (see Chapter 2) which may be expressed differentially.

Methanococcus voltae, for example, contains four hydrogenases, two of which contain at the active site, in addition to nickel and Fe, Se. The Se-containing hydrogenases are formed constitutively, whereas the Se-free enzymes are synthesized only under Se depletion (Table 3.1). Likewise, two species of *Methanococcus* are able to use organic compounds for methanogenesis. Both strains contain multiple hydrogenases, and it has been observed that in the presence of acetate one of the hydrogenase operons is repressed (Table 3.1). Moreover, transcript analysis of the putative [Fe] hydrogenase gene *hydA* in *Clostridium acetobutylicum* revealed a fermentation-directed control. The level of *hydA* mRNA is high in cells from acidogenic or alcohologenic phosphate-limited continuous cultures but low when the cells undergo solventogenesis (Table 3.1).

Hydrogenase isoenzymes are also common among the metabolically more versatile bacteria (see Chapter 2). For instance, H_2 metabolism and isoenzyme composition in enteric bacteria, including *Escherichia coli* and *Salmonella typhimurium*, appear to be differentially regulated under the two modes of anaerobic life, fermentation and anaerobic respiration (Table 3.1). Furthermore, biosynthesis of the individual isoenzymes appears to be controlled at a global level by the quality of the carbon source,

Table 3.1 Examples of hydrogenase regulation in response to environmental and physiological factors[a]

Organism	Type of hydrogenase	Relevant characteristics	Physiological function	Effectors and/or conditions affecting hydrogenase gene regulation[b]	Reference[c]
Alcaligenes hydrogenophilus	2 [NiFe] hydrogenases	Cytoplasmic NAD reducing Membrane-bound cytochrome b reducing	Energy conservation	H_2, carbon and energy source limitation	1
Anabaena cylindrica sp. PCC7120	[NiFe] hydrogenase	Putative membrane-bound cytochrome b reducing	H_2 recycling during N_2 fixation	$\underline{O_2}$, nitrogen limitation, heterocyst formation	2
Bradyrhizobium japonicum	[NiFe] hydrogenase	Membrane-bound cytochrome b reducing	Energy conservation H_2 recyling during N_2 fixation	H_2, $\underline{O_2}$, nickel, carbon and energy source limitation	3
Clostridium acetobutylicum	[Fe] hydrogenase	Putative cytoplasmic, ferredoxin linked	H_2 production during fermentation	Fermentation pathway, phosphate limitation	4
Escherichia coli	[NiFe] hydrogenase 1/2	Membrane-bound cytochrome b reducing	H_2 uptake under anaerobic conditions H_2 production during fermentation	Formate, molybdenum, nitrate, carbon source limitation, low pH	5, 6
Escherichia coli	[NiFe] hydrogenase 3	Membrane-associated component of the formate hydrogen lyase complex	H_2 production during fermentation H_2 uptake under anaerobic conditions	Anaerobiosis, carbon source limitation, phosphate limitation, molybdenum, nitrate, formate	7, 8
Methanococcus voltae	2[NiFeSe] hydrogenases [NiFe] hydrogenases	F_{420} reducing F_{420} non-reducing F_{420} reducing F_{420} non-reducing	Methanogenesis Methanogenesis Methanogenesis Methanogenesis	Constitutive Constitutive $\underline{Se}$ $\underline{Se}$	9

Table 3.1 Continued

Organism	Type of hydrogenase	Relevant characteristics	Physiological function	Effectors and/or conditions affecting hydrogenase gene regulation[b]	Reference[c]
Methanococcus mazeii	2[NiFe] hydrogenases	Membrane-bound cytochrome *b* reducing (Vht)	Methanogenesis	H_2/CO_2, methanol, <u>acetate</u>	10
		Membrane-bound cytochrome *b* reducing (Vho)	Methanogenesis	Constitutive	
Methanococcus barkeri	2[NiFe] hydrogenases	F_{420} reducing	Methanogenesis	H_2/CO_2, methanol, trimethylamine, acetate	11
Nostoc muscorum	2[NiFe] hydrogenases	Cytoplasmic NAD reducing	Energy conservation	Constitutive	12
		Membrane-bound cytochrome *b* reducing	H_2 recycling during N_2 fixation	Nitrogen limitation, light	
Pseudomonas hydrogenovora	[NiFe] hydrogenase	Membrane-bound cytochrone *b* reducing	Energy conservation	H_2	13
Ralstonia eutropha	[NiFe] hydrogenases	Cytoplasmic NAD reducing Membrane-bound cytochrome *b* reducing	Energy conservation	H_2, O_2, nickel, carbon and energy source limitation,	14
Rhizobium leguminosarum	[NiFe] hydrogenase	Membrane-bound cytochrome *b* reducing	H_2 recycling during N_2 fixation	O_2, symbiosis	15
Rhodobacter capsulatus	[NiFe] hydrogenase	Membrane-bound cytochrome *b* reducing	Energy conservation	H_2, O_2	16, 17
			H_2 recycling during N_2 fixation		
Rhodospirillum rubrum	[NiFe] hydrogenase	Membrane-bound component of the CO dehydrogenase complex	H_2 production during CO oxidation	CO, O_2	18

Notes

a Only those hydrogenases are listed which are investigated on the regulatory level.

b Effectors and/or conditions acting negatively are underlined.

c The list is restricted to the most relevant references: 1, Lenz et al. (1997); 2, Carrasco et al. (1995); 3 Durmowicz et al. (1998); 4, Gorwa et al. (1996); 5, Atlung et al. (1997); 6, Richard et al. (1999); 7, Rossman et al. (1991); 8, Rosental et al. (1995); 9, Sorgenfrei et al. (1997a); 10, Deppenmeier (1995); 11, Vaupel and Thauer (1998); 12, Axelsson et al. (1999); 13, Ohtuski et al. (1997); 14, Lenz and Friedrich (1998); 15, Brito et al. (1997); 16, Toussaint et al. (1997); 17, Dischert et al. (1999)); 18, Fox et al. (1996).

known as catabolite repression. This is exemplified by the case of E. coli, which contains four hydrogenase systems. Formate is the major effector directing the expression of genes coding for hydrogenase 3 in E. coli. This observation is entirely consistent with its role in formate hydrogen lyase-dependent H_2 evolution during fermentative growth (Table 3.1). The physiological function of the uptake hydrogenases 1 and 2 in E. coli is less well defined. Expression of the hya operon, coding for hydrogenase 1, is co-regulated with that of hydrogenase 3 which implies that an H_2-recycling process takes place during fermentative growth. The transcription factors that govern hya expression, however, are distinct from those instrumental in the control of the hydrogenase-3 operon (see Section 3.2). The hya operon is induced by formate and repressed by nitrate, its expression is strongly elevated by entry into the stationary phase pointing to a dependence on an alternative RNA polymerase σ subunit known as σ^S. In addition the transcriptional regulators AppY and ArcA influence the expression of hya. A fumarate respiration-linked H_2 uptake function is apparently assigned to hydrogenase 2 of E. coli. Consistent with this view is the fact that the hydrogenase 2 operon is controlled by the global regulator Fnr which directs the expression of genes involved in anaerobic respiration (Table 3.1).

A particularly interesting regulatory system has been uncovered in the cyanobacterium Anabaena cylindrica sp. strain PCC7120 (Table 3.1). Expression of the hydrogenase encoding hupL gene occurs during cellular differentiation to heterocysts which are specialized in N_2 fixation. Developmentally controlled gene rearrangments, resulting in the excision of a 10.5 kb DNA element from the hupL gene, encoding the large subunit of an NiFe uptake hydrogenase, lead to activation of hydrogenase gene expression. Coexpression of hydrogenase and nitrogenase genes has also been observed in other microbes such as Nostoc muscorum, Rhizobium leguminosarum and Bradyrhizobium japonicum (Table 3.1). Details of selected well-studied systems are depicted in Section 3.2.

Aerobic bacteria which often use H_2 as an alternative energy source express hydrogenase genes, with a few exceptions, when the substrate is provided (Table 3.1). How do these organisms recognize the presence of H_2, the smallest molecule on Earth? The underlying molecular mechanisms are subject of current research and will be discussed in Sections 3.2 and 3.3.

3.2. Diversity of hydrogenase operons and their transcriptional control

Analysis of the genomic arrangement of hydrogenase genes revealed that in most cases studied so far, the hydrogenase genes occur as tightly clustered functional units, or operons. These are also present also in methanogens as illustrated in Fig. 3.1. The proteobacterial species show complex hydrogenase gene clusters consisting of hydrogenase subunit genes and varying numbers of accessory genes (Figs 3.2–3.6) which code for functions involved in metal centre assembly, protein maturation and gene regulation (Chapter 4). Total genome sequences, now available for a few bacterial and archaeal species, revealed that the hydrogenase accessory genes may also be dispersed on the chromosome, e.g. in the cyanobacterium Synechocystis sp. PCC 6803, the hyperthermophilic bacterium A. aeolicus and the archaeon Methanococcus jannaschii (Bult et al. 1996; Deckertet al. 1998; Kaneko et al. 1996). It is too early to

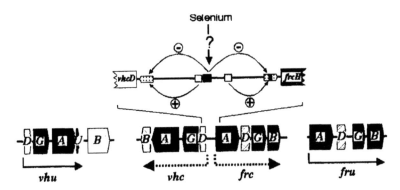

Figure 3.1 Se-controlled regulation of hydrogenase gene transcription in *M. voltae*. The intergenic regulatory region between the divergent *vhc* and *frc* operons is shown in a magnified form. Dotted boxes indicate the promoter regions. Open and solid boxes represent positive and negative regulatory elements, respectively. Hydrogenase structural genes are shaded in dark grey, accessory genes are hatched, genes involved in metallocentre assembly are dispersed over the chromosome and are not displayed. The functions of the *vhcD* and *vhuD* genes are unknown. Transcriptional start sites and the potential length of transcripts resulting from regulated and constitutive promoters, respectively, are indicated by dotted and solid arrows below the operons.

speculate if the difference in gene organization has any impact on hydrogenase gene expression and/or the activity of the resulting hydrogenase proteins. Therefore this section will focus only on those systems which have been well studied biochemically and genetically.

The arrangement of genes in an operon permits their coordinate expression by command of a principal regulator. Such a regulator normally binds to a promoter region, near the site of transcription initiation. Depending on the system, the regulator may exert a positive or a negative effect leading to induction or repression of transcription. The activity of the regulator may be modulated by low-molecular-weight compounds such as the substrate or a metabolite. For instance, the FhlA protein which controls the *hyc* (hydrogenase 3) operon in *E. coli* is modified by formate (Fig. 3.2) (Hopper *et al.* 1994). On the other hand, the regulators of hydrogenase operons might also be the target of a complex signal transduction cascade involving sensor protein(s) which recognize a given external or internal stimulus. These sensor proteins usually transmit the information by chemical modification of the regulator and thus direct its capacity to activate or to repress transcription (Section 3.3).

A major mechanism used by bacteria to respond to environmental changes is the so-called two-component regulatory system which uses phosphorylation/dephosphorylation as a means of information transfer (Hoch and Silhavy 1995). A standard two-component regulatory system consists of a response regulator and a sensor, a histidine protein kinase, which is able to autophosphorylate at the expense of ATP hydrolysis and to transfer its phosphoryl group to the response regulator. The sensor kinase may function also as a phosphatase, removing the phosphoryl group from the regulator. In these bifunctional cases, the direction of the sensor-mediated process is governed by the stimulus. Likewise, the phosphoryl group of the regulator may as

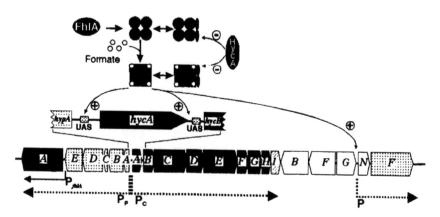

Figure 3.2 The formate regulon of *E. coli* hyc genes encoding structural components of the hydrogenase 3 are shaded in dark grey. Genes involved in metallocentre assembly (*hyp*) are shown in dotted boxes. Open boxes represent the cryptic *ascBFG* genes which form an operon for the degradation of β-glucosides. The hydrogenase-specific endoprotease gene *hycI* and the *hydN* gene are illustrated in hatched boxes. The regulatory genes (*hycA, fhlA*) and their respective products are marked in black. Not shown is *fdhF*, which encodes the seleno polypeptide of formate dehydrogenase H. FhlA binding sites (UAS) are indicated. Transcriptional start sites and the potential length of transcripts resulting from regulated and constitutive promoters, respectively, are indicated by dotted and solid arrows below the gene cluster.

well be hydrolysed by a distinct phosphatase, a potential extra component of a signal transduction system. Moreover, a possible extremely short half-life of the phosphorylated regulator is often observed (Hoch and Silhavy 1995).

Genes belonging to the superfamily of two-component regulatory systems have been identified in hydrogenase gene clusters of *B. japonicum* (Fig. 3.3), *Rhodobacter capsulatus* (Fig. 3.5) and *Ralstonia eutropha* (Fig. 3.6) (Dischert *et al.* 1999, Lenz and Friedrich 1998, Van Soom *et al.* 1999). Hydrogenase gene expression in *R. leguminosarum* is entirely adapted to symbiotic N_2 fixation and guided by the two master regulators of this control circuit, namely FnrN and NifA (Fig. 3.4) (Brito *et al.* 1997). The expression of the *hyc* operon in *E. coli* is coordinated with the expression of the formate hydrogen lyase complex (Sauter *et al.* 1992). Unlike standard two-component systems, the response regulator FhlA is not activated by phosphorylation but by binding formate as the effector molecule (Fig. 3.2). The two Se-sensitive operons in *M. voltae* are both positively and negatively controlled and the sites of regulation are well defined (Fig. 3.1) (Beneke *et al.* 1995); however, the trans-acting regulatory proteins remain to be elucidated. In the following, the regulation of six well-established hydrogenase systems is presented in more detail.

The methanogenic archaeon *M. voltae* harbours four hydrogenase operons (Chapter 2), two of which, *vhc* and *frc*, encode [NiFe] hydrogenases. The remaining two operons, *vhu* and *fru*, encode [NiFeSe] hydrogenases (Fig. 3.1, Table 3.1). The Se-containing hydrogenases are synthesized constitutively. Transcription of the *vhc* and *frc* operons only takes place under conditions of Se depletion, when the two Se-containing isoenzymes cannot be made in sufficient amounts (Berghofer *et al.* 1994).

A Free-living

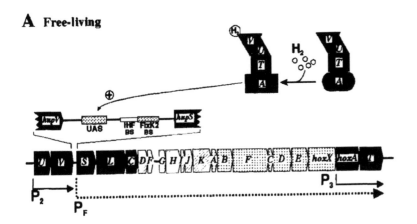

B Symbiosis

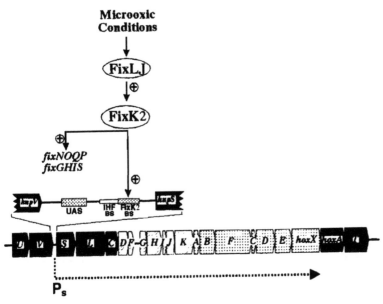

Figure 3.3 Hydrogenase gene expression in B. *japonicum* under free-living conditions (A) and during symbiosis (B). Hydrogenase structural genes (*hup*) are shaded in dark grey, accessory genes (*hup*) are hatched, genes involved in metallocentre assembly (*hyp/hox*) are shown in dotted boxes. The regulatory genes (*hox/hup*) and their respective products are marked in black. The *hupNOP* genes, of which *hupN* encodes a nickel permease (Chapter 4), are located 8.3 kb upstream of the *hupSL* genes and are not included in this figure. Putative FixK2 (FixK2 BS), IHF (IHF BS) and HoxA binding sites (UAS) are indicated. Transcriptional start sites and the potential length of transcripts resulting from regulated and constitutive promoters, respectively, are indicated by dotted and solid arrows below the gene cluster.

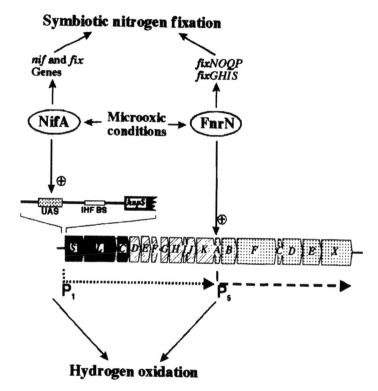

Figure 3.4 Symbiotic hydrogenase gene expression in *R. leguminosarum.* Hydrogenase structural genes (*hup*) are shaded in dark grey, accessory genes (*hup*) are illustrated in hatched boxes, genes involved in metallocentre assembly (*hyp*) are shown in dotted boxes. Putative IHF (IHF BS) and NifA binding sites (UAS) are indicated. Transcriptional start sites and the potential length of transcripts resulting from regulated and constitutive promoters, respectively, are indicated by dotted and solid arrows below the gene cluster.

A coordinated regulation of *vhc* and *fru* is facilitated by the genetic linkage of the two operons, which are separated by a relatively short 450 bp intergenic region. This region contains the promoters and sites implicated in positive and negative transcriptional regulation (Fig. 3.1) (Beneke *et al.* 1995). Mutational analysis has shown that a sequence located in the centre of the intergenic region is involved in negative regulation of both operons. Furthermore, there is evidence for an additional element within the promoter region of the *frc* operon which is involved in establishing the negative control of this gene group. Upstream of either operon, positive regulatory sites were identified that are necessary for full expression of the operons in the absence of Se. It is not yet understood how the signal 'Se deprivation' is transduced to the regulatory machinery. One possible explanation is that a negative regulator or a cofactor is a selenoprotein with a short half-life. Such a protein would not be actively synthesized when Se is depleted. An earlier hypothesis proposed that the

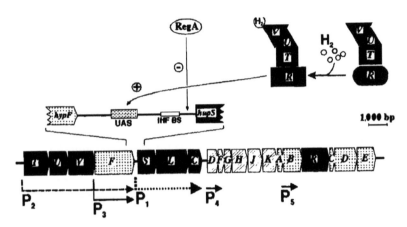

Figure 3.5 H$_2$-dependent hydrogenase gene expression in *R. capsulatus*. Hydrogenase structural *hupSLC* genes are shaded in dark grey, accessory *hup* genes are illustrated in hatched boxes, genes involved in Ni insertion are shown in dotted boxes. The regulatory *hupTUVR* genes and their respective products are marked in black. HupR (UAS) and IHF (IHF BS) binding sites are indicated. Transcriptional start sites and the potential length of transcripts resulting from regulated and constitutive promoters, respectively, are indicated by dotted and solid arrows below the gene cluster.

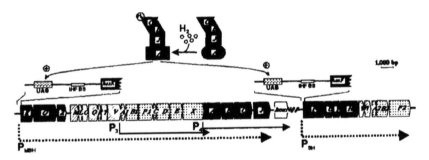

Figure 3.6 H$_2$-dependent expression of the hydrogenase regulon of *R. eutropha*. Hydrogenase structural genes (*hox*) are shaded in grey, MBH- and SH-specific accessory genes (*hox*) are illustrated in hatched boxes, genes involved in metallocentre assembly (*hyp*) are shown in dotted boxes. The regulatory genes (*hox*) and their respective products are marked in black. The *hoxN* gene encodes a nickel permease. Putative HoxA (UAS) and IHF (IHF BS) binding sites are indicated. Transcriptional start sites and the potential length of transcripts resulting from regulated and constitutive promoters, respectively, are indicated by dotted and solid arrows below the gene clusters.

smallest, Se-containing subunit (VhuU) of the Vhu hydrogenase acts as a regulatory factor. This assumption, however, is unlikely since a fusion of VhuU and the large subunit VhuG does not affect regulation (Pfeiffer *et al.* 1998).

The hydrogenase 3 of *E. coli* is part of the formate hydrogen lyase complex (FHL) whose corresponding genes are organized in the formate regulon (Fig. 3.2) (Rossmann *et al.* 1991). Hydrogenase maturation proteins are encoded by the *hypABCDE* gene complex which is co-transcribed with the regulator gene *fhlA*. FhlA belongs to a subclass of response regulators whose activity is modulated by binding of an effector molecule. In the case of FhlA the effector is formate (Hopper *et al.* 1994). The *hyc* operon genes predominantly code for structural components of hydrogenase 3 with the exeption of HycA and HycI which code for a negative regulator and a maturation endopeptidase, respectively. A third transcriptional unit encodes HypF, a maturation component, and HydN whose function is unknown. FhlA complexed with formate binds to the upstream activator sequences (UAS) and activates the σ^{54}-dependent promoters P_P, P_C and P (Fig. 3.2). P_{fhlA}, on the other hand, is a weak constitutive promoter dependent on σ^{70}-containing RNA polymerase. Formate production by pyruvate formate lyase is the main signal for the onset of anaerobiosis. Formate binds to the regulator FhlA (possibly a tetramer), which in turn activates transcription from the UAS motifs upstream of P_P, P_C and P (Hopper *et al.* 1994). The transcription activation leads to an increase in FhlA. This autogenous control of FhlA is counteracted by the activity of HycA, an anti-activator which may function by direct interaction with FhlA (Sauter *et al.* 1992). Expression of the *hyc/fhl* genes leads to the degradation of formate and consequently to the decrease of the effector concentration. Hence, the Fhl system is well balanced by the two positively acting components, FhlA and formate production, and their negatively acting counterparts, HycA and formate degradation.

B. japonicum, the N_2-fixing symbiont of the soybean plant, synthesizes its membrane-bound uptake hydrogenase under two extremely different life styles: under free-living microaerobic conditions, in the presence of H_2, and during symbiosis in the root nodules. In free-living bradyrhizobia, hydrogenase gene transcription is induced by molecular H_2 provided that the O_2 concentration is low and trace amounts of nickel are available (Fig. 3.3A) (Black *et al.* 1994). Under these conditions HoxA, a response regulator of the NtrC family, is the major transcriptional factor. HoxA binds to UAS and activates transcription at the σ^{54}-dependent P_F promoter. Transcription activation is facilitated by the IHF protein (Black and Maier 1995). The activity of HoxA is proposed to be modulated by the histidine protein kinase HupT and the H_2 receptor HupUV. P_2 and P_3 are presumably housekeeping promoters which provide a low-level expression of the regulatory *hupUV* and *hoxAhupT* genes to establish the H_2-sensing apparatus under non-inducing conditions (Black *et al.* 1994; Van Soom *et al.* 1999). During symbiosis it is supposed that the hydrogenase improves the efficiency of N_2 fixation by recycling the H_2 which is released by the nitrogenase as a side product. Under these conditions hydrogenase operon expression is integrated into a complex regulatory network which coordinates the processes involved in symbiotic N_2 fixation (Fig. 3.3B). Transcription activation at the P_S promoter depends on the Fnr-like FixK2 protein (Durmowicz and Maier 1998). Expression of the *fixK2* gene is controlled by the response regulator FixJ whose activity is modulated by its cognate O_2-sensing histidine protein kinase FixL.

Transcription of hydrogenase genes of *R. leguminosarum* bv. viciae is co-regulated with N_2-fixation genes (*nif* and *fix*) and controlled by two global activators, NifA and FnrN, in response to the microaerobic conditions inside the legume nodules (Fig. 3.4) (Brito *et al.* 1997; Guttierrez *et al.* 1997). This environment ensures the expression of hydrogenase genes. The hydrogenase activity enables the bacteroids to take up H_2 evolved by nitrogenase (Brito 1997). Two major promoters, P_1 and P_5, have been identified within the *R. leguminosarum* hydrogenase gene cluster. P_1 is a σ^{54}-dependent promoter responsible for the symbiotic activation of at least the *hupSLCD* operon, and likely for activation of the remaining *hup* genes as well. The symbiotic activation of P_1 is mediated by NifA, the N_2-fixation global transcription activator, and by the integration host factor (IHF). P_5 is an Fnr-type promoter responsible for the microaerobic expression of the *hypBFCDEX* genes independent of the symbiotic life style (Fig. 3.4). Unlike *B. japonicum*, *R. leguminosarum* is not able to synthesize its hydrogenase under free-living conditions since the transcriptional activator gene *hoxA*, which is located downstream of the *hypX* gene (not shown in Fig. 3.4) has been inactivated by accumulation of frameshift and deletion mutations (Brito *et al.* 1997).

The hydrogenase genes of the phototroph *R. capsulatus* are organized in several transcriptional units which are expressed in an H_2-dependent manner (Fig. 3.5). Two promoters have been unambiguously identified, P_1 upstream from the *hupSLC* structural gene operon and P_2 upstream from the *hupTUVhypF* operon (Elsen *et al.* 1996; Toussaint *et al.* 1997). The *hypF* gene can also be transcribed at a low level from its own promoter (P_3). The *hupDFGHJK* genes are separated from the *hupSLC* operon by a 200 bp intergenic region which contains a putative σ^{70}-dependent promoter (P_4). The *hupR* regulatory gene is transcribed together with the *hyp* genes from a promoter (P_5) localized in the *hypA* region (Fig. 3.5). The P_1 promoter which directs the transcription of the *hupSLC* operon is the σ^{70}-dependent type. Transcription activation requires the binding of two regulators. The positive regulator HupR binds to a specific palindromic sequence TTG-N_5-CAA localized at -160 to -150 nt upstream from the transcription start site (Dischert *et al.* 1999; Toussaint *et al.* 1997) and the histone-like IHF protein binds to an AT-rich sequence centred at nucleotide -87. The IHF protein of *R. capsulatus* is a global regulator which activates the *hupSLC* transcription without being essential (Toussaint *et al.* 1997). On the other hand, HupR is absolutely required to obtain transcription of the *hupSLC* operon. The P_1 promoter is activated in response to H_2 but negatively controlled by the global regulator RegA, which binds upstream from *hupS*, between the binding sites of IHF and RNA polymerase (Elsen *et al.* 2000). The P_2 promoter, although weakly expressed, is also regulated. It is almost inactive under autotrophic growth conditions. Thus, in *R. capsulatus*, hydrogenase synthesis is regulated in response to the presence of H_2 and organic substrates, through the activity of the H_2 sensor (HupUV) and the two-component signal transducing system (HupT/HupR) and in response to redox through the global regulatory system (RegB/RegA).

The genes for the membrane-bound (MBH) and the soluble hydrogenase (SH) of *R. eutropha* are arranged into two major operons on the indigenous megaplasmid pHG1 (Fig. 3.6). Both operons are preceded by σ^{54}-dependent promoters, designated as P_{MBH} and P_{SH}, respectively (Schwartz *et al.* 1998). Transcription initiation at the P_{MBH} and P_{SH} promoters depends on the transcription activator HoxA. Thus, the hydrogenase operons constitute a regulon. HoxA, which belongs to the NtrC subfamily of response regulators, receives signals from an H_2-dependent multicomponent signal transduction chain consisting of the H_2 sensor HoxBC and the histidine protein kinase HoxJ (see Section 3.2) (Lenz and Friedrich 1998). If H_2 is available, HoxA becomes activated. In the activated state HoxA binds at UAS and initiates transcription by contacting the σ^{54}-containing RNA polymerase. This protein–protein interaction leads to the formation of the open complex. Putative IHF binding sites

(IHF BS) in the MBH and SH promoter regions indicate that this process is stimulated by IHF (Zimmer et al. 1995). Two additional promoters of the σ^{70}-dependent type are located upstream of the *hyp* genes (P_3) and the regulatory genes (P_4), respectively, and provide a weak constitutive level of the corresponding gene products (Fig. 3.6). The Hyp proteins are not only required for cofactor insertion into the MBH and the SH but are also necessary for the activity of the H_2-sensing HoxBC protein (T. Buhrke and B. Friedrich, unpublished data). Thus, the low constitutive activity of the P_3 and P_4 promoters maintains the H_2-sensing signal transduction apparatus in a ready state under non-inducing conditions (Schwartz et al. 1998). This enables the cells to react instantaneously in the presence of molecular H_2. H_2-induced transcription from the P_{MBH} and P_{SH} promoters results in the synthesis of active MBH and SH (see also Chapter 2). Moreover, transcription from the strong MBH promoter augments *hox A* and most likely *hoxBCJ* expression under inducing conditions (Schwartz et al. 1998). Increased synthesis of HoxA in turn raises the transcription level from the P_{MBH} and P_{SH} promoters provided H_2 is present. This positive feedback loop leads to an adjustment of hydrogenase synthesis to the external concentration of molecular H_2.

3.3. Hydrogen sensing and signal transduction

3.3.1. R. capsulatus

To isolate genes and to study their function, it is common practice to isolate mutants with a given phenotype, e.g. mutants unable to synthesize hydrogenase. Complementation of the mutants by a gene bank then allows identification of the mutated gene(s). The study of the physiology of hydrogenase-deficient mutants (Hup⁻) gave clues on the role of *hup/hyp* genes in hydrogenase synthesis and in H_2 metabolism of *R. capsulatus*. To identify hydrogenase regulatory mutants, the promoter of the structural *hupSL* hydrogenase genes was fused with the reporter gene *lacZ*. Subsequently the *hupS::lacZ* fusion, carried on a plasmid, was introduced in the wild-type strain B10 and in various Hup⁻ mutants. The *lacZ* gene encodes the β-galactosidase enzyme whose activity can easily be measured colorimetrically. Activation of hydrogenase gene expression was correlated with an increase of β-galactosidase activity indicating that *hupS::lacZ* expression was under the same control as the chromosomal *hupS* gene. The use of mutant strains has been instrumental in assigning genes responding to environmental signals.

Molecular H_2 was one of the first environmental factors shown to stimulate the synthesis of hydrogenase and that of β-galactosidase in cells containing the *hupS::lacZ* fusion (Fig. 3.7). High gene expression was observed under conditions that favour nitrogenase synthesis, e.g. anaerobiosis, light, malate-glutamate medium or nitrogen-limited medium. The highest level was obtained with cells grown in darkness in the presence of H_2 and O_2 (Colbeau and Vignais 1992) or in cells grown autotrophically with CO_2 as carbon source and H_2 as reductant.

3.3.1.1. H₂-specific signal transducing system

The H_2 signal transduction cascade involves an H_2 sensor (HupUV), the regulatory soluble [NiFe] hydrogenase and a two-component regulatory system (HupT/HupR).

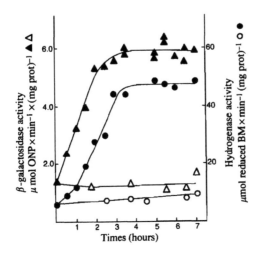

Figure 3.7 Increase in hydrogenase and β-galactosidase activities during growth with H_2 of B10 (pAC142) cells containing the hupS::*lacZ* fusion. ONP, o-nitrophenol; MB, methylene blue; prot, protein; solid symbols, H_2 added; open symbols, no H_2 added.

The response regulator HupR is a transcriptional activator necessary for H_2 induction of *hupSL* gene expression (Toussaint *et al.* 1997). HupR activity can be modified by phosphorylation of the aspartic residue in position 54 in the HupR protein. HupR activates *hupSL* transcription in the non-phosphorylated form, since HupR proteins mutated in the phosphoryl-accepting Asp54, or in which Asp54 has been deleted, have the same capacity to activate *in vivo* *hupSL* transcription in the absence of H_2, as wild-type HupR in the presence of H_2 (Table 3.2) (Dischert *et al.* 1999). The IHF protein (Toussaint *et al.* 1991) activates the *in vivo* transcription of the *hupSL* genes by bending DNA (Fig. 3.8) and facilitating the interaction of HupR, bound to upstream activating DNA sequences (UAS) of the promoter, with RNA polymerase bound at the transcription start site of *hupS*. The TTG-N_5-CAA binding site of HupR (Figs 3.9 and 3.10) is located upstream from the IHF site of the *hupS* promoter.

The HupT protein is a sensor histidine protein kinase. HupT$^-$ mutants have a higher hydrogenase activity than wild-type cells, i.e. they are derepressed for hydrogenase synthesis (Elsen *et al.* 1993). In other words, normally HupT exerts a negative control on hydrogenase synthesis unless H_2 is present (Fig. 3.11). The use of double *hupT, hupR* mutants has shown that HupT and HupR act in the same (H_2) signalling pathway (Toussaint *et al.* 1997). HupT and HupR communicate by transphosphorylation (Fig. 3.12) and form a two-component regulatory system. Since HupR is active in the non-phosphorylated form, it is proposed that HupT exerts its negative control on *hupSL* expression by phosphorylating the activator HupR (Dischert *et al.* 1999).

The presence of the inducer H_2 is not sensed by HupT but by the *hupUV*-encoded hydrogenase. Although very weak, the hydrogenase activity of HupUV could be detected. Figure 3.13 shows the hydrogen–deuterium (H–D) isotope exchange reaction

Table 3.2 Effect of mutations in H_2 signalling (HupR/HupTUV) or redox signalling (RegA/RegB) pathways on hydrogenase activity in *R. capsulatus*

Protein	Function	Mutated protein	H_2ase activity[a]		
			$-H_2$	$+H_2$	Features
		wt	7	45	H_2 dependent
HupUV	H_2-sensing H_2ase	Hup(UV)Δ	58	48	H_2 independent Increased activity
HupT	Histidine protein kinase Autophosphorylable H_2 signal transmitter	HupTΔ	118	95	H_2 independent Increased activity
HupR	Response regulator Transcription activator Phosphorylated by HupT-P	HupR (wt)[b]	1	32	H_2 dependent
		HupRΔ	1	1	loss of activity
		HupRD$_{54}$Δ	46	30	H_2 independent
		HupRD$_{54}$R	46	32	H_2 independent
IHF	DNA bending Global regulator	IHFα	4	17	H_2 dependent Non-essential activator
RegB	Histidine protein kinase Redox sensor	RegB	37	53	Global regulation
RegA	Response regulator Repressor Phosphorylated by RegB-P	RegA	34	59	Global regulation

Notes

a µmol methylene blue reduced $\times$ h^{-1} $\times$ mg protein^{-1}

b The HupR$^-$ mutant VBCl, in which the chromosomal *hupR* gene has been deleted (*hupR*Δ), was complemented with *hupR*-containing plasmids. HupR = wild-type protein; HupRD$_{54}$Δ = the codon for Asp$_{54}$ in the *hupR* gene has been deleted; HupRD$_{54}$R = the codon for Asp54 in the *hupR* gene has been replaced by a codon for arginine.

Strains were grown in mineral salt medium at 30°C aerobically in darkness with or without 10 per cent H_2. Hydrogenase was assayed directly in aliquots from the cultures at an OD$_{660}$ of ca. 1.5.

catalysed by HupUV in the presence of D_2 (Vignais *et al.* 1997, 2000) and the formation of H_2 upon addition of reduced methyl viologen at pH 4. These are two partial reactions typically catalysed by hydrogenases. As for HupSL, formation of active *hupUV*-encoded hydrogenase requires the participation of *hyp* gene products (e.g. HypF) (Colbeau *et al.* 1998). Like the HupT$^-$ mutants, the Hup(UV)$^-$ mutants are derepressed for hydrogenase synthesis and there is no response to H_2 in the mutants (Elsen *et al.* 1996) (Fig. 3.11). The low level of *hupUV* gene expression is in keeping with a signalling function of the HupUV protein. HupUV is an H_2-signalling hydrogenase, which acts in concert with HupT (the respective mutants have the same phenotype). It is still unknown how HupUV and HupT communicate in response to H_2.

Figure 3.8 Ribbon view of the *R. capsulatus* IHF protein bound to its DNA binding site upstream from the *hupS* gene (cf. Toussaint *et al.* 1994).

3.3.1.2. Global redox control by RegB/RegA

Global regulators capable of integrating signals from nitrogen-, carbon- and H_2 metabolism have recently been identified in photosynthetic bacteria. The global regulator RegA, an anaerobic activator of photosynthesis which is phosphorylated by the histidine protein kinase RegB, is involved in negative control of hydrogenase gene expression in *R. capsulatus*. The global redox control by RegB/RegA is superimposed onto the specific control of hydrogenase synthesis in response to H_2, mediated by the HupT/HupR system (Table 3.2). The RegB/RegA system can also activate the synthesis of nitrogenase by increasing the expression of the transcriptional activator NifA, promoted by NtrC. Thus, there is a genetic link between the synthesis of uptake hydrogenase and of nitrogenase in *R. capsulatus* (Elsen *et al.* 2000).

3.3.2. R. eutropha

3.3.2.1. A multicomponent signal transduction system controls H_2-dependent hydrogenase gene expression in R. eutropha

The facultative chemolithoautotroph *R. eutropha*, a member of the β subgroup of proteobacteria, harbours two [NiFe] hydrogenases which are involved in energy conservation from H_2 (Chapter 2). Since H_2 is available at very low concentrations in aerobic environments where these bacteria thrive, the complex synthesis of the hydrogenases (Chapter 4) has to be efficiently regulated in response to H_2. H_2-dependent gene expression in *R. eutropha* is mediated by a complex multicomponent signal transduction pathway (Fig. 3.14) (Lenz and Friedrich 1998) whose individual components are listed in Table 3.3.

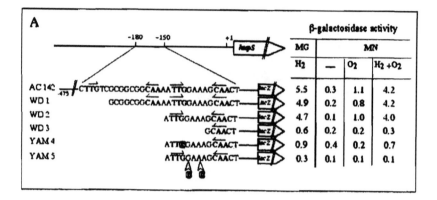

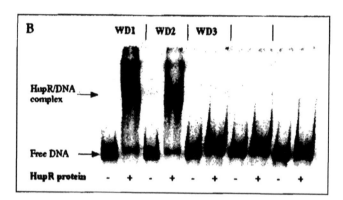

Figure 3.9 Identification of cis-regulatory sequences in the *hupS* promoter region. (A) *In vivo* experiments. DNA fragments encompassing the regulatory sequences of the *hupSL* promoter were progressively deleted, fused with the promoterless *lacZ* reporter gene (which encodes β-galactosidase), inserted in plasmids that were transferred to the wild-type strain B10. The β-galactosidase activities expressed in B10 cells grown in malate-glutamate (MG) or in malate-ammonia (MN) medium, in the presence or absence of H_2 or O_2, reflect the *in vivo* activity of the *hupSL* promoter. The table on the right shows that transcriptional activity is maximal in the presence of H_2 and requires the TTG-N5-CAA sequence, which is the binding site of the activator HupR. (B) *In vitro* experiments. The HupR protein was purified and used in electrophoretic mobility shift assays (EMSA). Only the DNA fragments containing the intact TTG-N5-CAA palindrome (WD1 and WD2, cf. (A)) form a complex with HupR. The complex HupR-DNA migrates more slowly in the electric field than free DNA (adapted from Toussaint *et al.* 1997).

R. eutropha harbours a third [NiFe] hydrogenase, denoted as the regulatory hydrogenase (RH), which enables the cells to sense the occurence of H_2 in the environment (Kleihues *et al.* 2000). The signal is subsequently transduced to a two-component regulatory system consisting of the soluble histidine protein kinase, HoxJ, and the response regulator, HoxA (Fig. 3.14). Although the molecular mechanism is yet unknown, preliminary experimental data indicate a complex formation between the RH and the kinase HoxJ (T. Buhrke and B. Friedrich, unpublished data). The communication

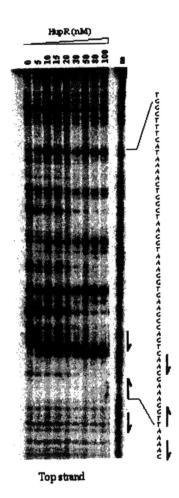

HupR (nM)

Top strand

Figure 3.10 Identification of the HupR binding site at the *hupS* promoter by DNase footprinting analysis. DNA footprinting is a technique which detects the binding of proteins to DNA by their protection of the DNA from attack by various chemical agents. The disappearance of bands in the vertical lanes reveals the increasing protection afforded by increasing concentrations of the HupR protein shown in nM at the top of each lane. The arrows on the sequence, on the right of the figure, show the HupR binding site.

between histidine protein kinases and their cognate response regulators is principally based on phosphotransfer reactions (Hoch and Silhavy 1995). A phosphotransfer between HoxJ and HoxA was recently demonstrated *in vitro* using purified components (M. Forgber, O. Lenz and B. Friedrich, unpublished data).

The response regulator HoxA is crucial for transcription activation at the σ^{54}-dependent SH and MBH promoters (see Section 3.2 and Fig. 3.6) (Schwartz *et al.* 1998). Both HoxA and the H_2-sensing RH act positively, since a knockout of the two proteins abolished hydrogenase gene expression and hence prevent the MBH and SH from being synthesized (Fig. 3.15, Table 3.3)(Lenz and Friedrich 1998). Genetic and biochemical

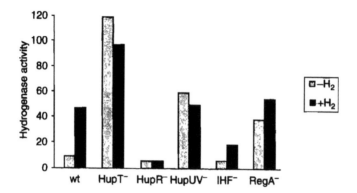

Figure 3.11 Hydrogenase activities in regulatory mutants of *R. capsulatus* affected in transcription of hydrogenase genes compared to the wild-type B10. Cells were grown aerobically (under air) in darkness in malate-ammonia medium at 30°C. Hydrogenase was assayed directly in aliquots from the cultures at an OD_{660} of approx. 1.5. The specific hydrogenase activity is given in μmol of methylene blue reduced per h per mg protein. The figure shows that the highest hydrogenase activities were observed in the HupT⁻ mutant and the Hup(UV)⁻ mutant, that those activities are independent of H_2, and that the transcription factor HupR is required for hydrogenase gene expression. In the RegA⁻ mutant, hydrogenase activity was derepressed but still capable of stimulation by H_2. It is concluded that HupT and HupUV exert a negative control on hydrogenase synthesis, which can be antagonized by H_2. The transcription factors HupR and IHF activate hydrogenase gene expression while the global regulator RegA inhibits it.

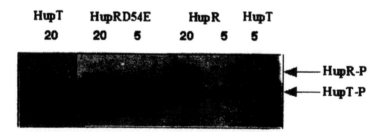

Figure 3.12 Phosphate transfer between the protein histidine kinase HupT and the response regulator HupR. HupT (7 pmol) was phosphorylated at 30°C with [γ-^{32}P]ATP for 5 min, before addition of wild-type (HupR) or mutated (HupR-D54E) HupR protein. Aliquots were withdrawn 5 min or 20 min after addition of HupR and analysed by SDS polyacrylamide gel electrophoresis. HupT20: HupT was phosphorylated for 20 min in the absence of HupR.

analysis revealed that the kinase HoxJ, on the other hand, has a primarily negative effect on hydrogenase gene expression. This negative control is released by the RH provided H_2 is available (Figs 3.14 and 3.15). The data obtained so far indicate that inactivation of HoxA results from its phosphorylation and that the non-phophorylated form of HoxA is active in transcription activation. This property puts the HoxJ/HoxA system

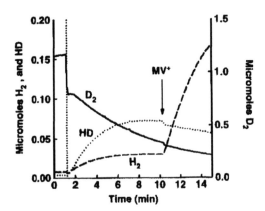

Figure 3.13 Time-course of H_2 and HD production and D2 consumption, in the D_2/H_2O system, catalysed by the *hupUV*-encoded hydrogenase. The HupUV activity was measured at pH 4 in extracts of the *hupSL* mutant JP91 containing plasmid pAC206, which encodes the *hupTUV* operon. The hydrogen–deuterium exchange was followed continuously in the aqueous phase by an on-line mass-spectrometric method. The reaction vessel, maintained under strict anaerobic conditions, was connected by a membrane inlet to the ion source of the mass spectrometer MM 8-80 (VG Instruments). The vertical dotted line indicates when the reaction vessel was closed and when the recording of concentrations of H_2 (....), HD (----) and D_2 (___) at masses 2, 3 and 4, respectively began. At the time indicated by the arrow, reduced methyl viologen (MV^+) was introduced in the medium.

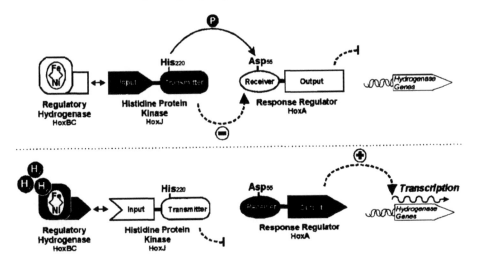

Figure 3.14 Model of H_2-dependent signal transduction in *R. eutropha*. Proteins which are regulatory active are highlighted in black. For details, see text.

apart from standard two-component regulatory systems (Lenz and Friedrich 1998). In compliance with this model is the observation that site-directed exchanges of the conserved phosphoryl acceptor residue Asp-55 in the receiver module of HoxA (Fig. 3.14) convert the system to an H_2-independent control. The HoxA variants

Table 3.3 Structural and functional properties of the hydrogenase regulatory proteins in *R. eutropha*

Protein	Size (kDa)	Type	Function	Mutation/amino acid exchange	Effect on hydrogenase gene expression
HoxB	36.5	RH, small subunit	[NiFe] hydrogenase active in H_2 oxidation and signal transduction	hoxBΔ	Abolished
HoxC	52.4	RH, large subunit		hoxCΔ	Abolished
HoxJ*	51.2	Histidine protein kinase	Sensor kinase, ATP-dependent auto-phosphorylation	hoxJΔ	H_2-independent
				HoxJG422S	H_2-independent
HoxA	53.6	Response regulator	DNA-binding protein, modified by phospho-rylation, activates transcription in response to H_2 and carbon supply	HoxAD55E	H_2-independent
				HoxAD55N	H_2-independent
				hoxAΔ	Abolished
RpoN	54.8	σ^{54}	Alternative σ factor of the RNA polymerase, required for HoxA-dependent activation of transcription	rpoN::Tn5	Abolished

HoxAD55E and HoxAD55N are still capable of activating MBH and SH gene transcription at a high level. Thus, these mutants are constitutive so far as the H_2-dependent regulation is concerned; however, they still respond to a global carbon catabolite control (Lenz and Friedrich 1998). This control also depends on HoxA as the principal regulator and enables the cells to turn off hydrogenase gene expression when fast growth-supporting organic nutrients are available.

3.3.2.2. The H_2 sensor of R. eutropha belongs to a distinct subclass of [NiFe] hydrogenases

The RH protein of *R. eutropha* is a dimeric [NiFe] hydrogenase consisting of a large subunit (HoxC) which harbours the [NiFe] active site and a small FeS-containing subunit (HoxB). This conformation resembles the prototypic [NiFe] hydrogenases as found in *Desulfovibrio gigas* (Kleihues *et al.* 2000). Nevertheless, the RH as well as the isologous proteins implicated in H_2 sensing in other bacterial species such as *B. japonicum* and *R. capsulatus* exhibit characteristic structural features (Fig. 3.16) which are representatively discussed for the RH of *R. eutropha*. The large HoxC sub-unit of the RH is devoid of a C-terminal extension and terminates at a histidine residue which is located at the cleavage site of a specific endoprotease which usually removes the C-terminal peptide after metallocentre insertion (Chapter 4). Thus, the maturation of the regulatory hydrogenase seems to be less complex. The small sub-unit HoxB lacks an N-terminal Tat signal peptide which generally directs the transport

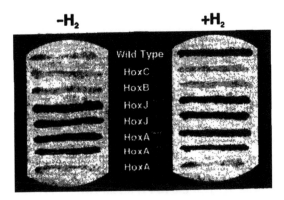

Figure 3.15 Hydrogenase synthesis in regulatory mutants of *R. eutropha*. Cells were grown on agar plates containing glycerol as the carbon source either in the presence or in the absence H_2. The cell material was transferred to filter paper and a triphenyl tetrazolium chloride-based hydrogenase activity staining was performed. Dark colour reflects activity of the MBH.

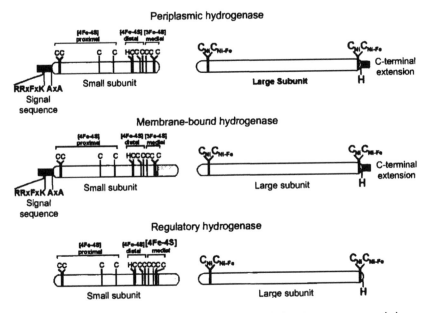

Figure 3.16 Primary structures of [NiFe] hydrogenases belonging to separate subclasses.

of periplasmic and MBH across the cytoplasmic membrane by the Tat system (Chapter 4). Since H_2 is a freely diffusible molecule and the cognate histidine protein kinase is of cytoplasmic nature (Lenz and Friedrich 1998), there is no need to anchor the RH to the membrane. Of great structural and functional importance is the C-terminal tail of HoxB. Such a tail is absent from the small subunits of periplasmic [NiFe] hydrogenases but usually present, although structurally distinct, in MBH (Fig. 3.16,

Chapter 2). Protein–protein interaction experiments revealed that this C-terminal domain of HoxB is necessary for the oligomerization of two RH dimers to form a tetramer and for contacting the histidine protein kinase HoxJ (T. Buhrke and B. Friedrich, unpublished results).

With the aid of a specifically designed overexpression system, the H_2 sensor of *R. eutropha* was characterized in more detail. The RH was identified as an Ni-containing hydrogenase which exhibits H_2-oxidizing activity in the presence of redox dyes (Fig. 3.17) (Kleihues *et al.* 2000). Both the hydrogenase activity and the regulatory function of the RH are Ni dependent. Fourier transform infrared spectroscopy (FTIR), electron paramagnetic resonance (EPR) and preliminary biochemical data revealed that the RH harbours an active site similar to that of prototypic [NiFe] hydrogenases including the two CN molecules and one CO molecule as ligands to the Fe (Pierik *et al.* 1998b). The catalytic activity of the RH is about two orders of magnitude lower than that of energy-generating hydrogenases. Furthermore, the RH activity is O_2 tolerant and insensitive towards CO (Pierik *et al.* 1998b). The unique structural and biochemical features of the RH assign this type of regulatory hydrogenase to a new subclass of [NiFe] hydrogenases (Kleihues *et al.* 2000).

3.4. Concluding remarks and perspectives

Hydrogenase gene regulation has to meet several requirements: (i) the environmental conditions to which the organisms are exposed, (ii) the physiological function of the hydrogenase in a given species, and (iii) the complex biosynthesis of [NiFe] hydrogenase which involves protein-assisted post-translational steps.

M. voltae responds to Se limitation by expressing two alternative Se-free [NiFe] hydrogenases. In this case Se acts as a regulatory signal of gene transcription, although the precise target is still unknown. *R. leguminosarum* takes advantage of hydrogenase activity during symbiotic N_2 fixation, hence it is conceivable that in this bacterium hydrogenase gene transcription is guided by the global N_2-fixation control and its major players NifA and FnrN. Unlike *R. leguminosarum*, *B. japonicum* uses its hydrogenase for energy generation also under free-living conditions. As a chemolithoautotroph it avoids the control by N_2, and adjusts hydrogenase gene expression to the availibility of H_2. This is recognized by *B. japonicum* and other aerobic H_2-oxidizing bacteria such as *R. capsulatus* and *R. eutropha* by a complex signal transduction chain consisting of an H_2-sensing [NiFe] hydrogenase, a histidine protein kinase and a response regulator.

The modular composition of [NiFe] hydrogenases allows their integration into various physiological pathways and frequently results in the occurrence of hydrogenase isoenzymes in a single cell. *E. coli* harbours four [NiFe] hydrogenases including H_2-evolving and H_2-consuming enzymes which are regulated separately. Hydrogenase 3 is a constituent of the formate hydrogen lyase system, its expression is governed by a σ^{54}-interacting response regulator whose activity is modulated by formate. The occurrence of multiple hydrogenases with distinct physiological functions in a single strain demands a high level of regulatory coordination between hydrogenase-specific and unrelated functions.

To obtain active hydrogenase positioned at a distinct cellular site the enzyme has to undergo multiple steps of protein-mediated maturation including [NiFe] cofactor

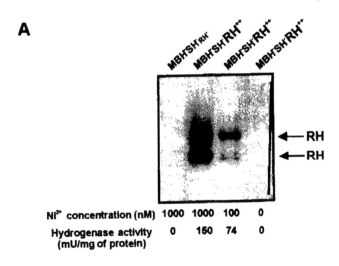

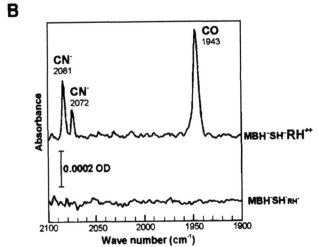

Figure 3.17 Catalytic and structural properties of the RH of *R. eutropha*. (A) Hydrogenase activity staining in gels, using phenazine methosulfate as the electron acceptor. Soluble extracts of cells devoid of the SH and the MBH, containing the RH at wild-type (RH$^+$) or overproduced level (RH^{++}) were separated by native PAGE. The cells were cultivated under various concentrations of NiCl$_2$. In parallel, hydrogenase activity was determined quantitatively by H2-dependent methylene blue reduction in soluble extracts. (B) FTIR spectra recorded for soluble extracts. The RH-overproducing strain shows the typical signals assigned to the diatomic ligands CN- and CO.

insertion, proteolysis and, if necessary, protein translocation. Thus, it is not surprising that the genes encoding the hydrogenase subunits are accompanied by hydrogenase-related accessory genes which form separate or extremely large transcriptional units. A fine tuning on the transcriptional and translational level is important to guarantee a balanced stoichiometric ratio of the various components.

We are at the very beginning of our understanding of the molecular background of hydrogenase regulation. The discovery of an [NiFe] hydrogenase which acts as an H_2 sensor, in concert with an unusual two-component regulatory system, has opened new insights into bacterial regulation and is an attractive model to be studied in more depth in the future. It is also desirable to learn more about the integration of hydrogenase control circuits into global regulatory networks which respond to redox, carbon, respiratory and nitrogen stimuli. Advances in functional genomics will hopefully offer novel tools to eludicate regulation of H_2 metabolism in organisms which are not yet accessible to genetic techniques.

Chapter 4

The assembly line

Robert Robson

With contributions from *August Böck, Marie-Andrée Mandrand-Berthelot* and *Long-Fei Wu*

4.1. Introduction

In recent years, biochemical, genetic and molecular biology studies have revealed that complex pathways are required in order that nascent hydrogenases mature into physiologically competent enzymes. This chapter covers the pathways which are required for the insertion of Fe, Ni and, in some enzymes, Se at the active site and for the transport of the enzyme into its correct physiological compartment. Some aspects of the maturation process, e.g. the formation and insertion of the bimetallic Ni-Fe site, are specific for hydrogenases but other aspects, e.g. the formation of [4Fe-4S] clusters and the membrane translocation pathway, are shared with other metalloproteins. This chapter does not address the biosynthesis of certain prosthetic groups, e.g. flavin adenine nucleotides (FAD, FMN), which though found in some hydrogenases are present in many other enzymes. Likewise there are many Fe-containing proteins in cells and Fe acquisition and metabolism has been reviewed extensively, and will only be mentioned herein. However, Ni is present in a few enzymes: hydrogenases, CO dehydrogenase and urease. Likewise Se occurs in relatively few enzymes and its insertion into proteins such as hydrogenases and formate dehydrogenase occurs via a remarkable co-translational mechanism. It is widely accepted that specific proteins known as chaperonins are required to control the folding of many proteins *in vivo*. There is much evidence to show that additional proteins, the so-called 'accessory proteins', are required to insert requisite metal centres in a manner which must be coordinated with the folding pathway. The complexity of these processes explains in part the usually high numbers of genes specifically required for hydrogenase activity. The complexity of the maturation processes for hydrogenases has implications for biotechnology. For example, to enhance expression of hydrogenase activity in any particular organism we must boost not only the production of the structural genes *per se* but may also need to enhance all those factors required for maturation.

4.2. General mechanisms of metal incorporation into proteins and enzymes

In the simplest of metalloproteins, spontaneous association can be observed *in vitro*. For example, in 1Fe-containing rubredoxins and small [2Fe-2S] and [4Fe-4S] ferredoxins reconstitution occurs under reducing conditions with high concentrations of Fe and sulfide. However, these examples are exceptional and it is unlikely that such

Table 4.1 Modes of metal insertion into protein

Mode	Examples
A. Spontaneous association	Zn^{2+}-dependent proteases
Ferredoxins	
B. Involvement of accessory proteins	
● Metal donor proteins	Cu-binding protein for N_2O reductase
● Chaperones	NifY for nitrogenase
	NarJ for nitrate reductase
● Synthesis of metal-sulfur centers	NifU and NifS for nitrogenase
	IscU and IscS for general [Fe-S] proteins
● Synthesis of metal-organic cofactors	Molybdopterin-cofactor in dehydrogenases
	FeMo-cofactor for nitrogenase
● Covalent insertion of metal-organic cofactors	Cytochrome *c* heme-ligase

high concentrations of Fe and sulfide exist *in vivo*. Furthermore, reconstitution is catalysed at least *in vitro* by sulfur transferases such as rhodanese (Bonomi *et al.* 1985). Hence, it is usually the case that metalloprotein/metalloenzyme assembly requires specific accessory proteins. We have become aware of the need for such accessory proteins either because structural proteins fail to contain any or the appropriate metal when overexpressed and/or because metalloenzymes structural gene operons in prokaryotes often contain additional genes in which mutations leads to inactive and often unstable enzymes. Some metal incorporation systems are listed in Table 4.1 and include systems for the insertion of metal-containing organic prosthetic groups, e.g. tetrapyrroles and pterins, metal donor proteins which present the metal-containing ligand to the enzyme, chaperones that control the folding of the metalloprotein, and relatively recently discovered systems for the synthesis of non-heme Fe or [Fe-S] clusters which are some of the most abundant metal centres in nature.

4.3. Metal acquisition

All organisms compete for metals in their natural environments. Such is the competition for metals that microorganisms and even plants may secrete a wide variety of high affinity and highly specific sequestrating agents and may possess several transport systems with different affinities and specificities to internalise those metals and their complexes. In some cases organisms can store or sequester these metals internally in storage proteins which also serves to nullify their potential toxicity.

4.3.1. Iron

Iron is essential for life and is required for many different types of iron-containing proteins. Microbes and other organisms go to extraordinary lengths to acquire Fe. Many microbes secrete specific and high affinity Fe chelators known as siderophores. More than 200 are known in bacteria alone (Neilands 1981). Siderophores overcome the problem of the low solubility of FeIII especially in oxidising environments, and microbes also use siderophores as a means of competing for iron since they make specific transport systems which specifically bind their own Fe-siderophore complexes and in some cases even the siderophore complexes of competitors. The importance of iron is further illustrated by the occurrence of eight different Fe transporters in *Escherichia coli* (Earhart 1996). Bacteria are unable to store many metals yet they can store Fe in ferritins and bacterioferritins (Andrews 1998).

4.3.2. Nickel

Microbes do not appear to produce specific chelating agents to sequester Ni from the environment; however, if they exist they are most likely to occur in obligate H_2 consuming organisms, e.g. methanogens. Ni^{2+} is known to be taken up on Mg^{2+} carriers. However, it is clear from studies in two organisms that organisms may possess highly specific Ni uptake systems. Both systems came to light through the study of mutants which require elevated levels of nickel in the medium to express normal levels of hydrogenase activity.

Nic$^-$ mutants of *Ralstonia eutropha* show low levels of the Ni hydrogenase, urease and lowered rates of nickel transport and require high levels of nickel in the medium for autotrophic growth (Eberz *et al.* 1989). The mutations were mapped to the *hoxN* locus on the megaplasmid which encodes a putative protein of 33 kDa which resembles a hydrophobic integral membrane protein and which probably is a high-affinity Ni-specific porter (Eitinger and Friedrich 1991).

In *E. coli*, mutants in a specific nickel uptake system were isolated as being pleiotropically defective in hydrogenase-linked fermentative and respiratory functions (Wu and Mandrand-Berthelot, 1986), a phenotype which was restored by the addition of 0.5 mM nickel to the medium. The mutants clustered in a locus called *nik* which has been cloned and sequenced and found to direct the synthesis of five proteins NikA, NikB, NikC, NikD and NikE that show high sequence identity to a family of transport systems named ABC (for ATP-Binding Cassette) transporters which function by coupling energy (released by the hydrolysis of ATP) to the transport process (Navarro *et al.* 1993). Mutants in the *nik* locus have intracellular concentrations of Ni^{2+} two orders lower than the wild type. When exogenous levels of Mg^{2+} are low, the mutation is suppressed probably because Ni is imported via the Mg^{2+} transporter. NikA encodes a soluble periplasmic metal-binding protein with a very high affinity for Ni^{2+} over $CO_2{}^+$, Cu^{2+} and Fe^{2+} (de Pina *et al.* 1995). Expression of the *nik* locus is induced under anaerobic growth conditions by the product of the regulatory gene *fur* and it is repressed by high extracellular Ni. This regulatory effect requires the *nikR* gene downstream of *nikE* which encodes a small protein of 15 kDa that potentially acts as a transcriptional repressor of the *nikABCDE* cluster (de Pina *et al.* 1999). The *nikR*-encoded protein is a member of the ribbon–helix–helix family of transcriptional factors (Chivers and Sauer 1999). It is functionally related to the ferric uptake regulation (Fur) protein that controls bacterial iron responsive promoters. Homologues of NikR have been found in several enteric bacteria and in many archaeal genomes which are known to possess one or several Ni metalloenzymes (Eitinger and Mandrand-Berthelot 2000). A scheme for Ni acquisition in *E. coli* is shown in Fig. 4.1.

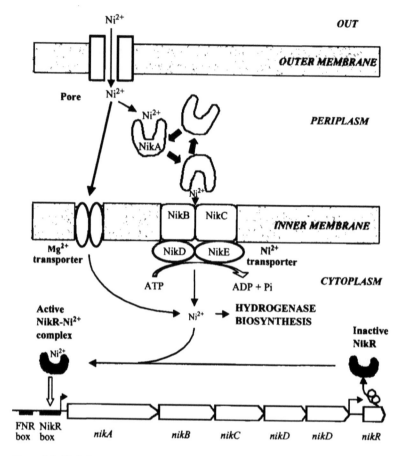

Figure 4.1 Nickel transport in *E. coli*. The lower part of the figure shows the *nik* gene cluster from *E. coli* which encodes a nickel specific ABC-type transporter and the repressor protein (NikR). The likely roles of the various *nik* gene products are indicated in the upper part of the figure. The transporter is encoded by the *nikBCDE* genes whilst a periplasmic nickel-binding protein is encoded by *nikA*. The system is expressed when nickel is low. High nickel represses expression via NikR and under these conditions nickel is transported via the Mg transporter (top left).

4.4. Accessory genes involved hydrogenase maturation

4.4.1. The role of housekeeping genes

Several genes have been identified as specifically required for the maturation of the NiFe(Se) hydrogenases. However, at present no genes have been identified which are specifically required for the insertion of Fe into the Fe-only hydrogenase (Chapter 2). Nevertheless given the complexity of the metal centres in these proteins, it is likely that cellular functions are required and these may have 'housekeeping' roles in the insertion of metals and maturation of other proteins. Indeed the *iscU*, *iscS*, *iscA*, *hscB*, *hscA*, *fdx* gene clusters which are conserved in many bacteria (Zheng *et al.* 1998;

Nakamura *et al.* 1999; Takahashi and Nakamura 1999) may be required for assembly and insertion of the [Fe-S] clusters in several different classes of hydrogenases although there is no direct evidence for this.

The discovery of genes involved in the maturation of hydrogenases arose from the characterisation of mutants in several organisms especially in *E. coli* affected in the expression of two or more hydrogenases or other enzymes. Early studies of mutants in hydrogenase in *E. coli* revealed a number of loci (*hydA*, hydB, hydF) as being required for hydrogenase activity (Sankar *et al.* 1985; Waugh and Boxer 1986; Sankar and Shanmugan 1988a,b). In *E. coli* one such locus (hydB) was originally mapped close to the genes for formate hydrogen lyase (the *fdh* or *fhl* genes) and the associated hydrogenase encoded by the *hyc* genes (Chapter 2). It was demonstrated that the gene cluster consisted of five genes called *hyp* (pleiotropic for hydrogenase) (Lutz *et al.* 1991). The genes *hypA*, *hypB*, *hypC*, *hypD* and *hypE* are probably co-transcribed and their properties and possible functions in the maturation of hydrogenase are listed in Table 4.2.

A second locus *hydA* which is required for all three hydrogenases in *E. coli* appears to contain an operon of two genes, hydN required for maximal activity of formate dehydrogenase and *hypF* required for activity of hydrogenases 1, 2 and 3 (Maier *et al.* 1996). Mutations in *hypB*, *hypD*, *hypE*, and *hypF* are pleiotropic for all three characterised hydrogenases, i.e. hydrogenases 1, 2 and 3 encoded respectively by the *hya*, *hyb* and *hyc* genes. The *hypB* mutant is phenotypically corrected by the provision of high levels (0.6 mM) of Ni to the medium (Waugh and Boxer 1986). However, mutations in *hypA* and *hypC* do not affect all three hydrogenases: HypA seems to be required for hydrogenase 3 only whilst a homologue HybF encoded within the *hyb* operon appears to function for the periplasmic hydrogenases 1 and 2. HypC appears to be a specific chaperonin for hydrogenase 3. Most if not all these *hyp* genes occur in all prokaryotes which contain NiFe hydrogenases but interestingly they do not seem to occur in the genome of *Thermotoga maritima*, an organism which appears to contain an Fe hydrogenase only. In most cases, especially amongst the Proteobactereaceae, e.g. *R. eutropha*, *Azotobacter* sp., *Rhizobium leguminosarum*, *Bradyrhizobium japonicum* (Chapter 2), the *hyp* genes are clustered in operons with a conserved gene order similar to that found in *E. coli* except that *hypF* is inserted

Table 4.2 Properties and putative functions of Hyp-proteins from *Escherichia coli* and other bacteria

Protein	Size (kDa.)	Property	Function
HypA	13.2	8.6% cysteine	Specific for Hyd3: HybF is homologue for Hyd1 and Hyd2
HypB	31.6	Homodimer, soluble	GTPase, Ni donor or storage
HypC	9.7	Soluble, oligomerises	Chaperone of large subunit of Hyd3
HypD	41.4	Soluble monomer, [3Fe-3S] or [4Fe-4S] cluster	HybG is homologue: pleiotropic
HypE	33.7	Soluble, monomer	Pleiotropic
HypF	81.5	Soluble, monomer. Possible acylphosphatase domain. No metal, putative zinc finger	Synthesis of CO and CN ligands from carbamoyl phosphate
HypX	62.3	Contains THF binding motif	Unknown

between *hypC* and *hypD*. However, the operon structure is not conserved in all bacteria. For example, quite remarkably, in *Methanococcus jannaschii* all the genes are scattered singly around the chromosome as discussed earlier (Chapter 2). It is also interesting to note that *R. eutropha* contains duplications of the *hypA* and *hypB* and two dissimilar *hypF* genes which are partially interchangeable as discussed further below (Wolf *et al.* 1998).

4.4.2. The hypA protein (HypA)

The *hypA* proteins are small proteins of between 109 and 121 residues. They are cysteine rich (e.g. nine cysteinyl residues in the *E. coli* example); only four cysteinyl residues are conserved throughout all examples. These are arranged into two -cys-X-X-cys- motifs which suggests that these proteins may be redox proteins potentially carrying an [Fe-S] cluster.

4.4.3. The hypB protein (the HypB family)

The *hypB* genes encode proteins of between 217 and 361 amino acid residues, which are quite highly conserved in different organisms and appear to consist of at least four domains in the eubacterial examples (Fig. 4.2), the first two of which appear to be absent in the archaeal examples. Domain 1 consists of a short region at the immediate N terminus and contains three invariate cysteinyl residues. Domain 2 is quite variable. It appears unexceptional in the case of the *E. coli* protein and contains only three histidinyl residues, but in the example from *B. japonicum* this domain is extraordinary in that it contains twenty-four histidinyl residues punctuated by aspartatyl or glycyl residues in a stretch of thirty-nine residues. Polyhistidinyl stretches are found in HypBs from many other organisms, e.g. *Rhodobacter capsulatus*, *Azotobacter* sp., *B. leguminosarum*. The multiplicity of histidinyl residues suggests a role in binding divalent metals, especially Ni and indeed artificial poly His-tag fusion proteins can be created by genetic manipulation that allow the purification of the required protein by Ni affinity chromatography (Rey *et al.* 1994). HypB proteins are known to bind and potentially store Ni^{2+} (Rey *et al.* 1994, Olson *et al.* 1997). Domain 3 of HypB contains a putative nucleotide-binding fold and the *E. coli* protein has been shown to hydrolyse GTP *in vitro* (Maier *et al.* 1993), an activity which is required for nickel insertion into hydrogenases *in vivo*. It is interesting to note that the HypBs in the Archaea, e.g. *M. jannaschii* and *Archaeoglobus fulgidus*, consist of the domains 3 and 4 only. Similar proteins are also located in gene clusters for other nickel enzymes, e.g. the UreG protein involved in Ni insertion into urease and CooC for maturation of CO dehydrogenase. In the latter case, a histidine-rich protein resembling domain 2 is encoded by CooJ.

4.4.4. The hypC proteins (the HypC family)

The *hypC* proteins are all relatively small proteins of between 75 and 108 amino acid residues which do not contain obvious metal-binding motifs. As will be discussed below (Section 4.5), these proteins may act as chaperones for hydrogenases during the process of maturation. This has been suggested from the studies of the interaction

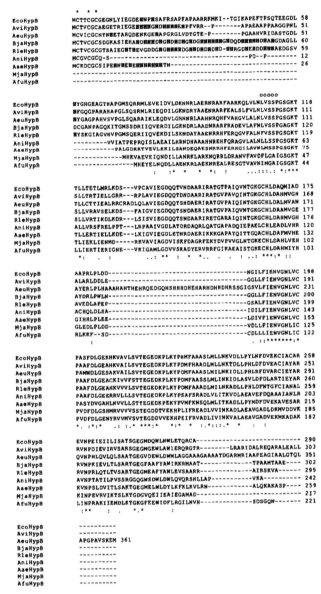

Figure 4.2 Multiple alignment of HypB proteins from Eubacteria and Archaea. Multiple alignment produced by ClustalW showing the remarkable variation in the histidinyl-rich regions between the HypB different proteins. Some interesting differences are the absence of the C-terminal histidine-rich domains in the archaeal proteins (Mja, *M. jannaschii*; Afu, *A. fulgidus*), the apparent lack of a histidine-rich domain in the *E. coli* protein (Eco) and the presence of the His-rich region at a different location in the protein in the example from *A. eutropha* (Aeu). Sources of other proteins in the alignment are as follows: Avi, *A. vinelandii*; Bja, *B. japonicum*; Rle, *R. leguminosarum*; Ani, *A. nidulans*; Aae, *A. aeolicus*. The Rossman fold of the putative nucleotide-binding domain is indicated by circles above the alignments.

between HypC and the large subunit, HycE of hydrogenase 3 from *E. coli*. Similar proteins also occur as accessory genes in some structural gene clusters encoding periplasmic hydrogenases, e.g. HoxL in *Azotobacter vinelandii* and *R. eutropha*, HupN and HupF in *R. capsulatus*, *B. japonicum* and *B. leguminosarum*. The *hybF* gene encoded in the gene cluster for hydrogenase 2 in *E. coli* may function for the maturation of both the periplasmic hydrogenases 1 and 2. Similar genes may be present in gene clusters for other metalloenzymes, e.g. *cooT* within the *coo* gene cluster for CO dehydrogenase, and *yrfC* in the fumarate reductase operon in *Proteus vulgaris*.

4.4.5. The hypD protein (HypD)

The *hypD* proteins are between 347 and 385 residues in length and show no significant identity to any other protein in the database. Furthermore the proteins does not seem to divide into any previously recognised domains or sub-domains. A homologue of HypD appears to be encoded by *hybG* within the cluster of genes for hydrogenase 2 in *E. coli*. Therefore the *hyb* operon appears to contain duplications of two *hyp* genes (see also HypC).

4.4.6. The hypE protein (HypE)

The *hypE* proteins are 302–376 residues long and appear to consist of three domains. Domain 1 shows sequence identity to a domain from phosphoribosyl-aminoimidazole synthetase which is involved in the fifth step in *de novo* purine biosynthesis and to a domain in thiamine phosphate kinase which is involved in the synthesis of the cofactor thiamine diphosphate (TDP). TDP is required by enzymes which cleave the bond adjacent to carbonyl groups, e.g. phosphoketolase, transketolase or pyruvate decarboxylase. Domain 2 also shows identity to a domain found in thiamine phosphate kinase. Domain 3 appears to be unique to the HypE proteins.

4.4.7. The hypF protein (HypF)

HypF proteins are relatively large proteins of between 729 and 806 residues which appear to contain several domains. Domain 1 corresponds to the N-terminal ~50 residues and shows sequence identity with acylphosphatases. This domain appears to be missing in the putative HypF from *Methanobacterium* sp. A second distinct domain occurs between residues ~100 and 200 and is characterised by four cys-X-X-cys motifs spaced over seventy-one residues and which resembles those found both in $Zinc^+$ finger proteins and [Fe-S] cluster binding proteins. The third domain which corresponds approximately to the C-terminal half of the protein is the only domain which is present apparently in the 394 residue long HypF1 protein from *R. eutropha*. However, *R. eutropha* contains a full length HypF (HypF2) in addition to HypF1 (Wolf *et al.* 1998). Originally, HypF proteins were thought to be involved in regulation given that $Zinc^+$ finger proteins act in transcriptional control; however, it is more likely that they are involved in maturation and that their effect on regulation may be indirect through their requirement for the maturation of nickel sensing hydrogenases (Colbeau *et al.* 1998).

4.4.8. The hypX protein (HypX)

A further gene, *hypX*, required for maturation of hydrogenase has been found in *R. leguminosarum*, *B. japonicum* and *R. eutropha* (Rey *et al.* 1996, Buhrke and Friedrich 1998) but does not seem to be present in *E. coli*. The *hypX*-encoded protein contains a region with sequence identity to N^{10}-formyltetrahydrofolate-dependent enzymes, e.g. phosphoribosylglycinamide formyl transferase which catalyses the fourth step in purine biosynthesis.

4.4.9. The hydrogenase specific endopeptidases (HybD, HyaD, HyI, HoxM, HupM, HupD)

Each NiFe(Se) hydrogenase structural gene cluster appears to include a gene encoding one of a family of polypeptides which range in length between 130 and 209 amino acid residues and which serve as highly specific endopeptidases for cleavage of the C terminus of the cognate hydrogenase large subunit. Mutations in these genes block cleavage of the C terminus of the large subunits and lead to inactive hydrogenases. In *E. coli*, *hycI* encodes the cognate endopeptidase for the *hycE*-encoded hydrogenase large subunit of the formate hydrogenase lyase complex (Rossmann *et al.* 1995). Purified *hycI* protein is a monomer and cleaves the HycE precursor when added to crude extracts of nickel-starved cells bearing a mutation in *hycI* but only when relatively high concentrations (400 µM) of Ni were added. Following cleavage, the protein was found to mature into an active enzyme in anaerobic conditions (Maier and Böck 1996). The crystal structure of HybD, the specific protease for hydrogenase 2 in *E. coli*, has been determined at the 2.2 Å resolution and it appears to be a distant member of the metzincin superfamily of zinc endoproteinases (Fig. 4.3). It is a twisted five-stranded β-sheet surrounded by three and four helices on each side respectively. A cadmium ion from the crystallization buffer was bound in a pseudo-tetragonal arrangement to Glu16, Asp62, and His93, and a water molecule. It was speculated that nickel may occupy a similar ligand environment during the cleavage event (Fritsche *et al.* 1999).

The cleavage site appears to be highly specific and requires that hydrogenase large subunits fold to a conformation which provides a ligand environment for nickel which may be similar to that of the native enzyme. Substitutions at any of the four conserved N- and C-cysteinyl residues block cleavage and also activity (Menon *et al.* 1993). Since purified HycI and HybD do not contain nickel (Theodoatou *et al.* 2000a), it appears that nickel must first bind to the hydrogenase subunit before cleavage occurs. In most enzymes, the site of C-terminal cleavage occurs between a histidinyl and a valinyl residue (Fig. 4.4). However, these residues are not entirely conserved as in the case of HycE where cleavage occurs between arginyl and methionyl residues. Some substitutions in the residues at the cleavage site are tolerated in hydrogenase 1 (Menon *et al.* 1993), and hydrogenase 3 (Theodoratou *et al.* 2000b) in *E. coli*. However, in HoxH, the large subunit of the NAD-reducing hydrogenase of *R. eutropha*, Ala for Pro replacement at residue 465, the first residue of the cleaved C-terminal peptide, permitted Ni-incorporation but blocked C-terminal proteolysis (Massanz *et al.* 1997). In HycE, two-thirds of the C terminus towards the cleavage site can be removed without blocking cleavage (Theodoratou *et al.* 2000b).

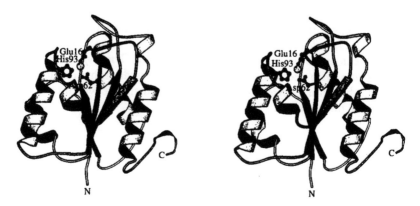

Figure 4.3 X-ray crystal structure of HybD, the specific protease for hydrogenase 2.

4.5. Pathway of maturation of NiFe(Se) hydrogenases

4.5.1. Outline of the maturation events

The pathway of maturation of the NiFe(Se) hydrogenases can be divided into several steps. Irrespective of the final location of the enzyme, the early events consist of co-translational insertion of selenocysteine (in the Se-containing hydrogenases), chaperonin assisted/controlled folding, insertion of the metal centres and subunit interactions. All these events probably occur in the cytoplasm even in those hydrogenases which are inserted into or translocated across membranes. Hydrogenase location is determined by the presence and absence of signal sequences usually on one subunit of the hydrogenase complex. The signal sequences contain the so-called twin-arginine motif which is recognised and exported by the relatively recently discovered *tat*-encoded secretion system which appears to be a generalised system for the secretion through membranes of folded proteins and protein complexes often with redox centres. A scheme outlining the role of the various *hyp* and other accessory gene products based on the proposed maturation pathway for hydrogenase 3 in *E. coli* and generalised to include potentially other NiFe(Se) hydrogenases is shown in Fig. 4.5.

4.5.2. Early events in maturation prior to C-terminal cleavage

The early events in the maturation of [Ni-Fe] hydrogenases in *E. coli* require the activities of the *hyp* gene products leading to the formation of a large subunit which is the substrate for the maturation endopeptidase discussed above. Maturation of the large subunit of hydrogenase 3 (L-SU) of *E. coli* occurs *in vitro* leading to the formation of an active enzyme complex (Maier and Böck 1996). Processing appears to be initiated by binding of the system-specific chaperone HypC to the uncleaved L-SU (pre L-SU) (Drapal and Böck 1998). The following events presumably require the synthesis of CO and CN as ligands and the incorporation of these together with the Fe and Ni atoms into the complex bimetallic site. The scheme suggests that the insertion of Fe precedes the insertion of Ni. The specific roles of most of the Hyp proteins are unclear at present. The HypB protein is not absolutely essential

```
AviHoxG    -ILRTLHSFDPCLACSTHVMSPDGQELTRVK
RcaHupL    -ILRTLHSFDPCLACSTHVMSAEGAPLTTVKVR
RleHupL    -ILRTIHSFDPCLACSTHVMSPDGQEMARVQVR
EcoHyaB    -ILRTLHSFDPCLACSTHVLGDDGSELISVQVR
AeuHoxG    -ILRTLHSFDPCLACSTHVMSAEGQELTTVKVR
EcoHybB    -VVRTIHSFDPCMACAVHVVDADGNEVVSVKVL
DgiHydB    -ILRTVHSYDPCIACGVHVIDPESNQVHKFRIL
DvuHydB    -ILRTVHSFDPCIACGVHVIDGHTNEVHKFRIL
DbaX       -VGRLVRSYDPCLGCAVHVLHAETGEEHVVNID
MvoVhcA    -MEMVIRAYDPCLSCATHTIGEEPKILSIHVCQGGKLIKTL
EcoHycE    -APLIIGSLDPCYSCTDRMTVVDVRKKKSKVVPYKELERYSIERKNSPLK
AvaHoxH    -VEAGIRAFDPCLSCSTHAAGQMPLHIQLVAANGNIVNQVWREKLGV

MvoVhuU    MVDETKLNLIEIVLRAYDPXYSCAAHMIVEDAEGNVVFEIVNDE

AeuHoxC    -VQHIVRSFDPCMVCTVH
```

Figure 4.4 C-terminal sequences of Ni-binding subunits in NiFe(Se) hydrogenases and C-terminal processing. The upper block of amino acid sequence alignments show the nascent C-terminal sequences of the large subunits of the NiFe hydrogenases from a range of different systems. The stars show the positions of the two cysteinyl (or selenocysteinyl) residues which provide ligands to the Ni. The arrow marks the position of the determined or presumed site of cleavage during the maturation process. AviHoxG, RcaHupL, RleHupL, AeuHoxG, DgiHydB, DvuHydB are the large subunits of the membrane-bound hydrogenases from A. vinelandii, R. capsulatus, R. leguminosarum, A. eutropha, D. gigas, D. vulgaris. EcoHyaB, EcoHybB, EcoHycE are the large Ni-bearing subunits of the NiFe hydrogenases 1 and 2 and the hydrogenase of the formate hydrogen lyase from E. coli. DbaX is the large subunit from the NiFeSe hydrogenase from Desulfobacter baculatus. AvaHoxH, the NiFe subunit of the bidirectional hydrogenase from Anabaena variabilis. AeuhoxC, cytoplasmic H$_2$-sensing NiFe hydrogenase from A. eutropha. The lower sequence MvoVhuU is the complete small peptide encoding the equivalent domain and cleavage site from the NiFeSe methyl viologen-reducing hydrogenase.

since mutations in *hypB* are overcome by the provision of high levels of Ni in the culture medium. Normally however, HypB probably acts to scavenge/sequester nickel present at low intracellular levels and probably to donate it in a GTP-hydrolysis-dependent step to a metal centre intermediate in the enzyme in which the N-terminal thiolates coordinate the iron and the C-terminal ones the nickel (Maier *et al.* 1995). Analysis of the various domains present in certain *hyp* proteins provides some clues that they may be involved in formation of CO and CN. There appears to be some resemblance between some of these proteins and enzymes involved in the fourth and fifth steps of purine biosynthesis. Domain 1 of HypX is most similar to a domain found in N^{10}-formyltetrahydrofolate-dependent enzymes involved in the metabolism of C$_1$ units, particularly phosphoribosylglycinamide formyl transferase which catalyses the fourth step in purine biosynthesis. Domains 1 and 2 of HypE show identity to enzymes involved which use or metabolise TDP, e.g. phosphoribosyl-aminoimidazole synthetase, which catalyses the fifth step in purine biosynthesis. HypF contains a domain similar to that found in acylphosphatases which suggests that a phosphorylated intermediate(s) may be involved. Recently, Paschos *et al.* (2001) have presented evidence that carbamoyl phosphate is the precursor of the -CN and -CO ligands. HypA and HypF both contain cysteinyl motifs which suggest that they contain potentially redox active metal centres. All the evidence suggests that the bimetallic site must be completed *in situ* before cleavage at the C terminus of the large subunit can occur.

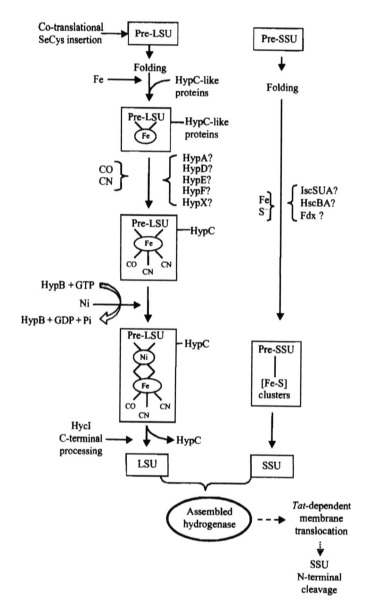

Figure 4.5 Postulated generalised pathway for maturation of NiFe(Se) hydrogenases. The pathway envisages that the assembly of the heterodimeric NiFe(Se) hydrogenase in which folding and maturation of the small subunit (SSU) occurs independently of the large subunit (LSU). The postulated involvement of various gene products is indicated at various stages, many of which are speculative at this time. The assembly of the dimer probably occurs after C-terminal cleavage of the pre-LSU and it is envisaged that the LSU is transported across cell membranes in the dimeric state coupled to the translocation of the pre-SSU after which the N-terminal signal sequence is removed. Various gene products are discussed in the text.

4.5.3. The role of the C-terminal cleavage of the large subunits of the NiFe(Se) hydrogenases

As discussed above, the C termini of the large subunits of the NiFe(Se) hydrogenases undergo proteolytic cleavage which can be detected as a slightly increased mobility of the polypeptides on SDS-PAGE gels. Cleavage is blocked physiologically by limiting intracellular nickel levels which has been achieved experimentally by growth of wild-type strains in nickel-deficient medium or by using mutants blocked in nickel uptake. Under such conditions unprocessed protein accumulates. Processing is initiated both *in vivo* and *in vitro* when nickel is restored so allowing the requirements and kinetics of the C-terminal cleavage to be determined. However, cleavage is also blocked by mutations in *hyp* and other accessory genes within structural gene clusters, and also in the structural genes *per se*.

In *A. vinelandii*, C-terminal cleavage occurs in crude extracts after restoration of nickel and requires hours rather than minutes, as is the case *in vivo*. It does not require *de novo* protein synthesis and surprisingly is not redox or O_2 sensitive. Also, it is not inhibited by well-established inhibitors of metallo- or serine protease families (Menon and Robson 1994).

From the behaviour of the various site-directed mutants of the C-terminal domains in HoxH of the NAD-reducing hydrogenase of *A. eutropha* (Massanz *et al.* 1997) and the HycE protein in *E. coli* (Theodoratou *et al.* 2000b), the primary role of the C terminus appears to be to block the interaction between subunits until the active site metals and ligands have been incorporated. This is required because, as in the case of the NiFe hydrogenase from *Desulfovibrio gigas*, the active site is probably buried at a subunit's interface. The C terminus might also serve to hold the structure in a ready conformation for metal insertion. It is not known if the C terminus folds to give a precise structure although secondary structure predictions suggest a β-sheet in the case of the *A. vinelandii* HoxG. However, there appears to be very little conservation of residues within the cleaved peptides which range in length from thirteen in HoxG from *A. vinelandii* to thirty-two residues in HycE from *E. coli* and in the latter example approximately two-thirds can be removed from the C terminus without affecting cleavage or further enzyme maturation.

4.6. Targeting hydrogenases

Hydrogenases are found in different cellular locations ranging from the cytoplasm, the periplasm, the cytoplasmic and thylakoid membranes and the hydrogenosomes present in a few eukaryotes.

4.6.1. N-terminal twin-arginine leader sequences

In several of those hydrogenases which are exported into the periplasm, it has been shown directly that their small subunits are cleaved at their N termini in a manner consistent with the N terminus acting as a signal sequence to direct export of the protein. In general, secretion or target signals are usually short and show distinct amphipathic structures comprising domains of net-positively charged residues flanking an essentially hydrophobic region. However, no consensus amino acid sequence emerges and individual amino acid substitutions do not have a marked effect on signal activity. However in the hydrogenases, the sequence which is removed is unusually long (~50 amino acid residues) and more closely resembles mitochondrial or chloroplast target signals. Also in the cases of periplasmic hydrogenase leader sequences a consensus sequence has been identified corresponding to -serine/threonine-arginine-arginine-x-phenylalanine-leucine-lysine- (Fig 4.6). The distinctive feature of this consensus sequence is the twin-arginyl motif by which such leader peptides are known. This consensus box is also common in signal sequences of other periplasmically located metalloenzyme complexes, e.g. trimethyl N-oxide (TMAO) reductase and nitrous oxide reductase.

4.6.2. Involvement of the tat genes

The importance of this consensus sequence was first shown when mutation of the first of the conserved argininyl residue to glutamate within the leader sequence of the [NiFe] hydrogenase from *Desulfovibrio vulgaris* was found to block export when fused to β-lactamase, a protein often used as a reporter to study protein translocation in bacteria (Nivière *et al.* 1992). Until recently it has been considered that all proteins are secreted across the cytoplasmic membrane by the *sec* gene encoded pathway which is well characterised for export of simple proteins that lack prosthetic groups, e.g. β-lactamase or alkaline phosphatase. However, the *sec* system appears to require that protein substrates be unfolded or in a molten state for export. However, when the structure of the periplasmic NiFe hydrogenase from *D. gigas* was revealed (Chapter 6), it was difficult to envisage how the *sec* system could be involved given that the enzyme is a dimer, that only one of the subunits carries an obvious leader sequence, and that the insertion of the metal centres requires cytoplasmic factors and possibly a nearly mature folded state of the protein. However, Santini *et al.* (1998) showed that mutants defective in essential components of the *sec* system were unaffected in their export of TMAO reductase. Earlier, twin-arginine motif containing signal peptides were recognised as important in the identification of proteins transported across the chloroplast and thylakoid membranes via a ΔpH-dependent mechanism (Chaddock *et al.* 1995) which seemed able to translocate folded proteins (Creighton *et al.* 1995). A component of the ΔpH-dependent thylakoid translocase, Hcf106, from Maize showed sequence identity to a gene that had recently been identified as affecting the activity and localisation of hydrogenase in *Azotobacter chroococcum* (Settles *et al.* 1997).

Proteins with sequence identity to Hcf106 were also found in the genomes of several bacteria including *E. coli* where two potential homologues *yigT* and *ybeC* occur at two different loci. Renewed analysis of the former loci (yigT, yigU and yigW) in *E. coli* revealed a cluster of not three but four potential genes renamed *tatA*, *tatB*, *tatC*, and *tatD* in which *tat* stands for twin-arginine translocation, which potentially encode proteins of 89, 171, 258 and 264 amino acid residues. TatA is predicted to have a structure consisting of an N-terminal transmembrane helix followed by a possible amphipathic helical domain. TatC is predicted to be an integral membrane protein composed of six transmembrane helices (Sargent *et al.* 1998). A strain of *E. coli* bearing mutations in both *tatA* and *tatE* (formerly *ybeC*) is defective in export of several periplasmic enzymes which bind a variety of redox cofactors including hydrogenases 1 and 2. This leads to the conclusion that the periplasmic hydrogenases are

```
DvuHydA    MKISIGLGKEGVEERLAERGVSRRDFLKFCTAIAVTMGMGPAFAPEVARALMGPRRPSVV 60
EcoHybA    --------MTGDNTLIHSHGINRRDFMKLCAALAATMGLSSKAAAEMAESVTNPQRPPVI 52
AviHoxK    -----MSRLETFYDVMRRQGITRRSFLKYCSLTAAALGLGPAFAPRIAHAMETKPRTPVL 55
AeuHoxK    -------MVETFYEVMRRQGISRRSFLKYCSLTATSLGLGPSFLPQIAHAMETKPRTPVL 53
BjaHupS    ----MGAATETFYSVIRRQGITRRSFHKFCSLTATSLGLGPLAASRIANALETKPRVPVI 56
RleHupS    -----MATAETFYDVIRRQGITRRSFTKFCSLTAASLGFGPGAATAMAEALETKERVPVI 55
EcoHyaA    -----MNNEETFYQAMRRQGVTRRSFLKYCSLAATSLGLGAGMAPKIAWALENKPRIPVV 55
RcaHupS    -----MSDIETFYDVMRRQGITRRSFMKSVRSPQHVLGLGPSFVPKIGEAMETKPRTPVV 55
AaeHupS    --------METFWEVFKRHGVSRRDFLKFATTITGLMGLAPSMVPEVVRAMETKPRVPVL 52
MmaVhtG    -MSTGMKNLTRTLESMDFLKMDRRTFMKAVSALGATAFLG-TYQTEIVNALEFAET-KLI 57
                  :       :  ** * *        :.   . :   ::         ::

DvuHydA    YLHNAECTGC------- 70
EcoHybA    WIGAQECTGCTESLLRA 69
AviHoxK    WLHGLECTCCSESF--- 69
AeuHoxK    WLHGLECTCCSESFIR- 69
BjaHupS    WMHGLECTCCSES---- 69
RleHupS    WMHGLECTCCSESF--- 69
EcoHyaA    WIHGLECTCCTESF--- 69
RcaHupS    WVHGLECTCCSESF--- 69
AaeHupS    WIHGLECTCCSESFIRS 69
MmaVhtG    WIHGSECTGCSE----- 69
```

Figure 4.6 Multiple sequence alignments of N-terminal sequences of NiFe(Se) hydrogenase small subunits. The alignments of the N-terminal sequences of the nascent NiFe(Se) hydrogenase small subunit using ClustalW clearly shows the conservation of twin-arginyl residues within the consensus sequence -arg-arg-X-phe-X-lys- considered to be important for export of hydrogenases via the *tat*- encoded secretion system. The residues around the N-terminal cleavage sites are not highly conserved consistent with cleavage being dependent on the Sec protease. Examples were as follows: membrane bound hydrogenases: DvuHydA, *D. vulgaris*; EcoHybA, *E. coli* (hydrogenase 2); AviHoxK, *A. vinelandii*; AeuHoxK, *A. eutropha*; BjaHupS, *B. japonicum*; RleHupS, *R. leguminosarum*; EcoHyaA, *E. coli* hydrogenase 1; RcaHupS, *R. capsulatus*; AaeHupS, *Aquifex aeolicus*; Mma VhtG, *Methanococcus mazeii*.

assembled in the cytoplasm in a potentially catalytically active form, and exported as such by the ΔpH-dependent twin-arginine (*tat*) translocase system, the details of which are under intensive research.

It is interesting to note that *R. eutropha* strain TF93 is pleiotropically affected in the translocation of redox enzymes synthesised with the twin-arginine leader peptides including the membrane-bound hydrogenase (MBH). In this strain MBH is active but is mislocalised to the cytoplasm but normal membrane localisation was restored after heterologous expression of the *tatA* gene from *A. chroococcum* in TF93 (Bernhard *et al.* 2000).

4.7. Conclusions, future directions and biotechnological implications

Maturation of members of the NiFe(Se) hydrogenase family is revealed as being a highly complex process which potentially involves as many as four phases. These include co-translational insertion of Se-Cys, the potential biosynthesis of the active site ligands CO and CN^-, the acquisition of metals and their presentation to the protein whose folding appears to be specifically controlled. These early events are followed by the C-terminal cleavage of the large subunit which apparently allows tertiary assembly and the subsequent targeting of the protein complex to its appropriate physiological location/compartment which may involve the novel *tat* gene encoded secretion system discovered only in recent years. Much needs to be learned about the specific roles of the various *hyp* genes. Are they involved in the biosynthesis of the CO and CN^- ligands? Do they form a scaffold on which some precursor of the active site is assembled before it is 'handed on' to the enzyme. What redox chemistry is involved in this process. Much needs to be learned about the mechanism of the *tat* system which is not confined to prokaryotes nor restricted for the membrane transloca-tion of hydrogenases.

Our knowledge of the assembly of the NiFe(Se) hydrogenases is incomplete and more or less confined to the Ni-containing subunits. Presumably the insertion of [Fe-S] and other centres into other subunits involves pathways which may be required more generally in the cell. For example, the *iscSUA-bscBA-fdx* genes might be required for manufacture/insertion of [Fe-S] clusters but there is no direct evidence for this. Our knowledge about the assembly of the Fe hydrogenases is even less com-plete. Possibly they require only housekeeping functions since until now no specific genes have been identified as being required for the assembly of these enzymes. This is probably because Fe hydrogenases have been found in organisms which are gener-ally not amenable to genetic analysis.

An important general implication of the need for complex maturation events is that if for biotechnological reasons it becomes necessary to overexpress hydrogenases, e.g. to increase production of H_2 either *in vivo* or *in vitro*, then it will probably not be sufficient to overexpress the structural genes *per se*. This was demonstrated in an early experiment where even though it proved possible to overexpress the subunits of Fe hydrogenase from *D. vulgaris* (Hildenborough) in *E. coli*, an active enzyme was not formed (Voordouw *et al.* 1987). In the case of the NiFe enzymes this may be even more problematical. There is no guarantee that increased levels of active enzyme would result from simply boosting the expression of all known maturation genes together with the cognate structural genes. It is clear that much work needs to be done on understanding the maturation processes.

Chapter 5

Hydrogenases and their activities

Richard Cammack

With contributions from *Simon P. J. Albracht, Fraser A. Armstrong, Boris Bleijlevens, Bart Faber, Victor M. F. Fernandez, Wilfred R. Hagen, E. Claude Hatchikian, Anne K. Jones* and *Harsh R. Pershad*

We now turn to the biochemistry of the hydrogenase molecule, and its catalytic activities. In order to discover how hydrogenase works, the standard biochemical approach is to study it in isolation. This means that the protein is extracted from the bacterial cells, purified and concentrated. The methods employed are standard for biochemical investigations but are complicated in some cases by the sensitivity of the enzymes to oxygen. In such studies it is always necessary to have a method of measuring the activity of the enzyme, known as an assay. There are many different assay methods for hydrogenase. Studies of the ability of hydrogenases to catalyse different types of reaction have proved to be rewarding, since they show that the enzymes can exist in a number of forms, with different catalytic properties. We will then discuss how these different forms were identified, and how the enzymes could be prepared in these different states for further examination. This has provided invaluable information about the catalytic activity of the enzyme.

An outline of the biochemical methods used for isolation and biochemical characterization of hydrogenases is given here. Further details are given in a review (Cammack *et al.* 1994).

5.1. Isolation of hydrogenase from cells: *Desulfovibrio gigas*

Ideally, we would like a source of enzyme in which the hydrogenase is present in large amounts. Unfortunately, although bacteria produce sufficient hydrogenase for their needs, this is not a large amount of protein, as the enzyme is highly active. So far, it has not been possible to overproduce a hydrogenase in a stable and active form by cloning of a hydrogenase-overproducing bacterial strain, which is a standard process in biotechnology for producing enzymes in bulk. This may be because the complex process that the bacteria use for assembly of the active sites is unable to cope with the large amounts of apoenzyme (Chapter 4). Thus, we have to culture large volumes of bacteria to extract a small quantity of pure material.

The purification of [NiFe] hydrogenase from *Desulfovibrio gigas* illustrates the general procedures required for the isolation of hydrogenase, developed from an original procedure of Hatchikian and LeGall (1978). Isolation of hydrogenase begins with the growth of *D. gigas* cells in a 300 l fermentor (Fig. 5.1). After 44 h of growth on lactate and sulfate, 260 g of cells is produced, and is recovered by centrifugation. The enzyme is extracted by breaking open the bacterial cells by pressure treatment, and then the solution is clarified by centrifugation. In the case of *D. gigas*, use can be

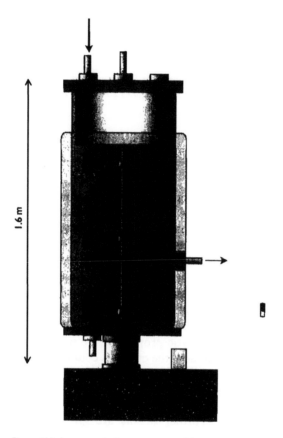

Figure 5.1 Large-scale fermentor used for culture of *D. gigas*, with tube of hydrogenase extracted from a 300 l batch. Height=1.6 metres.

made of the fact that the enzyme is located in the periplasmic space of the cell, between the inner and outer membranes. The outer cell wall of the cells can be disrupted by freezing and thawing in a slightly alkaline buffer containing the metal-ion chelator ethylenediamine tetra-acetate (EDTA), which releases the hydrogenase into solution. The inner cell membrane, which contains most of the other constituents of the cell, remains intact and can easily be removed by centrifugation (Fig. 5.2).

The enzyme is then purified by a sequence of chromatographic columns, which separate the proteins by their molecular size and charge. The steps of the purification procedure are summarized in Fig. 5.2. Methods for separating proteins are being improved, with the development of new chromatography media, but the principles remain the same. All proteins are polymers of the same twenty amino acids, but they differ in size, electrostatic charge and surface properties. The solution of proteins extracted from the bacterial cells is put through a series of separation procedures that sift the mixture of proteins on the basis of their molecular properties. These procedures have to be gentle, to protect the delicate molecular structure, but as selective as possible. The proteins are kept in a buffer solution that keeps the pH stable, and the purification is usually done at ice temperatures to avoid denaturation.

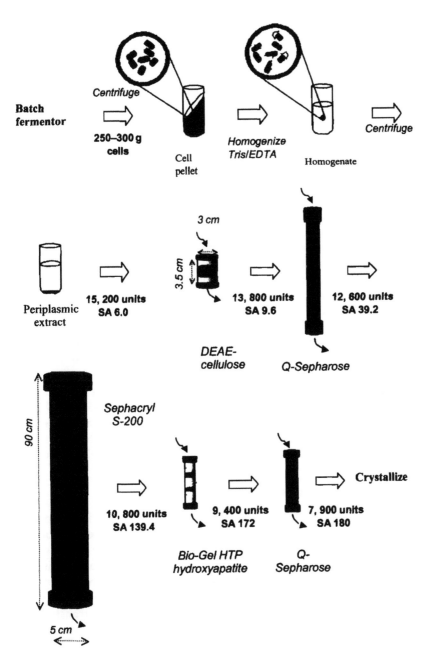

Figure 5.2 Flow diagram of purification of hydrogenase from D. fructosovorans. At each stage of fractionation, the enzyme activity and protein concentration are measured for each fraction. The fractions with the greatest specific activity (S.A., units/milligram protein) are taken onto the next stage.

At each stage during the purification procedure, samples of the solution are taken and assayed for enzyme activity, in order to select the fractions with high purity and quantity of enzyme. There is a trade-off between high purity and high yield of enzyme in the final product. For production of crystals, a very high purity and intact molecular structure are required. For other experiments, such as spectroscopy, it may not be necessary to remove other proteins that do not show up in the spectra, in which case the fractions are chosen for highest yield at the expense of purity. In general, the fresher the protein preparation, the better.

After the production of the solution of proteins, the purification of the protein requires six different steps. Silica gel powder is added to the solution and filtered off, to adsorb large quantities of cytochrome c_3. Several steps of column chromatography follow.

Ion-exchange materials with the diethylaminoethyl (DEAE) substituent, such as DEAE-cellulose, separate the proteins on the basis of molecular charge. DEAE is a chromatographic medium with a positive charge, which attracts molecules that have a negative surface charge, such as hydrogenase. The proteins are selectively removed from the ion-exchange columns by washing with salt solution.

Gel-filtration chromatography materials such as Ultrogel AcA44 separate molecules on the basis of molecular size. Larger protein molecules are unable to penetrate the small pores in this material and pass through the column more rapidly than the smaller protein molecules.

Hydroxylapatite column chromatography relies on the selective adsorption of the protein onto the surface of calcium phosphate. The final stage of purification, on a high-performance DEAE column, is carried out just prior to crystallization of the protein.

After the column chromatography steps, about half of the original enzyme activity is recovered, with a purification factor (increase of enzyme activity per milligram of protein) about sixty-fold. From the original 300 l of cell culture, 30–50 mg of hydrogenase are produced. The enzyme is judged to be pure by both polyacrylamide gel electrophoresis and analytical ultracentrifugation and it is suitable for crystallization (Chapter 6). The whole purification process involves six stages, and takes about a week.

5.2. Anaerobic techniques

For many manipulations of hydrogenases it is necessary to avoid the presence of oxygen, because it either interferes with the reductants being used, or poisons the enzyme. Some hydrogenases, notably the [Fe-] hydrogenases, have to be isolated in the absence of oxygen. In an aerobic atmosphere, this is difficult to achieve in the laboratory. In many cases, 'anoxic' conditions may suffice. Solutions are freed of oxygen by evacuation, or flushing with an inert gas such as nitrogen or argon. The solutions are kept in enclosed vessels purged with nitrogen or argon, and transferred with syringes though rubber septa. In practice, it is difficult to prevent the leakage of oxygen into the system through joints and plastic tubing. Reducing agents such as dithionite will remove oxygen efficiently but will also react to form hydrogen, which may interfere with the experiment. Purging the solution with knallgas (a non-explosive mixture of nitrogen and hydrogen) is another method of keeping the hydrogenase solution reduced. A more secure system is to use an anaerobic glovebox, in which the nitrogen atmosphere is continuously purified by passage through columns of oxygen-absorbing materials (Fig. 5.3).

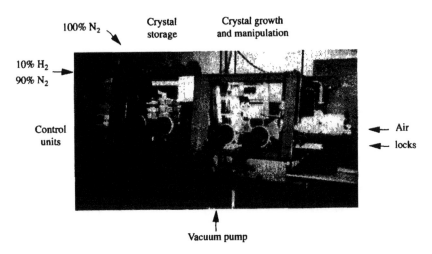

Figure 5.3 Anaerobic glovebox used for purification of oxygen-sensitive enzymes.

5.3. Assays of activity – H₂ production and consumption, isotope exchange

The quantity of an enzyme is generally measured by its ability to catalyse a reaction. A method of measuring activity is known as an assay. For hydrogenases, there are an exceptionally large number of different types of assay, since they catalyse different types of reaction. Sometimes we can learn more about the way that hydrogenases work, by measuring the rates at which the different reactions are catalysed under different conditions, varying the pH for example. Technical details of the different assay methods are described in the Methods in Enzymology series (Cammack *et al.* 1994).

Four general methods may be used for the assay of hydrogenases based on their ability to catalyse the following reactions.

5.3.1. Evolution and oxidation of H₂

Production of H₂

$$D_{red} + 2H^+ \leftrightarrow D_{ox} + H_2, \tag{1}$$

where the electron donor D is a low-potential compound such as cytochrome c_3 or methyl viologen.

Oxidation of H₂

$$H_2 + A_{ox} \leftrightarrow 2H^+ + A_{red}, \tag{2}$$

where the electron acceptor A may be either a low-potential compound, such as cytochrome c_3 or methyl viologen, or a higher-potential compound such as methylene blue or dichloroindophenol (DCIP). Hydrogenases in different states can be distinguished by their different reactivities with these different types of acceptors (Section).

Reactions may be followed by observing the oxidation of D or reduction of A, by spectrophotometry. A simple assay uses the dye methyl viologen, which turns blue on reduction. Alternatively, hydrogen production or consumption can be measured in various ways:

1   Amperometrically, using a Clark-type electrode. This is a platinum electrode surrounded by a gas-permeable membrane.
2   By gas pressure in a manometer.
3   By gas chromatography with a thermal-conductivity detector.

A practical difficulty should be mentioned here, which is also relevant to all processes that require exchange of H_2 between the gas and solution phases. Methods (2) and (3) depend on measuring the partial pressure of hydrogen in the gas phase over the solution. They depend on the assumption that the gas phase is in equilibrium with the solution, and this is not necessarily the case, unless there is vigorous mixing. The rate of exchange of gas molecules is surprisingly slow, particularly at low H_2 concentrations.

5.3.2. Deuterium or tritium exchange reactions with H$^+$, in the absence of electron donors or acceptors

For example, with deuterium gas ($^2H_2$ or D_2)

$$^2H_2 + 2\,^1H_2O \leftrightarrow {}^1H_2 + 2\,^1H^2HO \tag{3}$$

and

$$^2H_2 + {}^1H_2O \leftrightarrow {}^1H^2H + {}^1H^2HO. \tag{4}$$

This activity has been detected even in apparently dry samples of hydrogenase.

Of the isotopes of hydrogen, deuterium can be detected by mass spectrometric analysis of masses 2, 3 and 4 using a membrane-inlet mass spectrometer (Fig. 5.4).

As an alternative, exchange of ^1H_2O with tritium gas, $^3H_2$, or tritiated water with H_2, can be measured by radioactive counting. $^3H_2$ is counted with a gas ionization chamber, or ^3H_2O can be measured by solution scintillation counting.

5.3.3. Interconversion of ortho- and para-H$_2$

These two isomeric forms of H_2 are physically different because of the pairing of the nuclear spins on the hydrogen nuclei in the molecule. They can be physically separated by chromatography at very low temperatures. Generally to convert one isomer to the other requires a cleavage of the H-H bond. As with the hydrogen-isotope-exchange reactions, no external acceptor or donor is required.

$$oH_2(\uparrow \uparrow) + H_2O \leftrightarrow pH_2(\uparrow \downarrow) + H_2O. \tag{5}$$

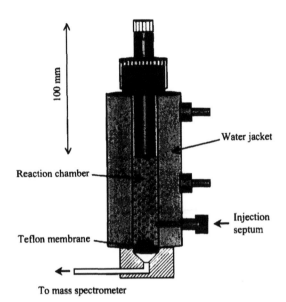

Figure 5.4 Reaction vessel for studying reactions of dissolved gases by mass spectrometry. The solution is agitated with a magnetic stirrer bar. The reaction space is limited at its lower end by a teflon membrane 12.5 mm thick, supported by a fritted steel disk above a cavity. This leads through a vacuum line to the ion source of a membrane-inlet mass spectrometer. (Adapted from P.A. Lespinat, Ph.D. thesis, Université Joseph Fourier, Grenoble 1988.)

The interconversion of *para-* and *ortho-*H_2 gas can be detected by the change in thermal conductivity.

5.3.4. Mechanistic implications of isotopic effects on rates

Isotope-exchange reactions can provide incisive information about the mechanisms by which hydrogenases catalyse their reactions. The ratios of products formed, and in the rates of reaction with different isotopes, can be measured. Different hydrogenases show significantly different proportions of $^1H_2$ and $^1H^2HO$ produced in reactions (3) and (4), and this effect has been used to identify the different types of hydrogenases in whole cells, without the need for purification (Berlier *et al.* 1987).

Kinetic isotope effects on the rates of whole or partial reactions of hydrogenases provide important clues to the internal workings of the enzymes. The isotopes of hydrogen, unlike those of the heavier elements, show significant differences in rates of reaction, owing to their different atomic masses. The reaction cycle of the enzyme depends on several steps in which hydrogen atoms are transferred from one site to another in the active site, and the rates of these transfers show significant kinetic isotope effects: the rates decrease in the order $^1H > {}^2H > {}^3H$. For example, the rate of photolysis of the Ni-C state of *D. gigas* hydrogenase is thirty times slower in deuterium oxide, ^2H_2O, than in water, ^1H_2O.

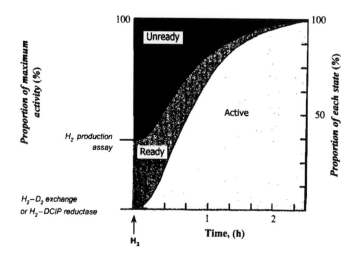

Figure 5.5 Activation of an [NiFe] hydrogenase such as that from *D. gigas*, by incubation with H_2, as measured by different assay methods. These are: production of H_2 with the low-potential donor methyl viologen; consumption of H_2 with the high-potential acceptor dichloroindophenol; and isotope exchange of $^1H_2$ with ^1H_2O. The proportions of the enzyme in three different states, unready, ready and active, are indicated by the shading.

5.4. Activation and activity states

A difficulty, in assays of the quantity and integrity of hydrogenase preparations, is that the activity depends considerably on the history of the sample. [NiFe] hydrogenases isolated under normal aerobic conditions do not display activity in the hydrogen-isotope exchange assay, even after prolonged removal of O_2. The same preparations show activity, although with variable kinetics, when assayed for H_2 evolution or H_2 uptake. All these activities are gradually stimulated by reductive treatments.

The apparently complex changes in activity of hydrogenases such as those from *D. gigas* have been interpreted in terms of interconversion between three states, designated the unready, ready and active states (Fernandez *et al.* 1985). This interpretation may be explained by reference to Fig. 5.5, in which the proportions of enzyme molecules in the various states are indicated during activation by H_2. The unready state (dark shading) is inactive in all assays, and requires prolonged reducing treatments to convert it to the active form. The ready state (lighter shading) is inactive towards H_2, and is inactive in the assays with electron acceptors of high redox potential such as DCIP. However, the ready state requires only brief reduction to become active; in H_2-production assay this reduction is carried out by the reducing substrate, and so the enzyme shows activity. The active state is fully active in all assays.

At present the reasons for the differences in activity are not completely clear, but are discussed in Chapter 7. A practical consequence is that in order to estimate the amount of enzyme, it is necessary to activate the enzyme fully before assay. This can be done by incubating the enzyme for a sufficient length of time under hydrogen, or, more rapidly, with a strong reducing agent such as dithionite + methyl viologen.

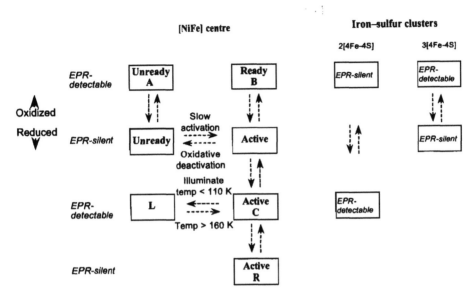

Figure 5.6 Activities and states of the [NiFe] centre and iron–sulfur clusters in the [NiFe] hydrogenase of *D. gigas*. Higher oxidation states are at the top, lower at the bottom.

The interconversions of the different states have been most extensively studied in the [NiFe] hydrogenases. A number of different terms have been employed to describe the states of the [NiFe] centre. Since EPR was the first spectroscopic technique used to observe the nickel in the enzyme, the states were labelled by reference to the EPR signals they gave. It is important to distinguish between the *state* of the centre, which is reflected in activity and many other properties, and the EPR signals that it displays. For example, the Ni-C active state can give rise to both 'split' and 'unsplit' EPR spectra, depending on whether the proximal [4Fe-4S] cluster is reduced or oxidized. The [NiFe] centre is the same in each case, but the spectrum is influenced by spin–spin interactions with the proximal cluster (Section 7.10). Figure 5.6 shows the different states and their interconversions. Various researchers have given different names to these states, somewhat confused by the use of the letter R to denote both 'ready' and 'reduced'. As will be seen in Section 7.1 in Chapter 7, this scheme has been expanded as additional diamagnetic 'EPR-silent' states have been revealed by FTIR spectroscopy. The characteristics of these states are described in Chapter 7.

5.5. Multiple hydrogenases in the same species

One method to observe the enzymes, when they are present, is to separate the proteins on non-denaturing polyacrylamide gel electrophoresis and detect them by an activity stain. This involves developing the polyacrylamide gel containing the protein bands in a solution containing H_2 and a tetrazolium indicator dye. The hydrogenase bands are stained by reduced tetrazolium. Care must be taken, since some early reports of apparent multiple hydrogenases were due to hydrogenase adhering to other proteins on the gel.

The presence of multiple hydrogenases, or isoenzymes, in one species of organism, caused confusion in the early investigations. For example, some methanogens and *Escherichia coli* may express four or more different hydrogenases, while some species of *Desulfovibrio* have hydrogenases of the [Fe]-, [NiFe]- and [NiFeSe]-types. The isoenzymes may be differently located in the cell, either in the cytosol, the cell membrane or the periplasm. We now realize that these hydrogenases are expressed under different conditions, for different purposes (Chapter 2). Some may reduce electron acceptors such as quinones, while others are used for H_2 production, and have electron donors such as ferredoxins or cytochromes.

5.6. Enrichment with isotopes

As will be explained in Chapter 7, spectroscopic methods are a powerful way to probe the active sites of the hydrogenases. Often spectroscopic methods are greatly enhanced by judicious enrichment of the active sites with a stable isotope. For example, Mössbauer spectroscopy detects only the isotope ^{57}Fe, which is present at only 2.2 per cent abundance in natural iron. Hydrogen atoms, which cannot be seen by X-ray diffraction for example, can be studied by EPR and ENDOR spectroscopy, which exploit the hyperfine interactions between the unpaired electron spin and nuclear spins. More detailed information has been derived from hyperfine interactions with nuclei such as ^{13}C, ^{17}O, ^{33}S, ^{61}Ni and ^{77}Se, in the active sites. In FTIR spectroscopy, replacement of the nuclei by heavier isotopes causes changes in the vibrational frequencies and this identified the ligands to the centre as -CO and -CN.

In contrast to radioactive isotopes, which are introduced in trace amounts, the stable isotopes in these experiments must be introduced as close to 100 per cent abundance as possible. This means that a high concentration of isotope is used and the naturally occurring isotope is excluded from the medium. Because of the complex way in which the enzyme is assembled, it is not possible to insert or exchange isotopes into the intact protein after it is formed. The exceptions are molecules that can diffuse into the active site, notably hydrogen isotopes from the water, and ^{13}CO. For other nuclei such as ^{57}Fe and ^{61}Ni, the only method of enrichment is by allowing the organism to grow on a medium in which the particular isotope is substituted for the naturally abundant one, and extracting the hydrogenase. Thus much painstaking (and expensive) work goes into the preparation of the materials used for the spectroscopic studies of the enzymes.

5.7. The hydrogen-consumption activity of the [NiFe] hydrogenase of *Allochromatium vinosum* in different redox states

Boris Bleijlevens

As explained in Section 7.4 the [NiFe] hydrogenase of *Allochromatium vinosum* can exist in various states, three of which are enzymically active. Activity is usually measured by following hydrogen consumption at 30°C with benzyl viologen ($E'_0 = -359$ mV) as electron acceptor. If performed with enzyme in the ready state (see Fig. 5.6 for an overview of all states), it only takes several minutes until full activity is

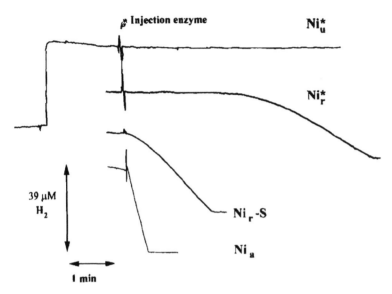

Figure 5.7 Activity of the various states of the [NiFe] hydrogenase from *A. vinosum* as determined with a Pt electrode at 30°C. The reaction was performed in 50 mM Tris/HCl (pH 8.0) in a volume of 2 ml. Oxygen was scavenged by adding glucose (90 mM) and glucose oxidase (2.5 mg/ml). Hydrogen peroxide was removed by catalase. When the system was anaerobic, an aliquot of H_2-saturated water was added, and a little later enzyme (5–10 nM) was injected. Benzyl viologen (4.2 mM) was used as electron acceptor.

observed (Fig. 5.7). Enzyme in the unready state cannot be activated under these conditions. Incubation for 30 min at 50°C under a hydrogen atmosphere is required to completely activate the enzyme in the latter state.

The active site in both ready and unready enzyme can be reduced with one electron to the Ni_r-S and Ni_u-S states, respectively. When carefully keeping the temperature low (2°C), the energy barrier required for activation will not be crossed and the enzyme remains inactive. The behaviour of these one-electron reduced states in activity assays is nearly the same as that of the oxidized states: the unready state still does not exert any activity at all under these conditions, but the ready state now shows only a short lag phase before getting active (Fig. 5.7). A curious observation was that although enzyme in the Ni_r-S state is easily activated when using benzyl viologen as electron acceptor, it did not activate when methylene blue (1.9 mM; $E'_0 = +11$ mV) was used. Active enzyme reacts immediately with methylene blue with a rate several times faster than that with benzyl viologen. Apparently, the Ni_r-S species still needs to undergo some redox-dependent structural change to become active. Active enzyme displays immediate hydrogen consumption (Fig. 5.7) independent of its initial redox state (Ni_a-S, Ni_a-C^* or Ni_a-SR).

The energy required to activate hydrogenase is in the order of 80 kJ/mol. Several possible processes have been suggested so far, e.g. a rearrangement of the spatial conformation of the active site or of the surrounding protein in order to allow substrate accessibility to the active site.

In conclusion it can be said that the nature of the activation process still remains obscure and may even include more than one process. The physiological relevance of the different inactive species also remains unclear.

5.8. Measuring redox potentials of hydrogenase

Wilfred R. Hagen and Richard Cammack

How do we measure the redox potential, E_m (or reduction potential, midpoint potential, half-wave potential) of the prosthetic groups in hydrogenase and its natural redox partners? The most commonly used technique is bulk titration. A solution of hydrogenase is chemically reduced, or oxidized, in small steps by substoichiometric additions of a reductant, e.g. sodium dithionite ($Na_2S_2O_4$), or an oxidant, e.g. potassium ferricyanide ($K_3Fe(CN)_6$). During the titration the potential of the solution, E, is constantly measured with a platinum wire connected via a DC voltmeter to a reference electrode. At certain potential values hydrogenase samples are drawn and the reduced fraction (or oxidized fraction, whatever is more convenient) of a prosthetic group is determined by spectroscopy, typically EPR (Fig. 5.8).

5.8.1. Mediated measurements of redox potentials

Redox proteins do not usually equilibrate with a platinum metal surface and, therefore, in order to establish thermodynamic equilibrium one must add a mixture of redox mediators to the solution. These are the same strongly coloured organic dyes

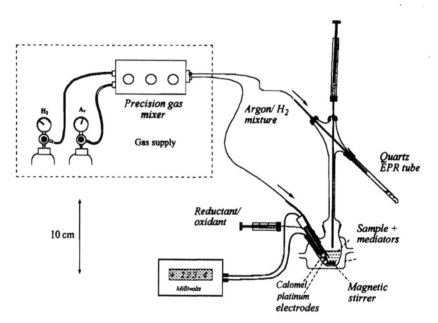

Figure 5.8 Electrochemical cell for measurement of redox potentials of hydrogenase by EPR spectroscopy.

that are frequently used in activity assays, e.g. benzyl viologen, methylene blue or dichloroindophenol. They would obscure the absorption spectra of the redox centres, and so UV/visible absorption spectroscopy is usually not an option for measuring the redox state of the protein. Infrared spectroscopy is still possible, since the mediators do not interfere with the characteristic absorption bands of the active sites of hydrogenases (see Section 7.8).

Let us, for example, determine the E_m of a [4Fe-4S] cluster in hydrogenase. The cluster has an EPR signal in its reduced form so we must rewrite the Nernst equation as

$$[red]/([red] + [ox]) = 1/\{1 + \exp[(nF/RT)(E - E_m)]\}.$$

The left-hand side of the equation is equivalent to the EPR amplitude as a fraction (or percentage) of the maximal EPR amplitude from the [4Fe-4S] cluster in fully reduced hydrogenase. Plotting EPR amplitude versus E gives a sigmoidal curve with half maximal intensity for $E = E_m$ (Fig. 5.9). This behaviour is typical of a one-electron reduction reaction in which one of the species being reduced is EPR detectable. A more complex behaviour is observed in the case where the reduced species undergoes a *second* redox step, which removes its spectroscopic signal. This is seen for the EPR signal of nickel in the active enzyme (Ni-C), as can be observed both in EPR (Fig. 5.10A) and FTIR spectroscopy (Fig. 5.10B). As will be seen later (Sections 7.2 and 7.8), the advantage of FTIR spectroscopy is that it can observe the [NiFe] site in all oxidation states. From the dependence of the EPR and FTIR signals on redox potential, it is possible to estimate the midpoint potentials of all of the redox steps in the reaction.

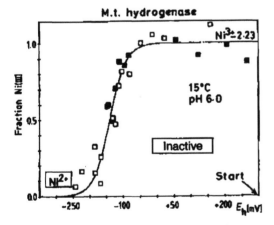

Figure 5.9 Example of a redox titration of nickel of hydrogenase from *M. marburgensis*. The amplitude of the Ni_u^* EPR signal is plotted against the measured redox potential. Half of the active sites in the enzyme solution is reduced at a redox potential (*midpoint potential*) of -140 mV (at pH 6). The $2H^+/H_2$ redox couple has an E_m of -354 mV at this pH. The line through the points is a theoretical line assuming a midpoint potential of -140 mV (Coremans *et al.* 1989).

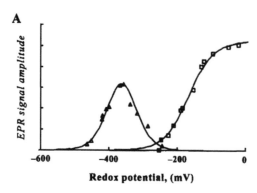

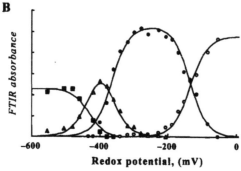

Figure 5.10 Redox titration of the Ni-C EPR signal in *D. gigas* hydrogenase, in the presence of mediators under partial pressure of H_2. (A) Titration monitored by EPR spectroscopy (data from Cammack *et al.* 1982, 1987). The data points were obtained by removing samples from a vessel as shown in Fig. 5.8. Data □ NiA signal; Δ NiC signal. (B) Titration monitored by FTIR spectroscopy (data from De Lacey *et al.* 1997). The spectra were recorded directly in a sealed optically transparent thin-layer electrode cell. Note that the oxidized and reduced species, which are undetectable by EPR, can be measured. Data: o $1946 \, cm^{-1}$ (NiB state); • $1914 + 1934 \, cm^{-1}$ (NiSR state); Δ $1952 \, cm^{-1}$ (NiA state); ■ $1940 \, cm^{-1}$ (NiR state).

Now the actual experiment: we need a titration cell that contains the enzyme and mediators. It has several side-arms; sealed with gas-tight septa; some are for the electrodes, others are for continuous flushing with wet argon gas and for additions of reductant and withdrawal of hydrogenase samples, both with gas-tight injection needles. The solution of 10–100 (μM) hydrogenase and redox mediators (each at 0.1–1 times the hydrogenase concentration) is stirred with a magnetic bar. An EPR sample is approximately 100 μl so, from 1.5 ml solution we can draw some twelve samples to put in twelve anaerobic EPR tubes to eventually get a twelve-point titration graph.

5.8.2. Titrations with H_2 pressure

There is something special about hydrogenase compared to all other enzymes: H^+ is always a substrate or a product, therefore it is impossible to study the enzyme in the absence of one of its substrates. This creates a unique problem for redox studies on active hydrogenase. When, in a reductive titration, the potential approaches that of the H^+/H_2 couple (-0.41 V at pH 7), the enzyme will produce an equilibrium partial pressure of H_2 in the solution. This will diffuse into the gas phase and be swept away by the flow of argon gas. As a result, the reducing power of added reductant will be converted into H_2, and the potential will remain constant (see e.g. Pierik *et al.* 1992). In other words, full reduction of the active site of hydrogenase is not possible in equilibrium experiments. One way out of this tricky situation is to titrate inactivated hydrogenase (e.g. in the presence of CO or at a low temperature), but that is, of course, not the real thing. Another possibility is reducing transiently in kinetic experiments, e.g. by rapid-mixing hydrogenase and reductant and rapid-freezing followed by spectroscopy, or in kinetic direct electrochemistry (see Sections 58.3, 7.6). The way to achieve equilibrium conditions is to use a gas phase that has the appropriate concentration of H_2 already in it (Fig. 5.8), ideally by means of a calibrated gas mixer (Cammack *et al.* 1987). The mixture of H_2 and Ar can then be adjusted to control the redox potential in the vessel.

5.8.3. Direct electrochemistry

Direct electrochemistry is another (somewhat less) widely applied technique to determine E_m of proteins. This approach requires an instrument called a potentiostat. No chemical reductant or oxidant is required. The potentiostat forces a potential on a working electrode versus a reference electrode. There is now also a third electrode, the counter electrode, to minimize current drawing through the reference electrode. The working electrode should directly reduce/oxidize the protein under study. Usually the experiment is done as cyclic voltammetry: the potential, E, is linearly varied between two limiting values, and the current through the circuit, used to reduce or oxidize the protein, is measured as a function of E. The resulting i–E plot shows a reductive (cathodic) and an oxidative (anodic) wave with peak potentials, E_{pc} and E_{pa}, ideally separated by $57/n$ mV. Ideally means that the electron transfer between protein and electrode is fast, and the rate-limiting step is diffusion of the protein to/from the electrode. We find the redox potential as $E_m = (E_{pc} + E_{pa})/2$. With carbon-based working electrodes the method nowadays works for most small (say, 5–25 kDa) electron-transferring proteins, e.g. the natural redox partners of hydrogenases: ferredoxin, flavodoxin or cytochrome. To save protein the cell volume can be made very small: 10–30 μl. The protein concentration is typically 0.3–3 mg/ml.

Unfortunately, diffusion-controlled direct electrochemistry very rarely works with enzymes. However, certain enzymes, including some hydrogenases, adsorb onto electrodes, and they may even develop electrochemical responses. Although these responses are more difficult to control experimentally, and to interpret theoretically, they may provide unique information. Such an experiment has been done with the soluble, monomeric [Fe] hydrogenase from *Megasphaera elsdenii* and *A. vinosum* (Pershad *et al.* 1999). From the complex response a value for the redox potential of the active centre of hydrogenase has been deduced for the first time: $E_m = -421$ mV at pH 7.0, i.e. a potential just slightly more negative than that of the H^+/H_2 couple (see Butt *et al.* 1997). More details of this approach will be given in Chapter 7.

5.9. The active ↔ inactive interconversion of an [NiFe]-hydrogenase at an electrode

Anne K. Jones, Harsh R. Pershad, Bart Faber, Simon P. J. Albracht, and Fraser A. Armstrong

[NiFe] hydrogenases can exist in a multitude of active and inactive redox states. It is known that inactive enzyme can be activated by reduction; conversely, active enzyme can be inactivated by oxidation (Section 5.4). In order to investigate the energetics, kinetics and $H^+:e^-$ stoichiometry of reductive activation and oxidative inactivation, we have examined the electrochemistry of *A. vinosum* [NiFe] hydrogenase. This technique has been developed for the study of the activity of electron-transfer proteins. The proteins are adsorbed directly onto the conducting edge of a graphite electrode (Fig. 5.11). Sensitive measurements are made of currents induced in response to a controlled, oscillating voltage. The voltammetric 'waves' in the current/voltage profile make it possible to identify the responses of redox centres in the protein, and measure their oxidation–reduction potentials. Moreover, for an enzyme, it is possible to observe 'catalytic waves' corresponding to the reduction or oxidation of substrates. This makes it possible to measure the rates of the catalytic reaction, without the complication of reactions with electron acceptors or donors.

In the case of hydrogenase, the substrate hydrons are always present and so at reducing potentials the enzyme will generate hydrogen. In solutions of H_2, the hydrogen-oxidizing activity can also be observed. Because measurements can be made over timescales of milliseconds to hours, it was possible to observe both the extremely rapid reaction of the enzyme with hydrogen, and the slow activation/deactivation processes at more positive potentials.

The enzyme adsorbed on the electrode showed large H_2 oxidation currents, with an activity greater than the catalytic activity with electron acceptors and donors (Pershad *et al.* 1999). This demonstrates that the electron transfer to [NiFe] active site, and reaction with H_2 are extremely efficient, and the factor that limits the enzyme-catalysed rate is the diffusion of the electron donors and acceptors to the enzyme.

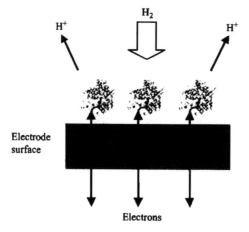

Figure 5.11 Diagram of hydrogenase molecules immobilized on an electrode surface.

The hydrogenase film on the electrode was very stable, and this allows the study of active/inactive interconversion under strict potential control. By comparing cyclic voltammetry and potential step chronoamperometry, we were able to integrate energetics, kinetics and $H^+:e^-$ stoichiometry of the reaction. The effects of pH on these processes could also be conveniently observed.

5.9.1. Energetics of reductive activation

Cyclic voltammetry can be used to monitor directly the hydrogen oxidation activity of the adsorbed hydrogenase film as a function of potential. Positive currents correspond to hydrogen oxidation and are proportional to the number of hydrogen molecules oxidized. When under enzyme control (hydrogen-saturated solutions and very slow scan rates), the voltammetry revealed a 'tunnel-diode' effect (Fig. 5.12). This means that over a certain potential range, the rate of H_2 oxidation decreases despite an increase in driving force for the catalysed reaction. This is a consequence of oxidative inactivation and the reverse, reductive activation (Section 5.4). This effect was also most pronounced at high pH.

The component waves in the voltammograms (a lower potential catalytic wave and a higher potential 'switch' wave associated with the activation/inactivation of the enzyme) can be seen in Fig. 5.13. The positions of their inflection points were obtained as local extrema in the first derivative with respect to potential (Fig. 5.13, inset). E_{switch} corresponds to the potential of the reductive reactivation process. Figure 5.13 shows that as pH is increased E_{switch} and E_{cat} both decrease. Furthermore, the pH dependence of E_{switch} could be fitted to a $1H^+:1e^-$ stoichiometry with an apparent pK value of 7.7 and a potential at the alkaline limit of $-166\,mV$.

These potentials are reasonably close to those observed from measurements of hydrogenase in solution (Table 5.1). (For *D. gigas* hydrogenase the midpoint potential

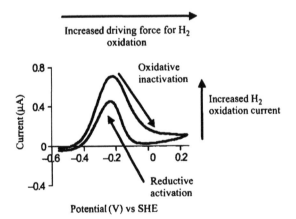

Figure 5.12 Typical cyclic voltammogram at 0.3 mV/s of a hydrogenase film adsorbed at a pyrolytic graphite edge electrode immersed in a pH 9.00 hydrogen saturated solution at 45°C (active–inactive interconversion is too slow to be reasonably studied at lower temperatures) and rotating at 1500 rpm. Under these conditions mass transport of hydrogen is not rate limiting.

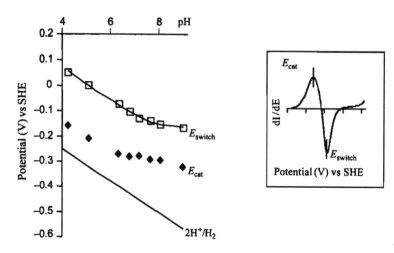

Figure 5.13   pH dependence of E_{switch} and E_{cat} at 45°C as determined from the derivatives of data such as those shown in Fig. 5.11. The solid line indicates the best fit to the E_{switch} data. Also shown is the potential of the $2H^+/H_2$ potential at 45°C and 1 bar hydrogen as calculated from the Nernst equation.

measured for reductive activation was -310 mV (Lissolo *et al.* 1984) and for oxidative deactivation was -130 mV (Mege and Bourdillon 1985). The midpoint potential for catalytic activity, corresponding to E_{cat}, was measured to be -360 mV (Fernandez *et al.* 1984).

Table 5.1   Redox potentials of *D. gigas* hydrogenase

Redox event	Em, pH 7 (mV)
Ni-A $\rightleftharpoons$ Ni-SU	-150
Ni-SR $\rightleftharpoons$ Ni-C	-270
Ni-C $\rightleftharpoons$ Ni-R	-390
Reduction of [3Fe-4S]	-35
Reduction of [4Fe-4S]	$-291, -340$
Reductive activation	-310
Oxidative deactivation	-133

5.9.2. Probing kinetics initiated by a potential step

The catalytic rate of hydrogenase adsorbed on the graphite electrode was measured by potential step chronoamperometry, in which current is monitored throughout a fixed sequence of potentials. This allows for direct observation of hydrogen oxidation activity at a particular potential over a period of time. Figures 5.14 and 5.15 show how chronoamperometry can be used to study the kinetics of reductive activation and oxidative inactivation respectively. A series of oxidative inactivation curves from several experiments like that shown in Fig. 5.14, showing the effect of pH on oxidative inactivation, are shown in Fig. 5.15. The kinetics of the reactivation process can be

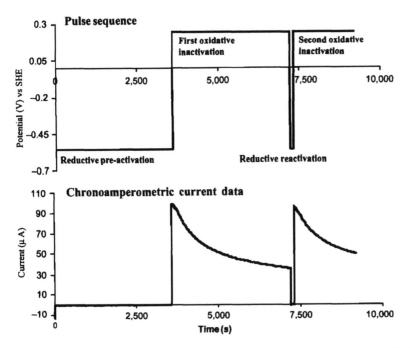

Figure 5.14 Pulse sequence and resulting data for a potential step experiment (hydrogen saturated solution, 45°C, 1,500 rpm and pH 6.94).

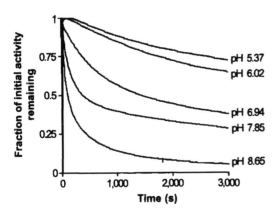

Figure 5.15 Fraction of activity of a hydrogenase film, as a function of time, for the first oxidative deactivation under the conditions described in Fig. 5.4.

further studied by varying the time of the reactivation excursion. One can see that as pH is raised, the rate of oxidative inactivation is increased.

They demonstrate several points:

1   *A. vinosum* hydrogenase films are extremely active as demonstrated by the initial hydrogen oxidation current of $100\,\mu A$. This corresponds to a turnover number of 10^3–$10^4 s^{-1}$.
2   Films are remarkably stable, losing no activity after 2 h.
3   The enzyme immobilized in a film can be inactivated by oxidizing potentials.
4   Complete reductive reactivation can be achieved (at 45°C) by a 2 min excursion to a reducing potential.

Chapter 6

Molecular architectures

Juan-Carlos Fontecilla-Camps, Michel Frey,
Elsa Garcin, Yoshiki Higuchi, Yaël Montet,
Yvain Nicolet and Anne Volbeda

6.1. Introduction

Many essential structural and functional features of hydrogenases have been derived from a wealth of various biochemical and spectroscopic methods. However, the knowledge of their atomic architectures have been obtained only very recently with the determination of the crystal structures of several hydrogenases belonging to both [NiFe] and [Fe] families (Table 6.1). These results have given a firm and unique structural basis to understand how these enzymes are actually working.

For molecules of molecular weight above 20,000 g/mol, X-ray diffraction remains the only experimental approach available to obtain detailed and reliable three-dimensional atomic models. The major steps of the method include the obtention of large and well-ordered crystals, their exposure to X-rays and collection of diffraction data and the 'phasing' of these data to obtain by Fourier analysis a three-dimensional view (or map) of the electron density of the molecule. Finally a three-dimensional atomic model of the protein is fitted like a hand in a glove within this map, using a 'kit' containing all the available biochemical and spectroscopic information (Table 6.2). The reliability of the final atomic model is of course dependent on the quality of the electron density map. This quality depends on the number of X-ray data per atom and on the resolution and accuracy of these data, which in turn are highly dependent on the size and quality of the crystals.

6.1.1. Growing crystals

The preparation of large (typically ~0.2–0.5 mm) and well-ordered crystals of hydrogenases, like those of many biomolecules, remains the most uncertain step of the X-ray crystallographic studies. Keeping in mind that a crystal is made of the periodic packing of structurally identical molecules (about 10^{13}–10^{15} molecules per typical crystal), a successful crystallisation experiment requires necessarily highly purified and structurally homogeneous material. Along these lines, hydrogenases are purified by chromatographic methods including preparative high performance liquid chromatography (HPLC) as a final step of the process. The purity and homogeneity of the samples are eventually probed by electrophoresis and HPLC. Moreover, since most hydrogenases are sensitive to oxygen, they should be maintained under anaerobic conditions to the very final step of the process and eventually be stored in argon or liquid nitrogen to avoid oxidative damages to their structural integrity. In the extreme

Table 6.1 X-ray structures of hydrogenases by the year 2001

Organism and functional state of the enzyme[a]	Class	Method[b]	Res.[c] (nm)	Reference	Year	PDB[d]
D. gigas unready	[NiFe]	MIR	0.28	Volbeda et al.	1995	1FRV
D. gigas unready		MR	0.25	Volbeda et al.	1996	2FRV
D. gigas active		MR	0.25	Garcin	1998[e]	f
D. vulgaris ready	[NiFe]	MIR	0.18	Higuchi et al.	1997	1H2A
D. vulgaris active		MR	0.14	Higuchi et al.	1999	1H2R
D. fructosovorans unready	[NiFe]	MR	0.27	Montet et al.	1997	1FRF
			0.18	Montet	1998[e]	f
Dm. baculatum active	[NiFeSe]	MR	0.22	Garchin et al.	1999	1CC1
C. pasterurianum	[Fe]	MAD	0.18	Peters et al.	1998	1FEH
D. desulfuricans	[Fe]	MAD	0.16	Nicolet et al.	1999	1HFE
D. desulfuricans	[NiFe]	MR	0.18	Matias et al.	2001	1E3D

Notes

a Desulfovibrio (D.), Clostridium (C.), Desulfomicrobium (D.); Unready: inactive oxidised forms, activated after a prolonged exposure to H_2; Ready: inactive oxidised forms, immediately active after exposure to H_2; Active: reduced forms (Chapter 7).
b Methods: multiple isomorphous replacement (MIR), molecular replacement (MR), multiple anomalous dispersion (MAD).
c Res.: resolution.
d Deposition code in the Protein Data Bank: http:www.rcsb.org.
e Thèse (Université Joseph Fourier, Grenoble).
f To be deposited.

Table 6.2 Structural features of the periplasmic [NiFe] hydrogenase of the sulfate-reducing bacterium *D. gigas*

Molecular weights (g/mol)	
Total	89,000
Large subunit, gene sequence	61,482 (551 a.a.)
Large subunit, mature sequence	59,450 (536 a.a.)
Large subunit, measured[a]	59,459 ± 6
Small subunit, gene sequence	28,324 (264 a.a.)
Small subunit, measured[a]	28,321 ± 3
Number of iron atoms	12
Acid-labile sulfides	12
[4Fe4S] $^{2+/1+}$ clusters	2
[3Fe4S] $^{1+/0}$ clusters	1
Nickel atom	1
Diatomic non-protein ligands	2 CN^-; 1CO
Other metal	Mg

Note
a By electrospray mass spectroscopy.

case of the [Fe] hydrogenase from *Clostridium pasteurianum*, which is particularly sensitive to oxygen, sodium dithionite is also added to the solutions.

Besides purity and structural homogeneity the obtention of crystals suitable for X-ray diffraction experiments depends on many other parameters including pH, temperature, protein concentration and the nature and concentration of the precipitant. It results that many crystallisation experiments and often large quantities of protein

Figure 6.1 Crystal of *D. vulgaris* hydrogenase.

(several tens of milligrams) are required to search for the optimal composition of the crystallisation medium.

Single crystals of both [NiFe] and [Fe] hydrogenases have been obtained (e.g. Higuchi *et al.* 1987, Fig. 6.1) by classical vapour diffusion, microdialysis or microcapillary batch diffusion methods, preferably under anaerobic conditions (Further reading: Ducruix and Giégé 1992). These crystals are equilibrated with glycerol and stored either under argon or, better, in a glove box under anaerobic conditions (oxygen concentration less than 1 ppm) or, best, in liquid nitrogen after 'flash cooling'. To determine the phases of the diffracted X-rays, for the [NiFe] hydrogenases from *Desulfovibrio gigas* and *Desulfovibrio vulgaris* (Miyazaki), crystals of various isomorphous heavy-atom derivatives were prepared: native protein crystals (which comprise around 50 per cent by volume of solvent) were soaked in solutions containing heavy-atom reagents at, typically, 5 mM concentrations.

A frequently asked question concerns the actual functional state of the crystallised protein. EPR analyses have shown, for example, that the enzyme present in aerobically prepared crystals of [NiFe] hydrogenase from *D. gigas* is mainly in its inactive, unready (Ni-A) state. However, it can be activated after prolonged incubation under hydrogen in the presence of methyl viologen (Nivière *et al.* 1987).

6.1.2. From X-ray diffraction to electron density

Except for 'in house' preliminary studies, the intensities of X-rays diffracted by hydrogenase crystals are now usually obtained with synchrotron radiation (Fig. 6.2) and detected by image plate or charge coupled device (CCD) detectors. To limit the damage induced by the powerful photon flux of synchrotrons, the crystals are usually mounted in a small loop, flash cooled in either liquid propane or nitrogen and stored

Booster   **Storage ring**

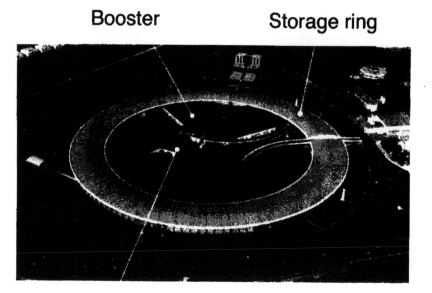

Linear accelerator

Figure 6.2 European Synchrotron Radiation Facility (courtesy *ESRF-Grenoble*).

at about 80 K. Subsequently, they are transferred and maintained at about 100 K in a stream of cold nitrogen gas during the X-ray diffraction data collection (Fig. 6.3).

To determine the first molecular architectures of both [NiFe] and [Fe] hydrogenases, the initial phases of the diffraction data were derived either by combining multiple isomorphous replacement (MIR; involving in each case, the preparation of several heavy atom derivatives) and anomalous diffraction methods (*D. gigas* and *D. vulgaris* [NiFe] hydrogenases) or by the so-called multiple anomalous dispersion method (MAD) (*C. pasteurianum* and *Desulfomicrobium desulfuricans* [Fe] hydrogenases). In the case of the [NiFe] hydrogenase of *D. gigas* and the [Fe] hydrogenase of *D. desulfuricans* these initial phases have been greatly improved and extended to high resolution by averaging the electron density of molecules related by non-crystallographic symmetry. For the other homologous [NiFe] hydrogenases the initial phases were obtained by molecular replacement (MR). This led to electron density maps with resolutions ranging from 0.29 to 0.14 nm (Table 6.1).

6.1.3. Building the atomic models

The three-dimensional atomic models are built manually within the electron density maps by interactive computer graphics and adjusted by interleaving crystallographic 'refinement' cycles. Due to the high quality and resolution of the electron density maps of hydrogenases, the fitting of the amino acid sequences and iron sulfur clusters is generally straightforward. By contrast the electron density of the catalytic site of both [NiFe] and [Fe] hydrogenases is more difficult to interpret. For example, the

Video   Cold stream nozzle   Synchrotron beam port

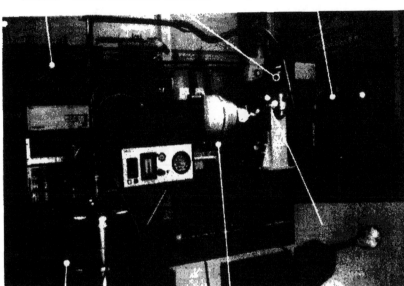

Liquid nitrogen tank CCD detector   Crystal

Figure 6.3 X-ray diffraction setup at ESRF (BM2) (courtesy *M. Roth IBS*).

electron density corresponding to the catalytic site of the *D. gigas* [NiFe] hydrogenase shows, besides two high peaks corresponding to a nickel and an iron atom, three protruding ellipsoidal features which can be modelled only with diatomic non-protein molecules (Fig. 6.4). In a similar way the electron density of the catalytic site of [Fe] hydrogenases can be modelled with two iron atoms, several diatomic ligands and a small molecule binding the two iron atoms through thiolates (Fig. 6.5). Since in both cases this complexity was completely unexpected, a great deal of effort and care had to be devoted to try to identify unambiguously the nature of metal ions and the non-protein ligands and to verify if they are intrinsic parts of the proteins rather than artefacts.

6.1.4. Identification of metal atoms by X-ray diffraction

Elsa Garcin

A small fraction of the X-ray radiation is absorbed and re-emitted by inner shell electrons. This 'anomalous' scattering, which becomes important for 'heavy' atoms, differs in phases and amplitudes from the 'normal' scattering and depends on both the nature of the atom and the X-ray wavelength. At wavelengths which are characteristic of each element, the anomalous intensity changes abruptly (absorption edge, e.g. at $\lambda = 1.743$ Å for iron and $\lambda = 1.487$ Å for nickel). Due to the high brilliance

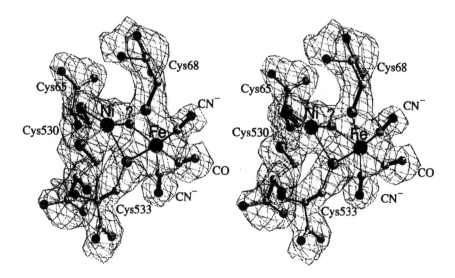

Figure 6.4 A stereoview of the six-fold averaged electron density map of the catalytic site of the *D. gigas* [NiFe] hydrogenase in the as-prepared oxidised form at 0.25 nm resolution (Volbeda *et al.* 1996).

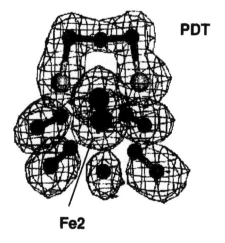

Figure 6.5 Electron density map of the catalytic site of *D. desulfuricans* [Fe] hydrogenase at 0.16 nm resolution (Nicolet *et al.* 1999). PDT: 1,3-propanedithiol.

and tunability of synchrotron radiation, it is possible to obtain accurate anomalous intensities at chosen wavelengths. This has been crucial to locate and identify unambiguously the metal ions present in the active site of *D. gigas* hydrogenase.

Atomic absorption spectrometry, EPR spectroscopy and inductively coupled plasma (ICP) analysis had shown that the *D. gigas* hydrogenase contains one nickel and twelve (± 1) iron atoms, eleven of which are distributed among the three [FeS] clusters. This strongly suggested that the 'remaining' twelfth iron atom could be one of the two metal ions revealed by the active site electron density. To verify this

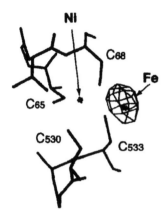

Figure 6.6 0.30 nm resolution double difference anomalous electron density map of the catalytic site of *D. gigas* [NiFe] hydrogenase, with X-ray data collected at both sides of the Fe absorption edge (Volbeda *et al.* 1996).

hypothesis X-ray data were collected at wavelengths close to the iron absorption edge and used to generate a double difference anomalous electron density map. As expected, strong peaks show up in this map at the [FeS] clusters, but also an additional one is found at one of the two metal sites of the catalytic centre (Fig. 6.6). This definitely confirms that this atom is iron.

The other metal that is close to Cys530 had been tentatively identified as a nickel. Cys530 in *D. gigas* hydrogenase is homologous to a selenocysteine in the [NiFeSe] *Desulfomicrobium baculatum* hydrogenase, which was demonstrated to be a ligand of the nickel by EXAFS studies and EPR measurements on [77]Se-enriched enzyme. Moreover, an anomalous electron density map obtained from X-ray data collected with a wavelength of 0.091 nm shows that the peak corresponding to the nickel atom is higher than that corresponding to the iron (Volbeda *et al.* 1996). This is in agreement with the expected values of the anomalous electron density of both these metal atoms at that wavelength. In the same way an Ni and Fe atom in the active site of the *D. vulgaris* hydrogenase have been unambiguously identified on anomalous electron density maps obtained from X-ray data collected at the Ni and Fe absorption edge respectively (Higuchi *et al.* 1997).

All these experimental results point unambiguously to a heterobimetallic catalytic centre in [NiFe] hydrogenases, which thus deserve their previously assigned name (Moura *et al.* 1988). Anomalous diffraction has been successfully used also to identify the selenium atom of the selenocysteine binding to the nickel atom in *Dm. baculatum* (Garcin *et al.* 1999).

6.1.5. Modelling the non-protein ligands

In the next chapter we will see how infrared studies on the inhibition of the *A. vinosum* [NiFe] hydrogenase by carbon monoxide led to the discovery of intrinsic diatomic non-protein moieties directly associated with the catalytic centres of [NiFe]

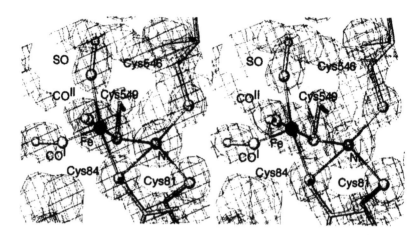

Figure 6.7 A stereoview of the electron density map of the catalytic site of the *D. vulgaris* [NiFe] hydrogenase in a reduced form at 0.14 nm resolution (Higuchi *et al.* 1999b).

and [Fe] hydrogenases. These moieties were subsequently identified by IR spectroscopy and chemical analysis as two CN⁻'s and one CO (Chapter 7). Consequently, the three characteristic ellipsoidal features of an electron density map of the *D. gigas* hydrogenase were modelled as two CN⁻'s involved in hydrogen bonds with protein atoms (one CN⁻ accepts two hydrogen bonds from the peptide NH and guanidinium group of Arg463 and the other one forms two hydrogen bonds with the peptide NH and the OH group of Ser486) and one CO completely surrounded by hydrophobic residues (Volbeda *et al.* 1996, Fig. 6.4).

The same approach has also been followed to locate the CN⁻'s and CO's at the active site of other [NiFe], [NiFeSe] or [Fe] hydrogenases (Table 6.1). However, in the particular case of the [NiFe] *D. vulgaris* hydrogenase (Higuchi *et al.* 1997, 1999a, 2000), one of the three diatomic ligands was modelled as SO on the basis of a higher value of the corresponding electron density and pyrolysis measurements (Fig. 6.7). More recently it has been proposed (Higuchi, personal communication) that, actually, the *D. vulgaris* crystals might contain a mixture of molecules with different non-protein ligands (i.e. SO or CO/CN⁻). Therefore, hydrogenase molecules might contain either two CN⁻ and one CO (Pierik *et al.* 1999) or one SO and two CO/CN⁻ ligands (Higuchi *et al.* 1997).

In the oxidised inactive [NiFe] hydrogenases, the electron density maps also show a small feature bridging the Ni and Fe sites. Due to the protein environment, this site has only room for a small species. In *D. gigas* hydrogenase this was assigned as a μ-oxo or hydroxo ligand on the basis of the electron density and EPR data of $^{17}O_2$ treated enzyme which had shown an Ni EPR signal broader than that of the $^{16}O_2$ treated enzyme (Chapter 7). In *D. vulgaris* hydrogenase, an Ni-Fe bridging feature was modelled as a sulfur species (Higuchi *et al.* 1997). This is in accordance with measurements of the amount of hydrogen sulfide liberated upon reduction of the enzyme (Higuchi *et al.* 1999a). Whatever its exact nature, no such bridging ligand is visible in reduced, active hydrogenases (Higuchi *et al.* 1999b; Garcin *et al.* 1999).

6.1.6. More metals

In [NiFe] hydrogenases a large peak of density is found in the vicinity of the C-terminal histidine of the large subunit. This peak has been modelled as a magnesium (*D. gigas*) or an iron ion (*Dm. baculatum*) (Garcin *et al.* 1999) following ICP analysis of the metal content of the respective proteins and anomalous X-ray dispersion data. In the *D. vulgaris* hydrogenase, an Mg atom was also modelled in the same location at the C-terminal end of the large subunit on the basis of the size and shape of its electron density, temperature factor and coordination (Higuchi *et al.* 1997).

6.2. Structures of [NiFe] hydrogenases

6.2.1. An overall view

With now four structures at hand, the *D. gigas* hydrogenase (Fig. 6.8), the first hydrogenase structure to have been determined (Volbeda *et al.* 1995), can be considered as a prototype of these enzymes (Fig. 6.9). The molecule appears as a globular heterodimer with a radius of about 3 nm. The large subunit contains the catalytic site. The small subunit holds the three [FeS] clusters which are disposed along an almost straight line, at ~1.2 nm from each other, with a [3Fe-4S] cluster located halfway between two [4Fe-4S] clusters. One of the [4Fe-4S] clusters is located at ~1.3 nm from to the active site and it is called proximal. The other [4Fe-4S] cluster, called distal, is found near the molecular surface with the imidazole ring of its unusual histidine ligand partially exposed to solvent.

The large and small subunits interact extensively with each other burying the catalytic site and the proximal [4Fe-4S] cluster at about 3 nm from the molecular surface. It is quite remarkable that the active site and the two most buried [FeS] clusters are located close to the almost planar subunit interface. The large subunit is anchored to the small subunit by about twenty-five side chains, of which several very conserved ones interact with the proximal [4Fe-4S] cluster, pointing to the role of this cluster as a direct partner of the catalytic site.

Interpretation of the electron density maps showed that the large subunit could not be modelled beyond His536 (Fig. 6.10), that is fifteen amino acids short of the 551 residues predicted by the nucleotide sequence (Table 6.2). At about the same time, the cleavage of this fifteen-residue stretch, which is performed by a specific protease, was reported to be an obligatory step for the maturation of the enzyme (Menon *et al.* 1993). It is also of interest to note that in all [NiFe] hydrogenase crystal structures this buried C-terminal histidine is ligated to a metal atom which is either a magnesium or an iron (see above).

The small subunit is composed of two domains. The N-terminal domain shows the characteristic architecture of flavodoxin with the phosphate moiety of the flavin cofactor occupying the binding pocket of the proximal [4Fe-4S] cluster. This N-terminal domain, including the proximal cluster, is found in all [NiFe] hydrogenases and is consequently an essential feature, both structural and functional, of these enzymes. By contrast, the C-terminal domain that binds the other [FeS] clusters is less organised and more variable in [FeS] cluster content and amino acid sequence.

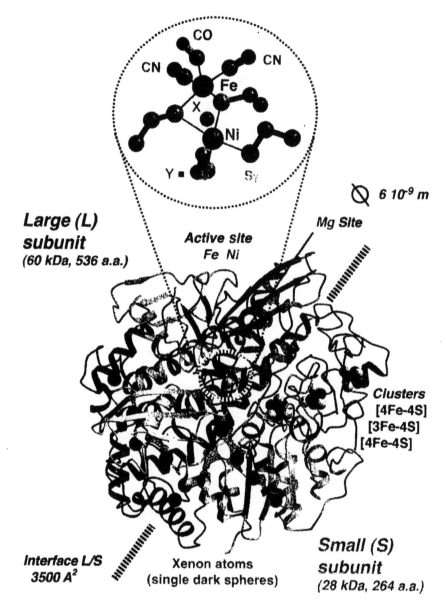

Figure 6.8 Schematic drawing of the three-dimensional structure of D. gigas [NiFe] hydrogenase
(PDB entry: 2FRV) and close up on the active site (top and Fig. 6.10). Ribbons and
arrows represent α-helices and β-strands, respectively. The large subunit (top-left)
contains the active site. The small subunit (bottom-right) includes the three [FeS]
clusters: (two [4Fe-4S] and one [3Fe-4S]). Single dark spheres indicate xenon atoms
diffused in the crystal structure of D. fructosovorans hydrogenase. These atoms probe
channels (grid) connecting the catalytic site and the external medium (Montet
et al. 1997). Note, in the active site, the three non-protein diatomic ligands to the iron
and the Xe site, bridging Ni and Fe, which is occupied by an O2⁻/OH⁻ species in the
most oxidised states and probably by a hydride in more reduced states. The vacant Y
Ni-coordinated site, close to one end of the hydrophobic channels, is a potential
molecular H_2 binding site.

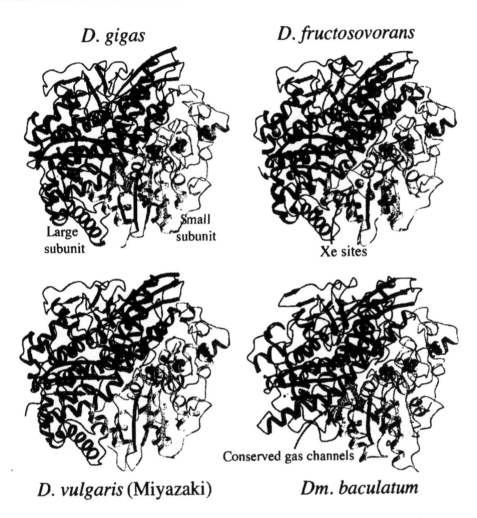

D. gigas

D. fructosovorans

Large subunit   Small subunit

Xe sites

D. vulgaris (Miyazaki)

Dm. baculatum

Conserved gas channels

Figure 6.9 Common architecture of [NiFe] hydrogenase crystal structures. [FeS] clusters, metal and xenon sites are shown as spheres. Included is also an averaged internal cavity map, calculated with an accessible probe radius of 0.1 nm, of *D. gigas* and *Dm. baculatum* hydrogenase.

6.2.2. Hidden at the centre of the molecule: The catalytic site

The catalytic site of the protein consists of a binuclear metallic centre with one nickel and one iron atom linked to the protein by four cysteic thiolates. In the oxidised form of the *D. gigas* [NiFe] hydrogenase, the nickel atom has three close and two distant ligands in a highly distorted square pyramidal conformation, with a vacant axial sixth ligand site, whereas the iron atom has six ligands in a distorted octahedral conformation. As mentioned above, three of the Fe ligands are diatomic molecules modelled as two CN⁻'s and one CO in *D. gigas*, *D. fructosovorans* and *Dm. baculatum* hydrogenases; also a μ-oxo species bridging the nickel and iron atoms was modelled in the oxidised forms of the former two hydrogenases (Table 6.1, Fig. 6.10). In the *D. vulgaris* enzyme, however, the iron ligands were modelled as two CO and one SO,

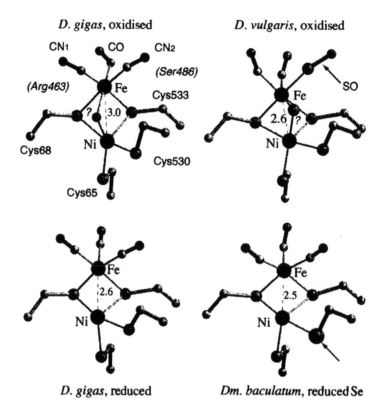

Figure 6.10 The catalytic site of [NiFe] and [NiFeSe] hydrogenases in oxidised inactive (top) and reduced active (bottom) states. Note the three non-protein diatomic ligands to the iron. The ? site bridging the Ni and Fe is occupied by an oxygen or sulfur species in the most oxidised states and probably by a hydride or molecular hydrogen in the most reduced states.

and the additional ligand bridging the nickel and the iron as a sulfur species. In all the reduced forms no significant density was found at this metal bridging position. Since it is difficult to detect hydride in electron density maps obtained by X-ray crystallography even at 0.14 nm resolution (Higuchi *et al.* 1999b), the possibility that the bridging site is occupied by a hydride ion, which would have mechanistic implications, cannot be excluded.

6.2.3. Accessibility of the [NiFe] active site

To 'shuttle' between the catalytic site located at the centre of the molecule and the surface, the components of the reaction, i.e. molecular hydrogen, protons and electrons, have to cover a distance of about 3 nm. To do so, they most probably use specific routes.

The spatial arrangement of the three [FeS] clusters in the molecule and the spectroscopic data suggest that they serve as relays for *electron transfer* between the active site and the surface, either following a through-space mechanism or through bonded orbitals with occasional through-space jumps. The proximal [4Fe-4S] cluster could directly exchange electrons with the active site, whereas the distal [4Fe-4S] cluster probably mediates, through its histidine ligand, the electronic exchanges between the hydrogenase and a functional partner, which could be a multihemic cytochrome. By contrast, the involvement of the medial [3Fe-4S] cluster is still debated due mainly to its redox potential which is far higher than those of the redox centres involved in the hydrogen activation. At any rate, reduction of the cluster potential through a site-directed mutation that transforms the [3Fe-4S] in a [4Fe-4S] centre had little incidence on the enzymatic activity (Rousset *et al.* 1998).

Protons move inside proteins through displacements of about 0.1 nm, which are mediated by rotational and vibrational movements of donor and acceptor groups. There exist several possible proton pathways connecting the active site and the molecular surface. These include histidines, carboxylate groups, which all have suitable pK_a's for transferring protons, and internal water molecules which, because they can simultaneously donate and accept two H bonds, can easily undergo H-bond exchanges. However, these putative pathways are waiting to be confirmed by site-directed mutagenesis studies.

One plausible pathway for protons within the large subunit is presented in Fig. 6.11. It starts with the nickel bound cysteic thiolate that is substituted by selenate in [NiFeSe] hydrogenases. This is H bonded to a completely conserved glutamic acid residue (Glu18 in *D. gigas*), which, in turn, is connected by a network of hydrogen bonds involving four structural water molecules, the C-terminal main chain carboxylate, an additional water molecule and another completely conserved glutamate (Glu46 in *D. gigas*) to a water ligand of the C-terminal Mg or Fe site, near the protein surface. The last water molecule is within H-bonding distance of two additional water ligands that are also H bonded to a third conserved glutamate (Glu321 in *D. gigas*). The pK_a of a water molecule may drop up to seven pH units when it is bound to a metal. Therefore metal-bound waters appear to be especially well suited for a role in proton transfer. This might provide an additional function for the metal site found at the C terminus of the large subunit, besides the obvious structural role and a possible involvement in the proteolytic processing of the C-terminal region of the immature large subunit.

6.3. Hydrogen channels

Yaël Montet and Patricia Amara

A crucial aspect of the diffusion of gases from the molecular surface to the active centres of enzymes is whether it requires random conformational fluctuations of the protein matrix or if specific channels within the protein are employed. We have investigated gas diffusion in Ni-Fe hydrogenases (Montet *et al.* 1997). Early proposals suggested that, as H_2 is a small molecule, it probably does not diffuse through any special channel. However, the analysis of the first X-ray structure of Ni-Fe hydrogenase showed that the active site is buried within the large subunit at approximately 3 nm from the surface, and this prompted us to consider the possibility of specific pathways for H_2 diffusion. A cavity map calculated with a probe radius of 0.1 nm, i.e. accessible for H_2, from the model of *D. gigas* hydrogenase, revealed a large network of

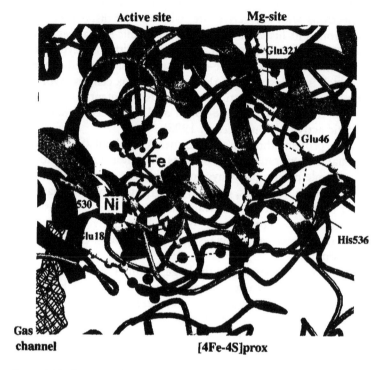

Figure 6.11 A zoom of *D. gigas* hydrogenase, showing the heart of the [NiFe] hydrogenase machinery. The [4Fe-4S]prox cluster is involved in *electron* transfer. A possible *proton* pathway, involving Glu18 and Glu46 of the large subunit and water molecules (grey spheres), between the catalytic [NiFe] and the C-terminal metal site is indicated with dashed lines. The end of a major internal hydrophobic cavity, which is part of a possible route for *molecular hydrogen*, is represented by a grid at the bottom-left. This end points towards a vacant Ni-coordination site.

mainly hydrophobic channels, connecting the active site to the molecular surface. This analysis also suggested different putative entrances at the protein surface and a unique entrance to the active site (Fig. 6.11).

In order to test whether these channels have specially evolved to facilitate the access of H_2 to the active site, we have investigated their possible role by using Xe binding in crystallographic experiments. Thanks to its ability to interact with proteins, xenon is a convenient choice to investigate the gas accessibility of protein interiors by X-ray crystallography, as it can be easily detected in difference Fourier electron density maps. Xenon binding experiments were carried out using crystals of Ni-Fe hydrogenase of *D. fructosovorans*, which shows 64 per cent amino acid sequence identity with *D. gigas*. X-ray diffraction data were collected on crystals that contained three independent hydrogenase molecules and were exposed to 9 bars xenon atmosphere. Ten major peaks were found in the averaged $(F_{Xe} - F_{native})$ map and they were attributed to ten xenon sites. All these sites are located in the channels calculated in *D. gigas* enzyme. Moreover their protein environment is highly hydrophobic.

Finally, we have simulated the diffusion of both xenon and hydrogen in the protein interior using molecular dynamics. First, in agreement with experiment xenon atoms were found not to be able to diffuse to the active site, in contrast to H_2 molecules. Second, remarkably, in all the obtained trajectories hydrogen molecules never diffused at random in the protein; rather, they always diffused through the calculated hydrophobic channels. In conclusion, even though some fluctuations of the protein medium might be necessary to enable the gas to pass through narrow channels connecting larger cavities, these hydrophobic channels seem to facilitate gas transfer and gas storage in the protein. Our study strongly suggests that they have a functional importance in gas transfer and play an instrumental role in gas metabolism.

6.4. Structures of [Fe] hydrogenases

6.4.1. An overall view

Two [Fe] hydrogenase structures have so far been determined: from C. *pasteurianum* (Cp) (Peters *et al.* 1998) and D. *desulfuricans (Dd)* (Nicolet *et al.* 1999). They have in common a large domain, which contains the catalytic site and three [4Fe-4S] iron sulfur clusters. The catalytic site and the closest (proximal) cluster are deeply buried inside the protein between two domains (or lobes), with access to a third, ferredoxin-like, domain that contains the two remaining (medial and distal) clusters. By contrast with [NiFe] hydrogenases the proximal [4Fe-4S] cluster is directly bridged to the binuclear active site by a cysteic thiolate (Fig. 6.12).

The two structures differ in that: (1) the cytoplasmic Cp hydrogenase is made of a single polypeptide chain. This includes the large active site and ferredoxin domains and two additional domains, each with one [FeS] cluster, which give the molecule a characteristic mushroom aspect; (2) the periplasmic Dd hydrogenase is made of two chains: the large one includes the active site and ferredoxin domains, whereas the small one is wrapped around the molecule and matches the C-terminal end of the Cp hydrogenase. This difference in quaternary structure is most likely due to the insertion of a signal peptide sequence in the gene of Dd at the N-terminal end of the corresponding small subunit polypeptide. This peptide is necessary for translocation of the enzyme into the periplasmic space (Hatchikian *et al.* 1999).

6.4.2. The active site

The core of the [Fe] hydrogenase catalytic site consists of a bimetallic centre with two iron atoms (termed Fe1 and Fe2) bridged by CO and a small molecule initially modelled in the Dd enzyme as a 1,3-propanedithiol on the basis of the electron density alone. Each iron atom is also liganded to two diatomic molecules, assigned to one CO and one CN^- (Fig. 6.13). It is quite remarkable that this catalytic site is bound to the protein through only one cysteic thiolate ligand to Fe1 and hydrogen bonds involving each of the CN^- ligands. Fe1 has six ligands in a distorted octahedral conformation whereas Fe2 has five ligands with an empty site in Dd and an additional putative water ligand in Cp. This site most likely binds hydrogen, as shown by the fact that CO, a competitive inhibitor of the enzyme, has been found to bind to it (Lemon and Peters 1999).

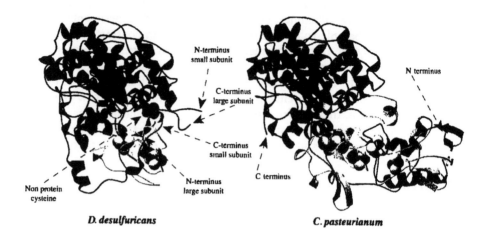

Figure 6.12 Schematic drawing of the three-dimensional structure of : (left) *D. desulfuricans* [Fe] hydrogenase, consisting of two subunits. The large one (light grey) contains the catalytic site and a ferredoxin-like domain containing the medial and distal [4Fe-4S] clusters (Nicolet *et al.* 1999). The small subunit (dark), homologous to the C-terminal region of *C. pasteurianum* hydrogenase, is wrapped around the large sub-unit. Ribbons and arrows represent α-helices and β-strands, respectively. (right) *C. pasteurianum* [Fe] hydrogenase, which consists of a single polypeptide chain. The large C-terminal domain (top) contains the catalytic centre consisting of a binuclear iron site bridged by a cysteic sulfur to the proximal [4Fe-4S] cluster. The N-terminal region (bottom) includes a [2Fe-2S] plant ferredoxin-like domain, a two [4Fe-4S] ferredoxin-like domain and a small exposed [4Fe-4S] domain (Peters *et al.* 1998). Ribbons and arrows represent α-helices and β-strands, respectively. (from Nicolet *et al.* 2000 with permission from Elsevier Science)

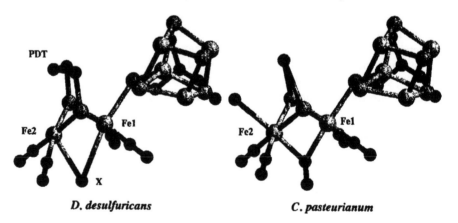

Figure 6.13 The catalytic site of: (left) the *D. desulfuricans* [Fe] hydrogenase. The *modelled* PDT: 1,3-propanedithiol (see text) bridges the two sulfur atoms bridging in turn Fe1 and Fe2 (medium-size gray spheres). Each iron binds two diatomic ligands: probably one CO (dark end) and one CN^- (gray end). The ligand bridging Fe1 and Fe2 has been modelled as an oxygen atom. (right) *C. pasteurianum* [Fe] hydrogenase (all ligands modelled as CO). The two sulfur atoms are bridged by a water molecule (black). (from Nicolet *et al.* 1999 with permission from Elsevier Science)

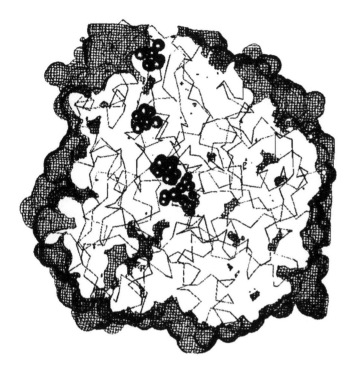

Figure 6.14 Gas access to the active site of *D. desulfuricans* [Fe] hydrogenase. A single channel (rep resented by a grid) connects the active site to the molecular surface. The internal extremity of the channel faces the vacant site of Fe2 (see Fig. 6.13). (from Nicolet *et al.* 1999 with permission from Elsevier Science)

6.4.3. Access

As in [NiFe] hydrogenases the catalytic site is deeply buried inside the molecule. Here again it appears that specific routes exist to mediate the transfer of electrons, protons and hydrogen between this site and the molecular surface. *Electrons* may be transferred by the iron–sulfur clusters spaced as in [NiFe] hydrogenase at typical intervals of about 1.2 nm. *Protons* might use several possible routes including an invariant cysteic thiolate, some negatively charged residues and water molecules. As in the case of [NiFe] hydrogenase *hydrogen* molecules could 'travel' through a hydrophobic channel (Fig. 6.14), the ends of which are facing the empty coordination site of Fe2 and the external medium.

Chapter 7

Spectroscopy – the functional puzzle

Simon P.J. Albracht

With contributions of *Patrick Bertrand, Boris Bleijlevens, François Dole, Bruno Guigliarelli, Wilfred R. Hagen, Randolph P. Happe, Wolfgang Lubitz, Michael J. Maroney, Christian Massanz, José J. G. Moura, Alice S. Pereira, Antonio J. Pierik, Oliver Sorgenfrei, Matthias Stein* and *Pedro Tavares*

Hydrogenase catalyses the simplest redox-linked chemical reaction in Nature, so one might assume that the task of solving the puzzle of how hydrogenases actually do this is a simple one. As the chapters in this book describe, the scientists involved did not anticipate the peculiarities discovered in these ancient enzymes. As already described in the previous chapters, we now know of two classes of enzymes which, when in the pure state, can activate H_2 without added cofactors.

7.1. [NiFe] hydrogenases

In the light of the structure of the enzyme (Chapter 6), we can see which parts of the amino-acid sequence of the protein have essential functions such as binding of the metal centres. Sure enough, these same amino acids are found in equivalent positions in the sequence of other [NiFe] hydrogenases; they are said to be conserved. The regions of sequence that contain these conserved residues are known as motifs. By looking for these motifs, we can learn about possible structure and function of hydrogenases for which we have sequences but no X-ray structures.

7.1.1. FTIR spectroscopy: The diatomic ligands

As previously described (Chapter 6), the [NiFe] hydrogenases minimally consist of a large and a small polypeptide (subunit). The active $NiFe(CN)_2(CO)$ site is located in the large subunit. The site was discovered by coincidence. As very often in science, one is looking for one thing and discovers another. This is how the importance of nickel for the enzyme was discovered in 1965 (Bartha and Ordal 1965), forgotten for fifteen years, and then rediscovered in the early 1980s (Friedrich *et al.* 1981; Graf and Thauer 1981). It was also by coincidence that, when I asked Prof. Woody Woodruff and Dr Kim Bagley in Los Alamos for their help to study the binding of the inhibitor carbon monoxide to hydrogenase from *Allochromatium vinosum* by monitoring the characteristic stretching frequency of the C-O bond by Fourier-Transform InfraRed (FTIR) spectroscopy, we discovered three additional absorption bands in the enzyme (Bagley *et al.* 1994). It took us quite some time and a lot of further experiments until we realized that these bands were not artefacts, but 'reported' on nearly every change in the active site (Bagley *et al.* 1995). We later identified the bands as due to CN^- and CO (Happe *et al.* 1997; Pierik *et al.* 1999) (see Section 8). It was through the X-ray structure that the Fe atom in the active site was discovered (Volbeda *et al.* 1995) and the CO/CN could be located (Chapter 6).

7.1.2. Amino-acid sequences

The large subunit

The Ni-Fe active site is bound to the large subunit by two CxxC motifs (x being any amino acid), one in the N-terminal region, and one very close to the C terminus. In spite of this, the length of the large subunit (as obtained from the encoding gene) can vary from 332 amino acids in the *Rhodobacter capsulatus* enzyme to 633 in the one from *Aquifex aeolicus*. In fact, only a few domains in the N-terminal and C-terminal regions seem to be conserved and the central part of the polypeptides can vary considerably. So it is the structure made up by the N- and C-terminal domains which is conserved throughout [NiFe] hydrogenases (see Fig. 7.1 for a schematic overview).

The small subunit

All [NiFe] hydrogenases contain at least one iron–sulfur (Fe-S) cluster with four Fe atoms and four sulfide ions, a [4Fe-4S] cluster, which is bound to the N-terminal part of the small subunit. As this cluster is not far (1.2 nm) from the active site, it is called the proximal cluster. Normally Fe-S clusters have the function of transferring electrons, one at a time, from place A to place B. The lengths of the small subunit can vary from 143 amino acids in the CO-induced enzyme in *Rhodospirillum rubrum* to 384 in *Helicobacter pylori*. The number and nature of Fe-S clusters vary as well, as can easily be inferred from the amino-acid sequences (Fig. 7.1). The majority of enzymes have, like the one from *Desulfovibrio gigas*, a [3Fe-4S] cluster which is further away from the active site (medial cluster), whereas a third cluster is near the surface of the protein (distal cluster; in the amino-acid sequence the order of binding of the clusters, from N to C terminus, is proximal, distal and medial, respectively; see Chapter 6 for details). The small subunits of H_2 sensors look like the *D. gigas* sequence, but have a Cys residue at the position of the Pro residue in the motif of the 3Fe cluster (and hence contain no 3Fe cluster). Other enzymes, notably the methyl-viologen reducing enzymes from Archaea, have the His ligand to the distal cluster replaced by a Cys residue and at the same time have the Pro residue near the 3Fe cluster replaced by a Cys residue, converting this cluster into a 4Fe one. In the F_{420}-reducing hydrogenases from Archaea the extra two clusters occur bound to two classical CxxCxxCxxxC motifs. Quite a few hydrogenases only contain the proximal cluster bound to a short version of the small subunit: the polypeptide ends shortly after the characteristic GCPP pattern, containing the fourth Cys ligand to the proximal cluster. The *Escherichia coli* hydrogenases 3 and 4 also contain only the proximal cluster, but have still a considerable stretch of amino-acid sequence after the GCPP pattern.

7.2. [Fe] hydrogenases

These contain Fe as the only transition metal and can be as simple as one polypeptide. Also these enzymes contain Fe-S clusters, but through the early work of the group of Veeger in Wageningen it became clear that the active site had to be something

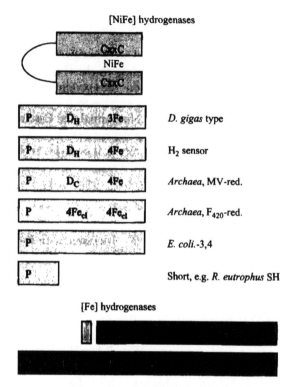

Figure 7.1 Schematic representation of the polypeptides and the prosthetic groups in the minimal functional modules of hydrogenases. Upper part: [NiFe] hydrogenases. The NiFe site is bound to two CxxC motifs in the N- and C-terminal regions of the large subunit. The length of the polypeptide connecting these two domains is greatly variable among the enzymes. All small subunits bind a proximal [4Fe-4S] cluster (P). Often additional clusters are present: a distal [4Fe-4S] cluster with a His residue as ligand (DH), a distal cluster like D_H, but with the His ligand replaced by a Cys ligand (D_C), a [3Fe-4S] cluster (3Fe), a [4Fe-4S] cluster with the same binding motif as the 3Fe cluster, except for a Cys residue replacing the otherwise conserved Pro residue (4Fe), or two classical cubanes bound to two CxxCxxCxxxC motifs ($4Fe_{c1}$ $4Fe_{c1}$). Lower part: [Fe] hydrogenases. There are two types which differ in the number of subunits and Fe-S clusters. The minimal functional module contains two subunits with the active site (6Fe) plus two classical cubanes ($4Fe_{c1}$) in the large subunit. Other enzymes consist only of one subunit with an extended N-terminal region containing an additional [4Fe-4S] cluster with a His residue as ligand ($4Fe_H$) plus a [2Fe-2S] cluster (2Fe). Other abbreviations: MV-red., methylviologen reducing hydrogenase; F_{420}-red., F_{420}-reducing hydrogenase; *E. coli*-3,4, hydrogenases 3 and 4 from *E. coli*; SH, soluble hydrogenase.

special. When the structural genes encoding the enzyme in *Desulfovibrio vulgaris* (the *D. vulgaris* enzyme has one extra subunit of 13 kDa which contains no Cys residue) were cloned and expressed in *E. coli*, an inactive protein was produced which otherwise behaved like the wild-type protein (and hence could be purified in exactly the same way without following activity as a marker) (Voordouw *et al.* 1987). This protein contained only two classical [4Fe-4S] clusters. Hagen *et al.* (1986) determined that the mature enzyme contained 14.4 Fe atoms and 12.3 acid-labile sulfur atoms per enzyme molecule. As the two cubane clusters each contain four Fe and four S^{2-}, Hagen proposed that the active site was a novel 6Fe cluster. In fact, after subtraction of the contributions from the cubane clusters, the active site in the *D. vulgaris* enzyme can be predicted to contain 6.4 Fe and 4.3 S^{2-}. The *Clostridium pasteurianum* hydrogenase[1] contains 20.1 Fe and 17.8 S^{2-} per molecule as determined by Adams *et al.* (1989). By Resonance Raman studies these workers found (W. Fu *et al.* 1993) that the enzyme contains three [4Fe-4S] clusters and one [2Fe-2S] cluster, in addition to the active site. As the Fe-S clusters require fourteen Fe and S^{2-}, this would leave 6.1 Fe and 3.8 S^{2-} for the active site in this enzyme. These numbers are in excellent agreement with the structure of the active site as determined by X-ray crystallography (Peters *et al.* 1998) (Chapter 6).

7.2.1. FTIR

As there is no similarity between the amino-acid sequences of [Fe] hydrogenases with those of [NiFe] hydrogenases, it was assumed that also the active sites would have little in common. Again, by sheer coincidence we found that [Fe] hydrogenases probably have an Fe(CN)(CO) group in the active site (Van der Spek *et al.* 1996). When testing whether the infrared bands in the *A. vinosum* enzyme (the origin of which was not yet known at the time) were unique, we (cooperation with W. R. Hagen) inspected many other (non-hydrogenase) redox enzymes with Fe-S clusters and/or Ni, but found no such bands there. It was a great surprise to find similar bands in samples of two [Fe] hydrogenases which were among the control samples. This discovery suggested that the architecture of the active sites in both classes of hydrogenases might be more alike than anticipated (e.g. an Fe-Fe site vs an Ni-Fe site). Also here the CN/CO bands appeared very sensitive to changes in the active site (Pierik *et al.* 1998a).

The crystal structures of two enzymes have been discussed in detail in Chapter 6. In the active site, one of the Fe atoms (called Fe1) from the Fe-Fe dinuclear site is linked to a [4Fe-4S] cluster via a Cys thiol group. In both crystal structures two non-protein thiol groups bridge between the two Fe atoms. In the *Desulfovibrio desulfuricans* enzyme, these thiols have been originally assigned to belong to a propane-1,3-dithiol molecule. Both structures predict two non-protein, diatomic ligands attached to each Fe atom. In addition, the *C. pasteurianum* structure shows an extra diatomic ligand bridging between the two Fe atoms. The Fe2 atom has an extra ligand ascribed to an OH^- or H_2O group. The accuracy of the crystal structures did not permit a characterization of the diatomic, non-protein ligands to the Fe-Fe centre.

A detailed FTIR study (Pierik *et al.* 1998a; see also Section 8) showed that the spectrum of the oxidized (inactive) *D. vulgaris* enzyme contains two infrared bands (at 2,106 and 2,087 cm^{-1}) which, in analogy to the [NiFe] hydrogenases (Happe *et al.* 1997; Pierik *et al.* 1999), were ascribed to the stretching frequencies of metal-bound CN^- groups, and two bands (at 2,007.5 and 1,983 cm^{-1}) due to end-on, metal-bound CO molecules (Fig. 7.2, upper trace). A fifth band at a considerably lower frequency (at 1,847.5 cm^{-1}) was attributed to a CO molecule bridging between two metals. Because the reduced, active enzyme (Fig. 7.2, lower trace) only showed one strong band in the ν(CO) frequency region, it was tentatively concluded that the oxidized enzyme was heterogeneous, resulting in two sets of $v(CO)/v(CN)$ bands. With the crystal structures of the di-iron site in hand (Fig. 7.2), however, a more straightforward interpretation of the FTIR and EPR data is possible. Both structures show the presence of two non-protein ligands to each of the Fe atoms with a size fitting that of a diatomic molecule (e.g. CN or CO). The FTIR spectrum (Fig. 7.2, upper trace) then immediately suggests the presence of one CN^- and one CO on each of the Fe atoms. Two cyanides on one Fe atom and two carbonyls on the other, seems less likely as vibrational coupling between two CO on the same metal usually leads to a large $(60–110\,cm^{-1})$ splitting between the two $v(CO)$ bands. This assumption remains to be verified by a 50 per cent enrichment in ^{13}C, as carried out with [NiFe] hydrogenases (Pierik et al. 1999). The C. pasteurianum Fe-Fe site structure (Fig. 7.2, upper structure), but not the D. desulfuricans one, shows an additional diatomic ligand bridging between the two Fe atoms. This is in good agreement with the $1,847.5\,cm^{-1}$ band in the active D. vulgaris enzyme, ascribed to a bridging CO molecule. FTIR spectra of the active D. vulgaris enzyme (Fig. 7.2, traces B and C), however, did not show such a band. A band of a bridging CO (at $1,811\,cm^{-1}$) was clearly detected in enzyme activated with H_2 and subsequently treated with CO.

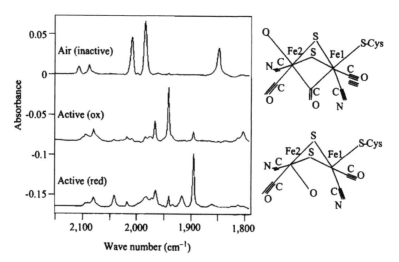

Figure 7.2 FTIR spectra and structures of the active site in [Fe] hydrogenases. Spectrum A (upper) from the aerobic, inactive D. vulgaris enzyme (Pierik et al. 1998a), shows two CN bands (around $2,100\,cm^{-1}$), two CO bands (around $2,000\,cm^{-1}$) and a bridging CO ($1,847\,cm^{-1}$). This agrees with a structure found in the C. pasteurianum [Fe]-hydrogenase-I enzyme under N_2 (upper structure; Peters et al. 1998). In spectra B and C (middle and lower) from the active D. vulgaris enzyme no band from a bridging CO is detectable. This agrees with the structure found in the D. desulfuricans enzyme under 10 per cent H_2 (middle structure; Nicolet et al. 1999). This agrees with the structure of the CO-treated C. pasteurianum enzyme (lower structure; Lemon and Peters 1999). In the structures 'S-Cys' indicates a thiol group from a Cys residue, which bridges between the Fe-Fe site and a [4Fe–4S] cluster; this Cys residue forms the only bond of the Fe-Fe site with the protein.

The X-ray structure of the CO-treated *C. pasteurianum* enzyme is now known (Fig. 7.2, lower structure) and perfectly explains the FTIR spectrum of the CO-treated *D. vulgaris* enzyme: two $v(CN)$ bands around $2,090\,cm^{-1}$, three $v(CO)$ bands between $1,950$ and $2,050\,cm^{-1}$ and one band from a bridging CO at $1,811\,cm^{-1}$. Nicolet *et al.* (1999) noted in the structure of the *D. desulfuricans* enzyme that two small ligands were in a position to form hydrogen bonds with main-chain N atoms, whereas the two other small ligands appeared to reside in hydrophobic pockets. Using our FTIR information of the *D. vulgaris* enzyme, they assigned the former to two CN^- ligands, and the latter to two CO ligands. Thus, the active site is depicted as in Fig. 7.2. At present it is not sure whether the bridging CO is an intrinsic, non-exchangeable ligand of the enzyme (see later).

7.2.2. Amino-acid sequences

For [Fe] hydrogenases, the positions of the binding sites for Fe-S clusters and the H cluster (6Fe), as deduced from the amino-acid sequences and the crystal structures, are depicted in Fig. 7.1. The eight Cys residues immediately preceding the conserved region binding the H cluster are grouped like those found in many proteins containing two classical [4Fe-4S] clusters. Remember that the dinuclear Fe site is only linked to the protein by one Cys residue. Some [Fe] hydrogenases, like hydrogenase[1] from *C. pasteurianum*, have an extended N-terminal region containing an extra [2Fe-2S] and [4Fe-4S] cluster. All Fe-S clusters serve as a path for electrons.

Both [NiFe] and [Fe] hydrogenases are often complexed to other redox-enzyme modules with various substrate specificities (dependent on the metabolism of the microorganism).

7.3. Two more enzyme classes dealing with H₂

The two classes introduced above are true hydrogen catalysts in the sense that the pure proteins accelerate both reactions of the equation $H_2 \leftrightarrow 2H^+ + 2e^-$. There are two more classes of enzymes which have H_2 as a reaction partner (see also Chapter 2). One class is formed by the nitrogenases, enzymes that reduce dinitrogen to ammonia. These enzymes contain complicated prosthetic groups based on iron and sulfide, often in combination with molybdenum or vanadium, and they produce H_2 during their activity cycle (at least one H_2 per N_2). Nitrogenase cannot, however, activate H_2 itself and is therefore (thus far) not considered a true hydrogenase. Another enzyme group is formed by the H_2-forming methylenetetrahydromethanopterin dehydrogenases found in certain methane-producing microorganisms. They do not contain any transition metal and the pure protein cannot activate H_2 either. Hydrogen binding and activation occurs on the specific redox partner N^5,N^{10}-methenyltetrahydromethanopterin ($CH\equiv H_4MPT^+$) once bound to the enzyme (Section 8.4).

7.3.1. Hydrogenases and oxygen

[Fe] hydrogenases are thus far only found in organisms growing under strictly anaerobic conditions. This makes it understandable that these enzymes are not tailored by Nature to be particularly resistant to oxygen and indeed this gas rapidly and irreversibly denatures most of these enzymes. The contrary is true for most [NiFe] hydrogenases. They are found in many organisms that can grow anaerobically as well as aerobically. Though inactivated by O_2, most of these enzymes are not denatured; they can be activated again via reduction, e.g. by their substrate H_2 (once O_2 is absent). Purple sulfur bacteria (e.g. *A. vinosum*) grow anaerobically during day time (producing ATP via their cyclic photophosphorylation system), but switch their metabolism to an aerobic one during night time. Hence, it is not so surprising that their hydrogenase survives contact with air. A few hydrogenases can even happily function in air; they seem not in the least perturbed by the presence of O_2! Also some H_2-sensor proteins can detect H_2 present in air.

7.3.2. Activation of H_2: A heterolytic process

The splitting of H_2 by hydrogenases is heterolytic (into H^- and H^+), rather than homolytic (into two H. radicals). The hydride is considered to deliver two electrons at a time to the enzyme. A [4Fe-4S] cluster in proteins can, however, accept only one electron. Other redox enzymes, e.g. flavoproteins, dealing with two-electron donors (like NADH) solve this problem by first accommodating both electrons onto the flavin, whereafter these electrons are transferred to an Fe-S cluster one at a time.

The minimal functional module in [NiFe] hydrogenases always contains the $NiFe(CN)_2(CO)$ site plus the proximal [4Fe-4S] cluster. The active site in [Fe] hydrogenases consists of the Fe-Fe site linked to a [4Fe-4S] cluster. Oxidation of the hydride is either an action of the dinuclear site alone, or a concerted action of this site plus the proximal cluster.

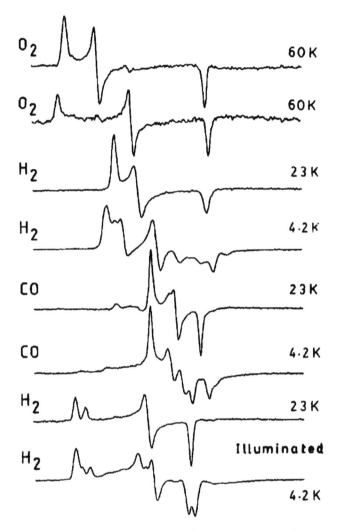

Figure 7.3 Overview of EPR signals from the active site of *A. vinosum* hydrogenase. From top to bottom: (i) Ni_u^* state; (ii) Ni_r^* state; (iii) Ni_a-C^* state; (iv) Ni_a-C^* signal split by the reduced, proximal Fe-S cluster; (v) Ni_a^*·CO state; (vi) Ni_a^*·CO signal split by the reduced, proximal cluster; (vii) Ni_a-L^* state; spectrum is an overlap of two different signals; (viii) Ni_a-L^* signal split up by the reduced, proximal Fe-S cluster. Spectra were taken from the thesis of J.W. van der Zwaan, Amsterdam, 1987.

7.4. Activity in relation to the redox states of the bimetallic centre in [NiFe] hydrogenases

7.4.1. Probing the behaviour of the NiFe(CN)₂(CO) site in an enzyme solution

Many [NiFe] hydrogenases (from anaerobic or semi-aerobic microorganisms) dissolved in aerobic buffer contain two unpaired electrons. One of these is located on the [3Fe-4S]

Figure 7.4 Space-filling representation of the $Fe(CN)_2(CO)$ model compound studied by Darensbourg *et al.* (1997). The cyclopentadienyl molecule is at the lower end of the figure. Prof. M. Y. Darensbourg is acknowledged for providing the coordinates.

cluster; the other is located in the active site. The oxidized proximal and distal clusters are diamagnetic (non-magnetic). Unpaired electrons can be detected by Electron Paramagnetic Resonance (EPR; see Section 5.4) spectroscopy. The EPR spectrum not only reveals the presence of such an unpaired electron, but can provide very useful information about the immediate environment of the unpaired electron as well as its actual concentration. An unpaired electron is often referred to as an unpaired spin and hence one often talks about the spin concentration. By a variety of tricks (the use of isotopes of Ni, Fe, S or Se with a magnetic nucleus) it has been established that the unpaired spin located in the active site is close to nickel and at least one of its sulfur ligands, but not to iron, so I will call it here the nickel-based unpaired spin (see Section 11 for an example of interaction with ^{77}Se). From an EPR point of view it can be described as due to a low-spin, trivalent nickel ion. The coordination geometry is rather asymmetric, but could be described as approximating to one of the following symmetries: tetragonally distorted octahedral, trigonal bi-pyramidal or square-pyramidal. EPR can thus be used to monitor changes at and around Ni (Fig. 7.3). The $Fe(CN)_2(CO)$ moiety is diamagnetic.

FTIR spectroscopy is ideally suited to specifically study the changes at the Fe site (for details see Section 7.8). In short, the CO-stretching frequency (and to a lesser extent also the CN-stretching frequency) is very sensitive to changes in the charge density on the Fe ion. It can be compared with the spectra of model compounds (see Fig 7.4, section 7.14).

7.4.2. Oxidized enzyme in air (inactive)

EPR spectra of aerobic [NiFe] hydrogenases are very similar, showing that the structure around nickel is highly conserved in all enzymes. In many enzymes the nickel-based spin can be in two, slightly different environments (coordinations). Often preparations are a mixture of two types of enzyme molecules. Both types are inactive. When oxygen is removed and hydrogen is provided, then one type of enzyme molecules shows full activity within minutes (at 30°C); this type is called ready. The other type is unready, since it takes hours before full activity is obtained (see Section 5.4 in Chapter 5 for further information). To distinguish between the two different states of

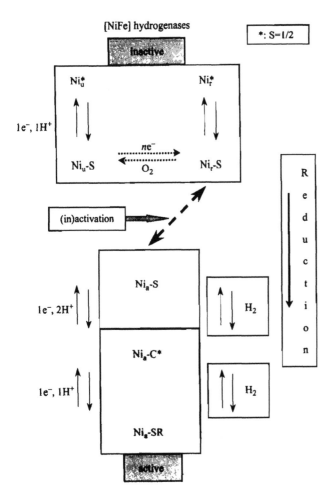

Figure 7.5 Overview of the several states of the active site in the [NiFe] hydrogenase from *A. vinosum*. Transitions can be invoked by redox titrations in the presence of redox mediators and involve both electrons (e^-) and protons (H^+) as indicated on the left side of the blocks. H_2, in the absence of mediators, can rapidly react only with enzyme in the active states (lower block), as indicated on the right side. Dotted arrows indicate a very slow reaction. Very similar states are found in the *D. gigas* enzyme.

the enzyme and their EPR signals from the nickel-based unpaired spin, they will be denoted here as the Ni_r^* and Ni_u^* states or signals (r for ready, u for unready, * for unpaired spin; see Fig. 7.5 for overview).

Experiments whereby reduced enzyme was reoxidized by O_2 enriched in ^{17}O (which has a magnetic nucleus) showed that an oxygen species ended up close to the Ni-based unpaired spin in the ready as well as in the unready state (Van der Zwaan *et al.* 1990). It could only be removed by full reduction and activation of the enzyme. The crystal structure (see Chapter 6) shows an oxygen atom close to nickel and it is

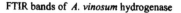

FTIR bands of *A. vinosum* hydrogenase

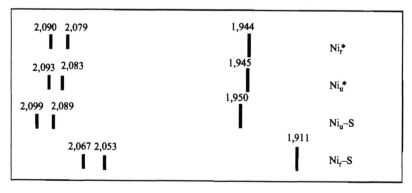

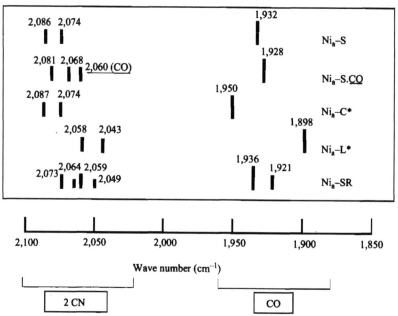

Figure 7.6 Overview of the CN and CO stretching frequencies observed for the several states of the *A. vinosum* [NiFe] hydrogenase. The symmetric and anti-symmetric stretching frequencies of the CN^- groups and of the stretching frequency of the CO molecule from the $Fe(CN)_2(CO)$ moiety in the active site are shown. The band of externally added CO is indicated in grey.

now assumed that as long as this oxygen species is there, the reaction with H_2 is severely hampered. EPR tells us that the Ni sites in the ready and unready enzyme are slightly different. The FTIR spectra indicate that the Fe sites in the Ni_r^* and Ni_u^* states are very similar. An overview of the FTIR-band positions of the several states of the *A. vinosum* enzyme is given in Fig. 7.6.

7.4.3. Enzyme with one electron added to the active site (inactive)

The active site in either the ready or unready state can accept one electron, as demonstrated with redox titrations (see Chapter 5). Since O_2 is a very greedy electron acceptor, one must take care that it is absent. Usually, redox titrations are reversible (even at 2°C), but the reduction of the *A. vinosum* enzyme in the ready state is irreversible at 2°C. We have shown that a reversible redox titration of the ready state is only obtained at elevated temperatures (>30°C). We do not yet understand the reason for these differences in behaviour of ready and unready enzyme.

Even after this initial reduction, the enzyme from *A. vinosum* remains inactive. When performed at 2°C and pH 6, the redox potential in such an enzyme solution can be lowered to -350 mV without any increase in activity; also no EPR signals of nickel-based unpaired spins are detectable. So the active site can shuttle between $Ni_{r,u}^*$ and the one-electron reduced state $Ni_{r,u}$-S (S stands for EPR silent), without activation of the enzyme. A proton accompanies the electron (apparently a strict charge compensation is obligatory, as the midpoint potential is dependent on the pH (-60 mV per pH unit). Although the changes in the EPR spectra suggest just a reduction of nickel in both cases, the FTIR spectra reflect clear changes at the Fe site for the ready enzyme, but not for the unready one (Fig. 7.6).

Upon reduction of the unready enzyme the electron/proton combination probably goes to the nickel site:

$$Ni^{(III)} - Fe^{(II)} + e^- + H^+ \rightarrow (H^+)Ni^{(II)} - Fe^{(II)}.$$

The infrared spectrum of the Ni_u-S state indicates even a slight decrease of charge density on Fe (a shift of the CN/CO bands to higher frequencies). In the infrared spectrum of the Ni_r-S state, however, all bands have shifted to considerably lower frequencies (the CO band shifts 43 cm^{-1}), indicating a greatly increased charge density on Fe. Two possible ways to explain this are e.g:

(i) The incoming electron and proton both go to the Fe site:

$$Ni^{(III)} - Fe^{(II)} + e^- + H^+ \rightarrow Ni^{(III)} - Fe^{(I)}(H^+).$$

Reduction of the low-spin $Fe^{(II)}$, which is non-magnetic, would create an unpaired spin. Its magnetic moment might easily couple to the nickel-based spin, thereby cancelling the total magnetism and so no EPR signal could be observed. The increased electron density on the iron would result in a large shift (50–100 cm^{-1}) to lower frequencies of the CN/CO bands; protonation of a thiolate ligand would largely reverse this shift. In model compounds protonation of a thiolate ligand to Fe can increase the stretching frequency of CO bound to Fe by 40 cm^{-1} (Sellmann *et al.* 1996). Usually reduction of $Fe^{(II)}$ is expected, however, to occur only at considerably lower potentials.

(ii) The added electron and proton both go to nickel, as in the case of Ni_u^*, but now the charge density on the Fe ion also increases due to a better electronic contact between the two metal ions in the ready state. The question is, why this would be so. The differences in the reversibility of the reactions for ready and unready enzyme also suggest that the effect and/or mechanism of reduction is different for both forms. Clearly more experiments are required to sort this out.

For *A. vinosum* hydrogenase, activation requires not only reduction, but an elevated temperature is essential as well. When enzyme in the Ni_r^* state is reduced at 30°C or higher, but not at 2°C, a rapid increase in activity is observed, with the development of a third EPR signal of a nickel-based unpaired spin, the Ni_a-C^* signal (a from active, C for third). Once in the active state, the active site in the enzyme displays properties quite different from those of the oxidized state: it now can exist in three different redox states (Fig. 7.5), one of which has an unpaired spin. The states are called Ni_a-S (one-electron reduced state of the Ni-Fe site in active enzyme), Ni_a-C^* for the intermediately reduced state and Ni_a-SR (R for reduced) for the most reduced state.

It is presently thought that the most obvious reason for the changed properties upon conversion from the unready to the ready state and then to the active state is the removal of the bound oxygen species (e.g. by reduction to OH^-, subsequent protonation to H_2O and removal of this molecule from the active site). As indicated above, reduction causes a slow Ni_u-S $\rightarrow$ Ni_r-S transition. It has been shown with the *D. gigas* enzyme (De Lacey *et al.* 1997) that at 40°C and the appropriate redox potential, the species with the $v(CO)$ of $1,914\,cm^{-1}$ (Ni_r-S in our scheme) prevails at high pH, whereas a species with $v(CO)$ at $1,934\,cm^{-1}$ (Ni_a-S) is the major form at low pH. It could well be that protonation of an OH^- bound to nickel in the Ni_r-S state forms a water molecule, which then leaves the active site (at 40°C, but not at 2°C; see also Section 5.7), whereafter the site becomes accessible for a rapid reaction with H_2.

7.4.4. The Ni_a-S $\leftrightarrow$ Ni_a-C^* reaction

Cammack *et al.* (1987) have shown that the reaction Ni_a-S $\rightarrow$ Ni_a-C^* in *D. gigas* hydrogenase is reversible, and the pH dependence of the potential indicated that it involved one electron and one to two protons. The redox potential must be set by adjusting the H_2 partial pressure in the gas phase as enzyme in the Ni_a-C^* state in the presence of redox mediators is reducing protons to H_2 (see Chapter 5). I have long held the opinion that the divalent nickel in the Ni_a-S state was further reduced to monovalent nickel. Recent experiments in my group (Section 6) have convinced me now that Ni_a-C^* rather represents trivalent nickel as initially suggested by other groups (Jacobsen *et al.* 1982; Teixeira *et al.* 1983). Hence the Ni_a-S $\rightarrow$ Ni_a-C^* reaction can be written as

$$Ni^{(II)} - Fe^{(II)} + e^- + 2H^+ \rightarrow Ni^{(III)} - Fe^{(II)}(H_2).$$

At this point it is not known if the proton taken up during reduction of Ni_r^* is still present. Note that the CN bands in the FTIR spectra do not shift, suggesting no changes in electron density on the Fe ion. The CO band, however, shifts $18\,cm^{-1}$ to higher frequency. Thus far there is no reasonable explanation for this behaviour; we do not yet know where H_2 (presumably as H^-/H^+) is precisely bound. As mentioned above, in the presence of mediating dyes, an H_2-producing reaction

$$Ni^{(III)} - Fe^{(II)}(H_2) + e^- \rightarrow Ni^{(II)} - Fe^{(II)} + H_2.$$

can occur as well, unless H_2 is present in the gas phase to counteract this.

The Ni_a-S $\rightarrow$ Ni_a-C^* reaction is also driven by H_2 alone, but then the reverse reaction is extremely slow (dotted arrow in Fig. 7.5). To explain the reaction with H_2 in the absence of mediators, involvement of an Fe-S cluster has to be assumed (see below).

In 1985 we discovered (Van der Zwaan *et al.* 1985) that the Ni_a-C^* state contains a water-exchangeable, bound hydrogen species not far from nickel. When in the frozen state, the bond of this species to the active site can be broken by illumination with white light at temperatures below that of liquid nitrogen resulting in the Ni_a-L^* state (see also Section 7.11). This means that the hydrogen species (H_2) in the reaction above is presumably bound (close) to the active site. In contrast to our initial assumption, we have now reasons to believe that nickel is not involved in this binding: the characteristic EPR signal of the Ni_a-C^* state can also be observed in irreversibly inactivated, aerobic enzyme and then is not sensitive to light (B. Bleijlevens and S.P.J. Albracht, unpublished experiments). We hence assume that the hydrogen species in the Ni_a-C^* state is bound close to the Fe site, presumably as a hydride with a nearby bound proton. Photodissociation of the hydrogen species leads to a large increase of electron density in the Ni-Fe site shifting all infrared bands from CN and CO to lower frequencies (see Fig. 7.6 and Section 7.8). An experimental approach to answer the question where the light-sensitive hydrogen species is located is ENDOR. This electron-nuclear double resonance technique has already been successfully applied to localize nearby protons in oxidized [NiFe] hydrogenases (see Section 7.12).

7.4.5. The Ni_a-C^* → Ni_a-SR reaction

In the presence of redox mediators, the Ni_a-C^* state can be further reduced to an EPR-silent state (by increasing the H_2 partial pressure) in a reaction requiring one electron and one proton. Assuming trivalent nickel in the Ni_a-C^* state, this reaction can be most simply interpreted as

$$Ni^{(III)} - Fe^{(II)}(H_2) + e^- + H^+ \leftrightarrow (H^+)Ni^{(II)} - Fe^{(II)}(H_2).$$

As the active enzyme is in equilibrium with H_2 also in the absence of mediators, a redox titration with He/H_2 mixtures has been performed as well (Coremans *et al.* 1992). The following reaction was observed:

$$Ni_a - C^* + 2H^+ + H_2 \leftrightarrow Ni_a - SR.$$

Consistent with the two-electron donor nature of H_2, the reaction behaved as an $n = 2$ Nernst redox reaction. It showed a pH dependence of $-66\,mV$ per pH unit, so again one proton was taken up for each electron. It is not known where all incoming protons are localized in the enzyme. The reaction shows that in addition to the light-sensitive hydrogen species bound to the active site in the Ni_a-C^* state, a *second* hydrogen can react at the active site and deliver its two electrons to the enzyme. We hence proposed that the active site of the *A. vinosum* enzyme has two sites where hydrogen can bind. If H_2 is completely removed, the Ni_a-C^* state persists for hours; this is unlike the situation in redox titrations in the presence of redox mediators. As the active site in the Ni_a-SR state has one electron more than that in the Ni_a-C^* state, an Fe-S cluster has to be involved in this reaction with H_2.

7.4.6. Involvement of Fe-S clusters in the redox reaction with H_2

From rapid-mixing rapid-freezing experiments (see Section 7.6), and also from XAS studies on hydrogenases such as the *A. vinosum* enzyme (Gu *et al.* 1996;

Davidson *et al.* 2000) (Section 7.13), it can be concluded that the formal valence state of Ni shuttles between 2+ and 3+. Hence, in the absence of mediators, we have to assume the involvement of one of the Fe-S clusters in the two-electron reaction of the *A. vinosum* enzyme with H_2:

$$Ni^{(III)} - Fe^{(II)} [4Fe-4S]^{2+} + H_2 \leftrightarrow Ni^{(II)} - Fe^{(II)} [4Fe-4S]^{1+} + 2H^+.$$

Only the proximal (P) and distal (D) clusters can be involved, as the 3Fe cluster has too high a redox potential. In [NiFe] hydrogenases, the proximal cluster is the only one which is strictly conserved in all enzymes. Hence, we assume that the proximal cluster is indispensible for the primary, rapid reaction of H_2 with the enzyme. Its short distance to the active site enables a clearly EPR-detectable spin–spin interaction of the unpaired spin of the reduced Fe-S cluster with the Ni-based spin (see Section 7.10). It is stressed here that when performing these reactions under equilibrium conditions, no change of the redox states of the Fe-S clusters can be observed (see below).

Why do the equilibrium reactions with H_2 not show a change of redox state of the Fe-S clusters?

This is what we consistently find by monitoring EPR and Mössbauer spectra (cooperation with Prof. E. Münck, Pittsburgh) of the Ni_a-S·CO, Ni_a-C* and Ni_a-SR states the *A. vinosum* enzyme produced by H_2 and/or CO at pH 8. A possible explanation is to assume that the individual enzyme molecules can exchange electrons. The best suited place for this is via the distal [4Fe-4S] cluster which is located on the surface of the protein. Such an exchange would occur on a one-electron basis and would be much slower (depends on the collision rate of the 90 kDa enzyme) than the reaction with H_2 (which is extremely fast and depends on the rate of diffusion of H_2 into the enzyme (Pershad *et al.* 1999)). Suppose that the Ni_a-C* state is initially formed with one Fe-S cluster in the oxidized state:

$$Ni^{(III)} - Fe^{(II)}(H_2)/[FeS]_P{}^+/[FeS]_D{}^{2+}.$$

Two such molecules could exchange an electron resulting in one molecule with two oxidized [4Fe-4S]$^{2+}$ clusters and the other with two reduces clusters:

$$Ni^{(III)} - Fe^{(II)}(H_2)/[FeS]_P{}^{2+}/[FeS]_D{}^{2+} + Ni^{(III)} - Fe^{(II)}(H_2)/[FeS]_P{}^+/[FeS]_D{}^+.$$

A simple reaction of the former molecule with H_2 would reduce its two clusters again. The final level of reduction of the Fe-S clusters would then only depend on the effective redox potential in the system. Under 1 per cent H_2 at pH 6, where the redox potential is about $-295\,mV$, part of the Fe-S clusters are oxidized. At this pH they only reduce under 100 per cent H_2. At pH 8 or higher, however, the redox potential of both 100 per cent H_2 and 1 per cent H_2 is low enough to keep all clusters fully reduced. So via intermolecular, one-electron transfer reactions the Fe-S clusters can presumably follow the current potential imposed upon the system by H_2 even in the absence of redox mediators. Of course, the presence of such mediators facilitates electron (re)distribution.

7.4.7. An enzyme with a blocked Ni_a-C^* → Ni_a-SR reaction

An interesting enzyme, from which we can probably learn more about the mechanism of H_2 activation, is the H_2-sensor protein. This protein is now found in several micro-organisms (see Chapter 3 for details). Recently the H_2 sensor from *Ralstonia eutropha* was studied with EPR and FTIR (Pierik *et al.* 1998b). Its active site is highly similar to the one in standard [NiFe] hydrogenases. Its enzymatic properties, however, are quite different. The sensor is two orders of magnitude less active, is always in the active Ni_a-S state (even in aerobic solution), is insensitive to O_2 or CO and can be reduced with H_2 to the Ni_a-C^* state, but not any further. Why this is so, is not understood yet, but it is consistent with the view that the rapid turnover of H_2 by hydrogenases showing an Ni_a-C^* signal is presumably not due to the Ni_a-C^* → Ni_a-C^* reaction. Rather, it might mainly involve the transition Ni_a-C^* + H_2 ↔ Ni_a-SR (Fig. 7.5).

7.5. Activity in relation to the redox states of the bimetallic centre in [Fe] hydrogenases

7.5.1. The oxidized D. vulgaris enzyme in air (inactive)

The active site in the aerobic, inactive *D. vulgaris* enzyme shows no EPR signals. The FTIR spectrum indicates the presence of a bridging CO, in addition to the end-on bound CN/CO pair to each iron atom (Fig. 7.2, trace A). We presently assume that both Fe atoms in the bimetallic site are low-spin ferric and that their $S = 1/2$ systems are anti-ferromagnetically coupled to a diamagnetic state.

7.5.2. Reduction and activation of the active site

The inactive enzyme can be activated by reduction. The underlying process is not yet known, but as indicated above, the apparent result is that in fully reduced enzyme the bridging CO is no longer discernable in either FTIR (Fig. 7.2, trace C) or the X-ray structure (Fig. 7.2, middle structure). When the active site in active enzyme is subsequently oxidized (by removal of H_2, or by addition of an oxidant), it shows the well-known $g = 2.10$ rhombic $S = 1/2$ EPR signal (see trace E in Fig. 7.7). We have ascribed this signal (Pierik *et al.* 1998a) to a low-spin $Fe^{(III)}$ centre, based on comparisons with a $[Ph_4P]_2[Fe^{(III)}(PS3)(CO)(CN)]$ model compound (Hsu *et al.* 1997), with very similar FTIR and EPR characteristics, as well as based on comparison with other low-spin $Fe^{(III)}$ compounds. The bimetallic site then presumably is an $Fe^{(III)}$-$Fe^{(III)}$ pair, obtained by oxidation from a fully reduced, diamagnetic $Fe^{(II)}$-$Fe^{(II)}$ pair.

Active, oxidized enzyme (with the $Fe^{(III)}$–$Fe^{(III)}$ pair) can rapidly react with H_2, as first demonstrated twenty-five years ago (Erbes *et al.* 1975). As only one electron can be accommodated by the $S = 1/2$ $Fe^{(II)}$–$Fe^{(II)}$ centre, the nearby [4Fe-4S] cluster is probably involved in this reaction as well:

$$Fe^{(II)} - Fe^{(III)} [4Fe-4S]^{2+} + H_2 \rightarrow Fe^{(II)} - Fe^{(II)} [4Fe-4S]^{1+} + 2H^+.$$

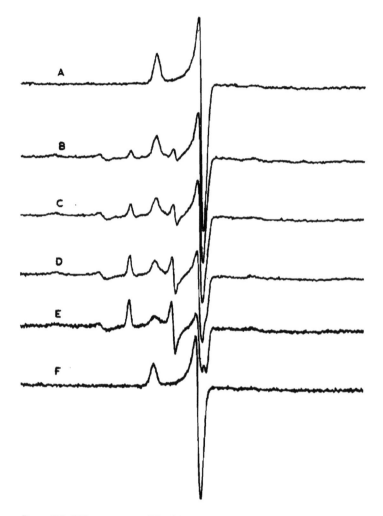

Figure 7.7 EPR spectra at 9K of *D. vulgaris* [Fe] hydrogenase. (A) H_2-reduced enzyme subsequently treated with CO; (B–E) Spectra recorded after various times of illumination
at 9K. (F) Spectrum recorded after warming up of the sample to 150K for 10 min.
Reprinted with permission from Patil, *et al.* (1988) and the American Chemical
Society.

This reaction is analogous to the rapid equilibrium reaction of H_2 with [NiFe] hydrogenases described above:

$$Fe^{(II)} - Ni^{(III)} [4Fe-4S]^{2+} + H_2 \leftrightarrow Fe^{(II)} - Ni^{(II)} [4Fe-4S]^{1+} + 2H^+.$$

In both cases the metal ion closest to the Fe-S cluster is being reduced. Hence, the
mechanism of H_2 oxidation may be very similar for the two enzyme classes. The main
change in the CO region of the FTIR spectrum during the reduction of the *D. vulgaris*

enzyme (Pierik *et al.* 1998a) (a shift of the main CO peak from 1,940 cm^{-1} to 1,894 cm^{-1}; Fig. 7.2, traces B and C) is in agreement with the reduction of an Fe$^{(II)}$–Fe$^{(III)}$ state to an Fe$^{(II)}$–Fe$^{(II)}$ one, both lacking a detectable bridging CO ligand.

Calculation of possible gas-access channels in the *D. desulfuricans* enzyme (Nicolet *et al.* 1999) indicates that there is a single channel connecting the surface of the protein with the vacant coordination site of Fe2 (see Chapter 6 for details). Hence it was proposed that Fe2 is the site of reaction with H$_2$. Carbon monoxide can bind to this vacant site: the active, oxidized Fe$^{(II)}$–Fe$^{(III)}$ state can easily react with CO. This results in an inhibited enzyme with a significantly changed EPR spectrum (the rhombic 2.10 signal is replaced by an axial 2.06 signal (Erbes *et al.* 1975; see traces E and F in Fig. 7.7). The bound CO in the *D. vulgaris* enzyme can be photolysed from the active site at low temperatures (Patil *et al.* 1988), resulting in the restoration of the rhombic *g* = 2.10 EPR signal (Fig. 7.7, traces A–E). The FTIR spectrum of the *D. vulgaris* enzyme (Pierik *et al.* 1998a) and the X-ray structure of the *C. pasteurianum* enzyme (Lemon and Peters 1999) in the CO-inhibited state point to the presence of a CO molecule bridged between the Fe atoms (Fig. 7.2). Experiments with ^{13}C-enriched CO (De Lacey *et al.* 2000) with the *D. desulfuricans* enzyme show that the bands at 2,016 and 1,971 cm^{-1} are due to the symmetrical and antisymmetrical stretch vibrations of two CO molecules bound to Fe2, the externally added CO molecule being one of them, whereas the 1,964 cm^{-1} band is from CO bound to Fe1 (Fig. 7.2). This experiment also indicated that the bridging CO is an intrinsic CO molecule.

7.6. Reaction of [NiFe] hydrogenases with H$_2$ and CO on the millisecond timescale

Randolph P. Happe

[NiFe] hydrogenase of the purple-sulfur bacterium *A. vinosum* can exist in various redox states (see Fig. 7.5). Until recently the redox transitions have been studied under equilibrium conditions only. I will shortly report here on the first rapid reactions of H$_2$ and CO with [NiFe] hydrogenases using the so-called rapid-mixing rapid-freezing technique. This technique enables the study of the reaction of an enzyme with its substrate and/or an inhibitor on a timescale down to 10 ms (milliseconds). The basic elements of

Rapid-mixing rapid-freezing technique

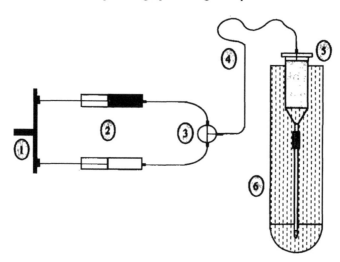

Figure 7.8 Rapid-mixing rapid-freezing apparatus. Two syringes (2) are filled with a solution of enzyme and substrate, respectively. By applying a large force (1) on the syringes, the solutions are driven into the mixer (3) whereafter the reaction starts. The reaction mixture flows through a reaction hose (4) and is then sprayed via a thin nozzle (5) in a funnel filled with cold isopentane ($-140°C$). This causes a rapid (5 ms) quenching of the reaction. The frozen powder is subsequently collected in an EPR tube attached to the funnel (6) and then is ready for EPR measurements. The funnel and EPR tube are held in a dewar with isopentane ($-140°C$). The reaction time can be varied by changing the length and diameter of the reaction hose.

the freeze-quench apparatus are depicted in Fig. 7.8. Because the reactions of hydrogenase are inhibited by oxygen, the experiments were carried out under anaerobic conditions by placing the freeze-quench apparatus in an Ar-filled glove box. The experiments were carried out by me and Winfried Roseboom in the group of Siem Albracht.

Hydrogenase of *A. vinosum* reduced by 100 per cent H_2 (0.8 mM) to the Ni_a-SR state was mixed in a 1:1 ratio with Ar-saturated buffer or with a buffer saturated with CO (Happe *et al.* 1999). The H_2 concentration is thereby halved, leading to an increase of the redox potential. Upon mixing with Ar-saturated buffer we found that the Ni_a-SR → Ni_a-C* + H_2 reaction was complete within 8 ms, implying a lower limit of $125 s^{-1}$ for the rate of this reaction. This rate is higher than that of the hydrogen-production reaction ($\approx 50 s^{-1}$) as assayed in standard activity measurements. Thus, the reaction as studied here is kinetically competent for functioning in hydrogen production by the enzyme. In time, there was no further change of the Ni_a-C* signal.

Mixing with CO-saturated buffer revealed that the reactions with CO were influenced by light. Hence all experiments were carried out in the dark. Within 8 ms an Ni_a-C* signal as strong as after mixing with Ar-saturated buffer was obtained. In the present case, however, this signal was not stable but disappeared within 40 ms (formation of the Ni_a-S·CO state as earlier studied by FTIR; see Fig. 7.6). The reactions show that CO does not bind to the Ni_a-SR or the Ni_a-C* states, but that the presence of CO leads to the rapid formation of the Ni_a-S·CO state.

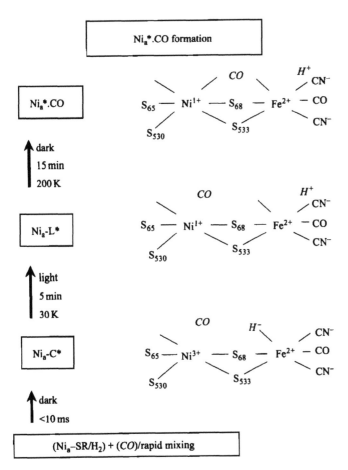

Figure 7.9 Light-dependent binding of CO to Ni at the level of the Ni_a-C^* state. After mixing of H_2-reduced enzyme with CO-saturated buffer in the dark, the Ni_a-C^* state is formed within 10 ms. Carbon monoxide does not bind to Ni due to its high valence state. When illuminated at 30 K the Ni_a-L^* state is formed, where the charge density at the Ni-Fe site has greatly increased (see also the shift of the FTIR bands in Fig. 7.6). Upon raising the temperature to 200 K in the dark, the nearby CO now can bind to the electron-rich Ni. The Ni_a^*.CO species has been earlier characterized by our group.

It was previously shown by our group that CO can bind to the active site at the level of both the Ni_a-S and the Ni_a-C^* states. Here we discovered, however, that CO can very rapidly (within 10 ms) reach the active site, but cannot bind to enzyme in the Ni_a-C^* state in the absence of light. Binding occurs only upon illumination. This could even be demonstrated in a frozen enzyme solution (at temperatures below −80°C) by a cycle of illumination and dark adaptation. Based on these findings we conclude that CO does not bind to Ni_a-C^* because of the low electron density on Ni (Fig. 7.9).

Thus, we assume that Ni is trivalent in this state, as suggested earlier by other workers. We also assume that a hydride is bound at or near the Fe ion in this state. Upon illumination the bond between the hydride and the Fe is broken, whereby the two electrons of the hydride are transferred to the bimetallic active site. The major part of the electron density moves to Ni. It is only then that CO, when given enough thermal motion (200 K), can bind.

These first rapid-mixing rapid-freezing studies on an [NiFe] hydrogenase have provided us with a new insight into the reactions with H_2 and CO with the enzyme. It also allowed us to make important conclusions concerning the formal redox states of the Ni ion in the three redox states.

7.7. Modifications in the H_2-activating site of the NAD-reducing hydrogenase from R. eutropha

Christian Massanz

An alternative approach to find out which ligands of the active site in hydrogenases are of importance for an optimal activity, is to study enzyme with point mutations. Several groups are actively working in this field and a detailed example is provided in this section.

The cytoplasmic NAD-reducing hydrogenase (SH) of the bacterium R. eutropha is a heterotetrameric enzyme, which contains several cofactors (Friedrich et al. 1996; Thiemermann et al. 1996). The Ni-containing subunit is called HoxH. This subunit plus the small subunit HoxY form the strictly conserved hydrogenase module with the Ni-Fe centre and a proximal [4Fe-4S] cluster. HoxF and HoxU represents the Fe-S/flavoprotein moiety which is closely related to a similar moiety in NADH:ubiquinone oxidoreductase. The SH has been subject to molecular biological techniques in order to study its modular structure, mechanism and biosynthesis.

A variety of deletions were introduced into the four SH structural genes to obtain mutant strains producing monomeric and dimeric moieties of the enzyme (Massanz et al. 1998). A monomeric HoxH form was generated which was proteolytically processed at the C terminus like the wild-type polypeptide. While the hydrogenase dimer displayed D_2/H^+-exchange and H_2-dependent dye-reducing activity (Chapter 5), the monomeric form did not mediate the activation of H_2 although a Ni-Fe centre was incorporated into HoxH. HoxFU dimers were formed and displayed NADH-oxidoreductase activity, while the monomeric subunits were proteolytically degraded rapidly in the mutant cells. Mixing the hydrogenase and the NADH-oxidoreductase modules in vitro reconstituted structure and catalytic function of the SH holoenzyme demonstrating that both modules were formed in a native state.

To investigate the biological function of the C-terminal extension of HoxH (Thiemermann et al. 1996), two amino-acid replacements were introduced by site-directed mutagenesis at the proteolytic cleavage site (Massanz et al. 1997). C-terminal proteolysis of HoxH was blocked by replacing the alanine, the first residue of the extension, by a proline. This prevents the oligomerization of the subunits to the tetrameric holoenzyme. In the second mutant, the C-terminal extension of HoxH was eliminated by substituting the alanine for a translational stop codon. Although this mutant subunit was able to form the oligomeric holoenzyme, it was almost devoid of Ni and contained only traces of H_2-activating functions. Additional experiments with this mutant strain, in collaboration with Simon Albracht (Amsterdam), showed that the $Fe(CN)_xCO$ moiety of the bimetallic centre is able to

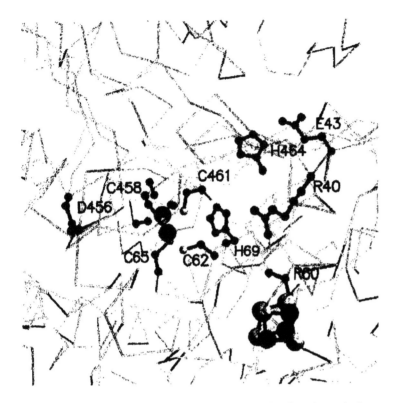

Figure 7.10 Structural and functional model of the NAD-reducing hydrogenase of *R. eutropha*, showing the positions of residues modified by mutagenesis.

incorporate independently from Ni into the SH. Furthermore, divalent metals like Zn and Co are apparently able to stabilize the tetrameric enzyme; whether they compete with Ni and Fe ions for binding at the active-site pocket is under investigation (unpublished results). These results showed (i) that a HoxH precursor with a C-terminal extension is not competent for subunit oligomerization and (ii) that the C-terminal extension of HoxH is necessary to direct a specific insertion of the complete Ni-Fe centre into the SH.

An important functional aspect in hydrogenase catalysis is the disposal of protons and the release of electrons. A set of histidines, an arginine and a partially exposed glutamate have been proposed for the transfer of protons from the active site of [NiFe] hydrogenases to the solvent medium (Chapter 6). The characteristics of mutant SH proteins, affected in the corresponding Glu43, His69 and His464 of HoxH (in the conservative N- and C-terminal domains; see Fig. 7.10), are in accordance with this hypothesis (Massanz and Friedrich 1999). After insertion of non-polar valine or leucine residues at this position, the Ni-containing tetrameric SH proteins showed only residual hydrogenase activity. D_2/H^+-exchange activities reached various levels from the wild-type protein. Since this activity depends on proton exchange, H^+-transferring functions seem not to be completely blocked. Replacement of Arg40 blocked H_2-dependent oxidoreductase activity completely; however, the enzyme is still functional for D_2/H^+ exchange. Apparently a proton exchange was uncoupled from electron transfer. At least H_2 was heterolytically cleaved at the modified active site, but electrons could not be transferred to an acceptor molecule. May be this strictly conserved, positively charged residue is directly involved in electron transfer between the Ni-Fe centre and the proximal [4Fe-4S] cluster or it is involved in a second proton chain which is essential for hydride oxidation.

7.8. FTIR spectroscopy of hydrogenases

Antonio J. Pierik

Like EPR spectroscopy has proven a useful tool to monitor nickel in [NiFe] hydrogenases, FTIR spectroscopy has fulfilled a key role in the identification and characterization of the rest of the active site of hydrogenases. FTIR spectroscopy entered the hydrogenase field through a cooperation of Albracht with Woodruff and Bagley. They observed that the [NiFe] hydrogenase from *A. vinosum* exhibited not only light-sensitive bands from externally added carbon monoxide but also vibrational bands from species near or coordinated to the active site (Bagley *et al.* 1994, 1995). The position of the bands (one strong band at 1,898–1,950 cm^{-1} and two 12–16 cm^{-1} separated, small bands at 2,044–2,093 cm^{-1}, see Fig. 7.2 for some representative spectra) was seen to reflect the redox state of the active site.

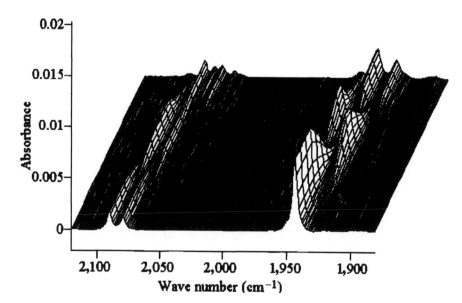

Figure 7.11   Reduction of A. *vinosum* hydrogenase in an FTIR cuvette by illumination in the presence of deazaflavin. An Ar-flushed solution of enzyme in the ready state, supplemented with deazaflavin and EDTA, was illuminated with white light for periods of about 4 min. After each illumination a spectrum was recorded. Reduction proceeds from the front to the back. Using the overview in Fig. 7.6, one can easily identify the several states of the enzyme by looking at the ν(CO) frequency. Adapted from (Pierik *et al.* 1998a).

Each state had its own characteristic set of three infrared bands from intrinsic non-protein groups, which could thus directly be used to define the redox state of the active site (see Fig. 7.6) (Bagley *et al.* 1995). Combined with the observation of a second metal ion with three non-protein ligands in the X-ray crystallographic structure of the [NiFe] hydrogenase of *D. gigas* (Volbeda *et al.* 1995), it soon became clear that the three infrared bands corresponded to the ligands in the active site. Steadily the full picture emerged when it was realized that the unidentified low-spin ferrous ion earlier observed by Münck and Albracht (Surerus *et al.* 1994) in Mössbauer studies on the *A. vinosum* enzyme might be the same as the iron ion close to the nickel (Volbeda *et al.* 1995), whereby the CN/CO ligands provide a strong enough ligand field to keep Fe low spin (Happe *et al.* 1997). A screening of seven hydrogenases, two of which were [Fe] hydrogenases, and a variety of Fe-S and Ni-containing proteins, gave further support that the occurrence of small non-protein ligands was unique for the active site of metal-containing hydrogenases (Van der Spek *et al.* 1996).

The breakthrough came when the nature of the infrared bands was unequivocally revealed by studies on *A. vinosum* [NiFe] hydrogenase isolated from cells grown on ^{13}C or ^{15}N-enriched media (Happe *et al.* 1997). Identification of the two small bands as the stretch vibration of cyanides was straightforward, since they both shifted upon ^{15}N and ^{13}C enrichment. However, the large band only shifted upon ^{13}C enrichment. Since the extent of the ^{12}C to ^{13}C shift was in perfect agreement with oxygen as the second atom of a diatomic oscillator, and since no other elements could have produced such shift, the large band was assigned to carbon monoxide. Chemical analysis of CO and CN, released upon denaturation, was in agreement with an $NiFe(CN)_2(CO)$ group in the active site. By wavelength dispersive anomalous diffraction the metal ion, to which the diatomic ligands were bound, was determined to be iron. On careful inspection, the electron densities originally modelled as three water molecules (Volbeda *et al.* 1995) were seen to be too dense and ellipsoidal for water and were assigned to diatomic molecules (Volbeda *et al.* 1996). Hydrogen-bonding interactions between (conserved) arginine and serine residues and two of the diatomic ligands (i.e. the CN moieties) allowed structural differentiation of the three diatomic ligands, which were otherwise beyond the level of resolution.

The assignment of the two small 2,044–2,093 cm^{-1} bands to two cyanides (Happe *et al.* 1997) had to be formally phrased in a different way. Partial enrichment with ^{15}N, leading to one C^{14}N and one C^{15}N in the same molecule, in vibrational model compounds (Lai *et al.* 1998) as well as the *A. vinosum* enzyme (Pierik *et al.* 1999) demonstrates that the two bands do not correspond to the individual cyanides but to the symmetric (in-phase) and asymmetric stretch modes of two vibrationally coupled nearly equivalent cyanides at an NC-Fe-CN angle greater than 60° (see also Volbeda *et al.* 1996). The in-phase stretch mode corresponds to the high frequency band.

After the elucidation of the origin of the infrared bands, FTIR spectroscopy has become an established tool for the study of hydrogenases. Specialized FTIR experimental setups, including e.g. optically transparent thin-layer electrodes (De Lacey *et al.* 1997) or deazaflavin-based light-induced photoreduction (Pierik *et al.* 1998a) (Fig. 7.11), allow the activation and redox chemistry of the active site to be monitored with superior convenience. The application is not limited to purified hydrogenases: low concentrations (10 μM) of the sensory hydrogenase of *R. eutropha* could be detected in cell-free extracts (Pierik *et al.* 1998b). Particularly for the investigation of the biosynthesis of hydrogenases and their diatomic ligands, this seems of great potential.

7.9. Using EPR and Mössbauer spectroscopies to probe the metal clusters in [NiFe] hydrogenase

Pedro Tavares, Alice S. Pereira and José J. G. Moura

EPR and Mössbauer spectroscopies have been successfully used to characterize iron–sulfur clusters. Hydrogenases are no exception. Here, we will describe the knowledge gained from applying these spectroscopies to the study of [NiFe] hydrogenase.

The magnetic features of iron–sulfur clusters are relevant probes to infer structural, redox and catalytic properties. Now, structures are best determined by methods such as X-ray crystallography (Chapter 6) or multidimensional NMR, which give detailed three-dimensional representations of proteins and their cofactors. However, they seldom give a full picture of the active states of enzymes. The knowledge of the three-dimensional structure is a starting point for mechanistic discussions. Also, structural determinations are limited by crystal availability (X-ray) or protein size (NMR). Such was the case of hydrogenases, where NMR seems to be excluded by the fact that hydrogenases are large multimeric enzymes and X-ray crystallography only in 1995 solved the first structure of an [NiFe] hydrogenase (Volbeda *et al.* 1995).

Twenty years ago, the first spectroscopic studies were initiated on hydrogenases, and Mössbauer was first applied in 1982 to characterize the iron clusters present in *D. desulfuricans* hydrogenase (Krüger *et al.* 1982). Different enzymes have been studied since through a spectroscopic approach. In the remainder of this section, we will present as an example of these studies the work done in the *D. gigas* [NiFe] hydrogenase (Huynh *et al.* 1987; Teixeira *et al.* 1989).

D. gigas hydrogenase can be considered a case study for the application of Mössbauer and EPR spectroscopies in conjunction with redox titration methodologies. H_2 (the substrate/product of the reaction) was used to control the redox state of the enzyme by varying the partial pressure of the gas. By doing so, several samples of the enzyme were obtained in different oxidation states and investigated in parallel both by Mössbauer and EPR spectroscopies.

In the most oxidized state (circa 0 mV), EPR spectroscopy shows a signal ($g = 2.02$) characteristic of [3Fe-4S] clusters in the +1 state. In agreement, the low temperature (4.2 K), Mössbauer spectrum shows the correspondent typical magnetic component. However, this component only accounts for approximately 27 per cent of the total iron. The remaining 73 per cent, invisible to EPR spectroscopy, can be assigned to two diamagnetic $[4Fe-4S]^{2+}$ clusters.

Further decrease in redox potential, -200 mV, only affects the [3Fe-4S] cluster properties (one electron reduced in the $[3Fe-4S]^0$ state). The EPR spectrum now shows a $g = 12$ feature due to a $S = 2$ spin system. Mössbauer data clearly illustrates this point, ruling out the possibility of a [3Fe-4S] to [4Fe-4S] cluster interconversion. The two [4Fe-4S] clusters remain in the diamagnetic state, indicating that they have a lower redox potential.

Raising the partial pressure of H_2 gas results in the reduction of the [4Fe-4S] clusters (circa -400 mV). The EPR spectrum changes dramatically. Three effects can be seen: (i) an extremely broad signal (extending approximately 300 mT) appears; (ii) concomitantly the $g = 12$ signal decreases in intensity and a broad signal develops at even lower fields; and (iii) some changes occur in the 'Ni signal' (see below). The Mössbauer data indicate that, at potentials below -250 mV, the $[3Fe-4S]^0$ remains at the same oxidation state. So, the disappearance of the $g = 12$ signal cannot be the result of further reduction of this cluster. The extremely broad signal was attributed to the reduced [4Fe-4S] clusters (Cammack *et al.* 1982). The assignment is fully supported by Mössbauer spectroscopy, which demonstrates that there are two reduced [4Fe-4S]$^{1+}$ clusters. Moreover, these clusters were found to be spectroscopically distinguishable and to have different redox potentials. Finally, the change detected in the $g = 12$ signal suggests that the reduced [3Fe-4S]^0 is sensitive to the redox state of the other metal centres, i.e. the [3Fe-4S]^0 clusters feel a paramagnetic centre developing in its vicinity. The exposure to substrate alters drastically the magnetic behaviour of the centres: in the oxidized state the metal clusters behave as isolated entities. On reaction to H$_2$, the [3Fe-4S]^0 interacts magnetically with a reduced 4Fe centre, as well as the Ni-Fe site with another reduced 4Fe cluster, as could be predicted by the structural arrangement revealed by X-ray crystallographic structure.

As described above, the combination of EPR and Mössbauer spectroscopies, when applied to carefully prepared parallel samples, enables a detailed characterization of all the redox states of the clusters present in the enzyme. Once the characteristic spectroscopic properties of each cluster are identified, the determination of their midpoint redox potentials is an easy task. Plots of relative amounts of each species (or some characteristic intensive property) as a function of the potential can be fitted to Nernst equations. In the case of the *D. gigas* hydrogenase it was determined that those midpoint redox potentials (at pH 7.0) were −70 mV for the [3Fe-4S]$^{1+}$ → [3Fe-4S]^0 and −290 and −340 mV for each of the [4Fe-4S]$>^{2+}$ → [4Fe-4S]$^{1+}$ transitions.

What about the hetero-dinuclear Ni-Fe centre?

EPR spectroscopy was a helpful tool to look into the enzyme active site, as well as its possible redox states. *D. gigas* hydrogenase belongs to the group of oxygen stable hydrogenases. After aerobic purification these enzymes are inactive. In this state, the EPR spectrum shows a rhombic signal denominated *Ni-signal A* (also designated as Ni$_u^*$, unready). Upon deoxygenation, the enzyme becomes active, and the EPR spectrum changes. A new rhombic signal (*Ni-signal B*, also designated as Ni$_r^*$, ready) is observed. At this stage the enzyme does not require activation steps to initiate its performance. Decrease of the solution potential leads to the disappearance of these signals (midpoint redox potential circa −150 mV at pH = 7.0). The resulting redox state is EPR silent. It is generally agreed that both Ni-signal A and Ni-signal B correspond to the Ni(III) oxidation state, and that the EPR-silent state results from a mono-electronic reduction to the Ni(II) state. Below −250 mV, two Ni-related EPR signals appear. They were designated *Ni-signal C* (also designated as Ni$_a$-C*) and 'signal $g = 2.21$'. Detailed EPR studies indicate that the signals represent two distinct states of the enzyme with the Ni-signal C species being a precursor of the $g = 2.21$ state. These signals maximize around −360 mV, and eventually disappear at lower potentials. Some controversy still remains in respect to the oxidation state of the Ni ion. Some speculation was made that this signal is related to an Ni(III)-hydride species, due to D/H isotopic effects, light sensitivity and proton ENDOR measurements (Franco *et al.* 1993; Albracht 1994). The $g = 2.21$ signal was indicated to arise from the magnetic interaction between the Ni-C species and another paramagnet in the enzyme (most probably the closest reduced [4Fe-4S] cluster, see Section 7.10). The release of a proton from the Ni-hydride species, with the transfer of one electron to the Ni ion, would again generate an Ni(II) EPR-silent state.

So far, biological Mössbauer spectroscopy is only sensitive to iron (or to be more precise to the ^{57}Fe isotope) which rules out the use of this technique to directly probe the Ni site. Unfortunately, the iron site seems to remain always in a low-spin state making its identification by Mössbauer very difficult since it will be probably unresolved from the [3Fe-4S] and [4Fe-4S] cluster components. Also, each iron atom in the protein contributes equally to the Mössbauer spectrum; the Fe site of the Ni-Fe centre is only 8.3 per cent of the total absorption.

An unusual absorption of a lone Fe atom, not belonging to the Fe-S clusters, was detected in the reduced enzyme from *A. vinosum* (Fig. 7.12). The nature of this Fe atom was a mystery, however. It was only when X-ray anomalous dispersion

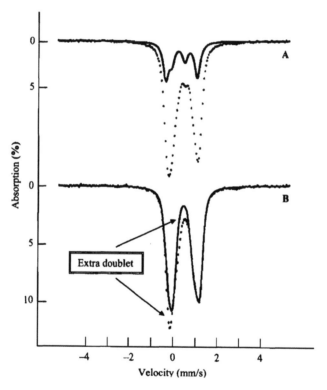

Figure 7.12 Analysis of the zero-field Mössbauer spectrum at 50 K of H_2-reduced *A. vinosum* [NiFe] hydrogenase. (A) Dotted line: experimental spectrum; solid line: theoretical spectrum for a [3Fe-4S]^0 cluster. (B) Dotted spectrum: experimental spectrum in A minus the theoretical spectrum or the reduced 3Fe cluster; this was expected to represent the spectrum of the two reduced [4Fe-4S] clusters. Solid line: theoretical spectrum for two reduced 4Fe clusters. The additional absorption in the experimental spectrum (arrows) comprised a doublet with a surprisingly small isomer shift (0.05–0.15 mm/s) and accounted for about 8 per cent of the total absorption, i.e. one Fe out of twelve (adapted from Surerus et al. 1994). The present interpretation is that this doublet represents the Fe atom in the Ni-Fe site.

analysis was carried out on crystals of native *D. gigas* hydrogenase, that a lone Fe atom was identified near the Ni atom in the hetero-dinuclear Ni-Fe centre (Volbeda *et al.* 1995, 1996). An independent proof that iron was involved in the structure of the catalytic site came from another nuclear resonance technique. A recent report (Huyett *et al.* 1997) presented a ^{57}Fe Q-band pulsed ENDOR study of ^{57}Fe-enriched *D. gigas* hydrogenase in the inactive Ni-signal A form. The nickel EPR spectra disclosed a signal from a ^{57}Fe cation with a small isotropic coupling of A(^{57}Fe) of circa 1 MHz, indicating that the cluster contains a non-heme iron atom in the highly unusual low-spin ferrous state.

7.10. Spin–spin interactions in [NiFe] hydrogenases

François Dole, Bruno Guigliarelli, Patrick Bertrand

Depending on their magnitude, spin–spin interactions may affect the magnetic characteristics of a system to different extent: in a polynuclear cluster, the strong exchange coupling between the metal sites determines the energy and the spin state of the low-lying energy levels. In contrast, the weak intercentre spin–spin interactions that appear in multicentre proteins can only alter the shape of their EPR spectra (Section 7.10.1).

A first point related to the occurrence of spin–spin interactions in [NiFe] hydrogenases concerns the nature of the Ni-A, Ni-B, Ni-C and Ni-L signals (also designated as the Ni_u^*, Ni_r^*, Ni_a-C* and Ni_a-L* signals, respectively): do these S = 1/2 species arise from the Ni centre alone or do they result from the exchange coupling between the two metal sites of the Ni-Fe centre? This point was recently addressed by studying the temperature dependence of the EPR signals exhibited by these various forms in the *D. gigas* enzyme (Dole *et al.* 1997). No departure from the Curie law could be detected, which rules out the existence of low-lying excited levels due to the exchange coupling between the Ni and Fe sites. From these results and other data obtained in magnetization experiments carried out on EPR-silent species, it was concluded that the Fe atom is low-spin ferrous in all forms of the enzyme, which very likely excludes a redox role for this site.

Let us consider now intercentre spin–spin interactions. Although the Ni and [3Fe-4S] centres are both paramagnetic in the Ni-A and Ni-B forms, no effects due to spin–spin interactions are observed in the EPR spectra given by the *D. gigas* and *D. vulgaris* Miyazaki enzymes. This means that the distance between the two centres is larger than 10 Å, in agreement with the arrangement displayed by the crystal structures. However, in some [NiFe] hydrogenases like that of *A. vinosum*, concomitant changes appear in the EPR signals exhibited by the Ni and [3Fe-4S]$^{1+}$ centres when the enzyme is fully oxidized. These spectral changes, which are frequency dependent, have been ascribed to spin–spin interactions between the Ni centre and an unidentified paramagnet. Since the strong interactions needed to account for these spectral changes are not related to the activity of the enzyme, their origin is still unclear.

The system made of the Ni-Fe centre and of the proximal [4Fe-4S] centre constitutes the basic functional unit of all [NiFe] hydrogenases. It is therefore especially interesting to investigate its properties by analysing the spin–spin interactions between these centres. Applying this strategy to the Ni_a-C* form of the *D. gigas* enzyme yielded

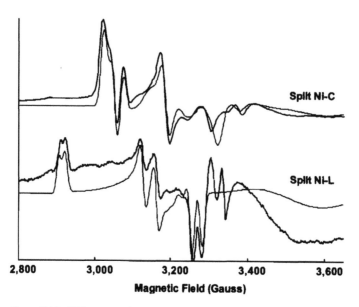

Figure 7.13 EPR spectra (heavy lines) and simulations (thin lines) of the split Ni_a-C* in reduced
D. gigas hydrogenase; split Nia-L* signal from illumination of Ni-C.

a relative arrangement of the Ni and $[4Fe-4S]^{1+}$ centres which is very similar to that
displayed in the crystal structure of the inactive Ni-A form, which demonstrates that
this part of the protein is not altered by the activation process. This study has also
provided valuable information about the location of the mixed-valence pair and the
orientation of the magnetic axes in the proximal $[4Fe-4S]^{1+}$ centre. The Ni_a-C* species
is light-sensitive, being converted upon illumination below $100 \, K$ into another species
called Ni-L. This is manifested by a significant increase in the anisotropy of the unsplit
EPR signal, which demonstrates that the structure of the Ni centre has been somewhat
altered. Since this process exhibits a kinetic isotope effect in D_2O, it probably involves
the dissociation of a bound proton. The split signal is changed by this process as well
(Fig. 7.13 upper spectrum). A detailed study carried out in the *D. gigas* enzyme has
shown that the orientations of the magnetic axes of the Ni_a-C* and Ni_a-L* species are
similar, which suggests that the proton involved in the photoconversion process is not
directly bound to the Ni atom. Besides, since the spectral changes observable in the
split Ni_a-L* signal are mainly due to the vanishing of the exchange coupling between
the Ni centre and the proximal $[4Fe-4S]^{1+}$ centre (Fig. 7.13, upper spectra), this proton is likely bound to the terminal cysteine facing this [4Fe-4S] centre (Fig. 7.14).

7.10.1. Intercentre spin–spin interactions

These interactions appear in multicentre proteins when neighbouring centres are
simultaneously paramagnetic, and give rise to a splitting of the resonance lines in the
EPR spectrum. Since they mask the individual characteristics of the centres, intercentre spin–spin interactions are often viewed by spectroscopists as complicated

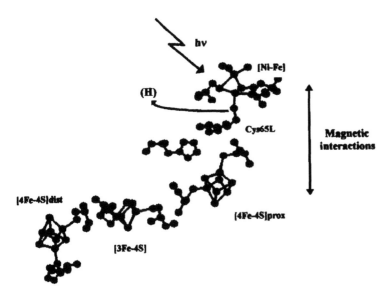

Figure 7.14 Arrangement of the NiFe centre in *D. gigas* hydrogenase. Vertical arrow indicates the possible interactions observed on illumination.

phenomena which preclude the full characterization of proteins. Their effects are, however, very sensitive to the relative arrangement of the interacting centres, so that their quantitative study can provide detailed structural information. The spin–spin interactions between two centres A and B comprise two contributions.

The dipole–dipole coupling is a through-space interaction which decreases as $1/r^3$, whose magnitude is about $10^{-3} cm^{-1}$ at $10 Å$. This term alone produces a symmetric splitting of the resonance lines whose magnitude depends on the orientation of the intercentre axis with respect to the external magnetic field, so that an anisotropic splitting is observed in a frozen solution spectrum.

The exchange coupling, which arises from the overlap of the magnetic orbitals, is a through-bond interaction. Its magnitude can reach several hundred cm^{-1} in polynuclear clusters where the interaction is mediated by a short bridge of covalent bonds, but it decreases very rapidly at large distances. Its main component can be written $-2J S_A.S_B$, where J is the so-called exchange parameter. This component alone gives rise to an asymmetric and isotropic splitting of the resonance lines.

The combined effects of the dipolar and exchange interactions produce a complex frequency-dependent EPR spectrum, which can however be analysed by performing numerical simulations of spectra recorded at different microwave frequencies. When centre A is a polynuclear centre, the value of its total spin $S_A = \Sigma S_i$ is determined by the strong exchange coupling between the local spins S_i of the various metal sites. In this case, the interactions between A and B consist of the summation of the spin–spin interactions between S_B and all the local spins S_i (Scheme II). The quantitative analysis of these interactions can therefore yield the relative arrangement of centres A and B as well as information about the coupling within centre A.

7.11. Biochemical properties of the selenium-containing hydrogenases

Oliver Sorgenfrei

The F_{420}-non-reducing hydrogenase (Vhu) has been anaerobically purified. The purified enzyme had an exceptionally high specific hydrogen-oxidation activity of 40,000 U/mg (Sorgenfrei *et al.* 1993a). This value is about 50- to 100-fold higher than the activity usually observed for [NiFe] hydrogenases. It has been shown that the purified hydrogenase consists of three subunits of 46, 31 and 3 kDa. The identification of the very small subunit VhuU on the protein level was especially interesting, since the polypeptide was predicted to contain selenium, which was known to be a ligand to nickel in the active site of the selenium-containing hydrogenase of *D. baculatum* (He *et al.* 1989). Indeed the existence of this subunit could be demonstrated on a peptide resolving gel (Sorgenfrei *et al.* 1993a). Via mass spectrometry it was furthermore shown that this peculiar subunit contains selenocysteine. The mass determination and the additional amino-acid sequence analysis of the complete subunit also showed unambiguously for the first time that the C-terminal processing observed earlier in other hydrogenases (Menon *et al.* 1991; Gollin *et al.* 1992; Kortlüke *et al.* 1992; Fu and Maier 1993) takes place behind a highly conserved histidine residue.

Knowing that this small subunit contains at least part of the nickel binding site, it was chemically synthesized as a sulfur analogue. It was shown that this synthetic peptide had a nickel binding capability (Pfeiffer *et al.* 1996). The structure of the isolated subunit containing zinc instead of nickel was determined using NMR spectroscopy. It resembled the structure of the homologous part of the *D. gigas* hydrogenase as determined by X-ray diffraction analysis (Volbeda *et al.* 1995). The surface of the peptide was mainly hydrophobic, suggesting that it could be located inside the protein in native hydrogenase. Indeed, the homologous part in the hydrogenase of *D. gigas* is buried inside the molecule (Volbeda *et al.* 1995).

The second selenium-containing hydrogenase (Fru) from *Methanococcus voltae* has been purified aerobically (Muth *et al.* 1987; Halboth 1991) and anaerobically (Sorgenfrei *et al.* 1997a). The purified enzyme consists of three subunits of 45, 37 and 27 kDa mass corresponding to the fruA, fruB, and fruG genes, respectively (Halboth 1991). The amount of protein and the specific activity obtained by anaerobic purification was significantly higher than that obtained by aerobic purification suggesting that at least part of the hydrogenase is irreversibly damaged by exposure to oxygen.

7.11.1. EPR analysis of the Ni-Fe site of the Vhu and Fru hydrogenases

The selenium-containing hydrogenases from *M. voltae* (Fru and Vhu) have been studied with EPR spectroscopy. (Anaerobically purified the active enzymes exhibited a rhombic EPR signal with its lowest g value at 2.01 (Fig. 7.15, upper panel), indicating that the unpaired electron is in an orbital with mainly d_{z^2} character.) The species responsible for this signal is light sensitive at low temperatures (Sorgenfrei *et al.* 1993b, 1997b). After illumination a signal was obtained which had its lowest g value at 2.05 (Fig. 7.15), indicating that the unpaired electron in this species is no longer in an orbital with d_{z^2} character, but rather in an orbital with most likely $d_{x^2-y^2}$ character. This is a common feature of [NiFe] hydrogenases in an active state (Albracht 1994).

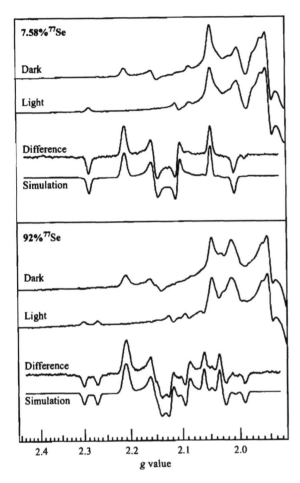

Figure 7.15 EPR spectra at 35 K of the Se-containing, F_{420}-non-reducing [NiFe] hydrogenase from *M. voltae*. Upper panel: (dark) enzyme with natural-abundance Se (7.58 per cent [77]Se) frozen under a few per cent H_2 in the dark (Nia-C* state); (light) after illumination at 35 K (Nia-L* state); (difference) difference spectrum dark minus light (four-fold enlarged); (simulation) simulation of the difference spectrum. Lower panel: enzyme 92 per cent enriched in [77]Se. Other conditions as in the upper panel, except for the difference spectrum which was three-fold enlarged. (Adapted from Sorgenfrei *et al.* 1993b.) The two-fold splitting from the I = 1/2 nucleus from [77]Se can be seen in many of the peaks in the lower panel.

It is generally believed that the light-sensitive changes are due to the cleavage of a metal–hydrogen bond, since the reaction velocity is dependent on the hydrogen isotope which is present (Van der Zwaan *et al.* 1985; Medina *et al.* 1996). Sorgenfrei *et al.* (1993b) demonstrated that incorporation of [77]Se, which has a nuclear spin of I = 1/2, into the F_{420}-non-reducing hydrogenase leads to an anisotropic splitting of the EPR signal prior to illumination and to a nearly isotropic splitting after illumination (Fig. 7.15, lower panel). The same characteristics were shown for the F_{420}-reducing hydrogenase, showing that it is not a peculiarity of the small subunit containing F_{420}-non-reducing hydrogenase (Sorgenfrei *et al.* 1997b). The existence of splitting caused by the ^{77}Se nucleus in both states was not easy to reconcile with the notion that the unpaired electron was in the d_{z^2} or $d_{x^2-y^2}$ orbital in the 'dark' and 'light' state, respectively. However, it could be explained by assuming that the electronic z-axis flips over upon illumination.

EPR measurements with oxidized, inactive, selenium-containing hydrogenase from *M. voltae* showed that there is a very small interaction, if any, between the unpaired electron and the ^{77}Se nucleus (Sorgenfrei *et al.* 1997b). In contrast, hydrogenase from *Wolinella succinogenes* enriched in ^{33}S showed a considerable hyperfine effect (Albracht *et al.* 1986). Since the nucleus of ^{77}Se is a much stronger magnet than the ^{33}S nucleus, it is obvious that the sulfur which causes the hyperfine interaction in the *W. succinogenes* enzyme is not homologous to the selenium in the *M. voltae* enzymes. The EPR signal of the oxidized hydrogenases indicates that the unpaired spin is in an orbital with largely d_{z^2} character. Thus the symmetry of the orbital is the same as in the 'as isolated' state prior to illumination. Nevertheless, the hyperfine interaction was only clearly seen in the 'as isolateded' state. Consequently it was concluded that the electronic z-axis is different in both states (Sorgenfrei *et al.* 1997b).

Carbon monoxide is a competitive inhibitor of [NiFe] hydrogenases. If active hydrogenases are treated with CO, the EPR properties change (Van der Zwaan *et al.* 1986; Cammack *et al.* 1987). It has been shown that in CO-treated F_{420}-non-reducing hydrogenase from *M. voltae* the CO and the selenium are ligands to the nickel at opposite positions (Sorgenfrei *et al.* 1996). Previously it has been argued that CO or hydrogen molecules could bind to the same position to the nickel (Van der Zwaan *et al.* 1986). The results obtained with the F_{420}-non-reducing hydrogenase from *M. voltae* confirmed that the CO indeed binds to the nickel, but that the hydrogen species more likely binds to the iron (Sorgenfrei *et al.* 1996), which is also part of the active site.

7.11.2. A model explaining the spectroscopic characteristics of the active reaction centre

The results can be explained with a model as depicted in Fig. 7.16. It is assumed that in the Ni_a-C^* state H_2 initially binds as a bridging ligand between the nickel and the iron in the active site (Fig. 7.16-I). After heterolytic cleavage the hydride could be bound to the iron and the proton to a bridging thiolate ligand, thereby weakening the nickel thiolate bond (Fig. 7.16-II). The unpaired electron would be in an orbital with d_{z^2} character in the direction of the nickel–selenium axis. Such a complex would explain the results obtained with the 'as isolated' enzyme (Ni_a-C^* state, electron in the d_{z^2} orbital, ^{77}Se splitting and no detectable splitting due to hydrogen; Sorgenfrei *et al.* 1993b, 1996).

According to this model the iron–hydride bond is cleaved upon illumination and the hydride, together with the proton, leaves the complex. This would lead to re-establishment of the nickel thiolate bond which was weakened in the former state, explaining the changes in the EPR spectrum observed after illumination (Fig. 7.16-III). After a flip of the electronic z-axis the selenium could in this state interact with the unpaired electron in an orbital with $d_{x^2-y^2}$ character. To get EPR-active, CO-treated

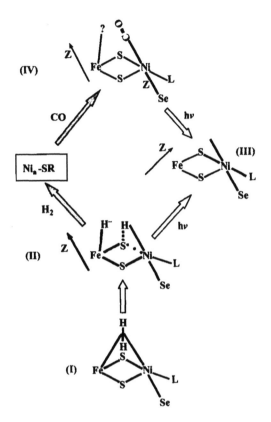

Figure 7.16 Working hypothesis to explain the EPR-spectroscopic properties of the [NiFe] hydrogenases from *M. voltae*. Thick black lines denote orbitals with d_{z^2} or $d_{x^2-y^2}$ symmetry bearing the unpaired electron. The direction of the electronic z-axis is indicated by a thin, solid black arrow. The question mark stands for an unknown ligand. See text for further details. (Adapted from Sorgenfrei *et al.* 1997.)

samples the enzyme is first incubated under hydrogen leading to an EPR-silent state known as Ni_a-SR (Sorgenfrei *et al.* 1996). Treatment of such enzyme with CO could give a complex as shown in Fig. 7.16-IV. This proposal is in line with the observation that CO and selenium bind on opposite sites of the nickel. No experimental evidence is available on the ligand of the iron in this state. Illumination of this complex could lead to breakage of the nickel–CO bond and a complex like the one shown in Fig. 7.16-III is obtained, as expected from the very similar EPR spectra of the illuminated forms of the hydrogen-loaded and CO-treated enzyme (Sorgenfrei *et al.* 1996). This model explains the spectroscopic characteristics determined for the active Ni-Fe centre. Interactions of this centre with other redox groups are not taken into account, although one has to be aware that an interaction with a [4Fe-4S] cluster has been observed (Section 7.10) (Cammack *et al.* 1985; Teixeira *et al.* 1989; Van der Zwaan *et al.* 1987). Direct evidence for this interaction has been obtained with the Fru hydrogenase (E. C. Duin, O. Sorgenfrei, J. Koch, N. Ravi, W. Roseboom, R. Hedderich, E. Münck and S.P.J. Albracht, unpublished results).

The model presented in Fig. 7.15 is not in agreement with the crystal structure of the inactive, oxidized hydrogenase from *D. gigas* (Volbeda *et al.* 1995). In the model derived from the X-ray diffraction analysis there is no space opposite the selenium atom. One way to explain this discrepancy is to assume that the enzyme undergoes a conformational change upon reductive activation. This is, however, not observed in the X-ray structure of two different, active enzymes (Garcin *et al.* 1999; Higuchi *et al.* 1999b). Another possible explanation would be that the structure of the active site in the hydrogenase from the sulfate reducer differs from the structure present in hydrogenase from a methanogen. A crystal structure is required to resolve this problem.

7.12. ENDOR spectroscopy

Wolfgang Lubitz

7.12.1. Basic principles

EPR has proved to be a sensitive tool for the detection and characterization of paramagnetic metal centres in biological systems (Lippard and Berg 1994). This method is mostly applied to frozen solutions, to stop molecular tumbling, which averages out the anisotropy of the g and hyperfine (hf) tensors. An alternative is to study single crystals, from which both the magnitude *and* orientation of the g and hf tensors are available, thereby providing detailed information about the geometrical and spatial structure. In hydrogenase research, EPR provided information about the iron–sulfur clusters and the various paramagnetic states of the Ni (NiA, NiB, NiC, NiL, etc.) (for a review see Albracht 1994). Recently, the first experiments on NiA and NiB have been reported for single crystals of the [NiFe] hydrogenase from *D. vulgaris* Miyazaki F (Gessner *et al.* 1996; Trofanchuk *et al.* 2000).

The value of EPR in establishing the electronic structure of the complex and obtaining details on the composition and geometry of the coordination sphere relies on the detection and analysis of the g- and hf-tensor components. In EPR only the g components and the hfc of the central metal (nuclear spin $I \neq 0$) can usually be resolved. However, in many metal complexes, like the Ni centre in hydrogenases, the unpaired electron is significantly delocalized onto the surrounding ligands leading to hyperfine coupling with magnetic ligand nuclei (^1H, ^{14}N, ^{13}C, ^{17}O, ^{33}S, etc.). These small interactions often cannot be resolved by EPR owing to the large EPR linewidths.

A way out of this dilemma is to perform an NMR-type experiment on the paramagnetic centre. NMR provides an inherently higher resolution than EPR, since only one line pair at frequences $\nu^{\pm}$ is obtained for each nucleus (or set of magnetically equivalent nuclei). The lines appear, to first order, symmetrically spaced around the Larmor frequency ν_n of the respective nucleus; for nuclei with $I = \frac{1}{2}$ and a hfc A the resonance condition is

$$\nu_{NMR}^{\pm} = \left| \nu_n \pm \frac{A}{2} \right|$$

Unfortunately, such NMR experiments are very insensitive because of the small population differences of nuclear energy levels and the very broad lines encountered in paramagnetic species. Therefore, a double resonance experiment must be performed in which the NMR transitions are detected via a simultaneously irradiated EPR line. Due to this detection scheme the sensitivity is increased by several orders of magnitude. The experiment is called Electron Nuclear DOuble Resonance (ENDOR). ENDOR can be performed in the continuous wave (cw) or pulsed mode (both EPR and NMR frequencies pulsed), for reviews (see Kurreck et al. 1988; Kevan and Bowman 1990). For complicated spin systems the method exhibits a drastically increased hyperfine resolution as compared with EPR so that even very small hfc's of distant and/or weakly coupled nuclei (e.g. of the protein matrix) can be detected. As an NMR experiment ENDOR directly identifies the type of nucleus (isotope) under consideration.

ENDOR experiments can be performed in liquid solution, in which only the isotropic hfc's (A_{iso}) are detected. They are proportional to the spin density at the respective nucleus. From the assigned isotropic hfc's a map of the spin density distribution over the molecule can be obtained. In frozen solutions and powders the anisotropic hf interactions can also be determined. Furthermore, the method allows the detection of nuclear quadrupole couplings for nuclei with $I \geqslant 1$. For dominant g anisotropy as found in many metal complexes the external magnetic field can be set to several specific g values in the EPR, thereby selecting only those molecules that have their g tensor axis along the chosen field direction. In such orientation-selected spectra only those hf components are selected that correspond to this molecular orientation ('single crystal-like ENDOR').

By far the most extensive information about the hf interaction is obtained by performing the ENDOR experiments on single crystals. Such measurements provide magnitude and orientation of the electron nuclear hyperfine couplings of the various magnetic nuclei from which details about the electronic structure and the geometrical arrangement of the spin centre are obtained with high precision.

7.12.2. Applications in hydrogenase research

Early ENDOR studies performed by Hoffman and coworkers on [NiFe] hydrogenase from D. gigas and Thiocapsa roseopersicina have focused on the substrate hydrogen that is expected to bind to the NiC state (Fan et al. 1991; Whitehead et al. 1993). The authors were able to show that exchangeable protons exhibiting rather large hfc's are present in this state (A ≈ 17–20 MHz). The respective resonances disappear upon illumination and conversion to the NiL state (Whitehead et al. 1993). The process was found to be reversible. This, for the first time, showed that the photoprocess and annealing involves dissociation and recombination of a hydrogenic proton supporting the idea that the NiC site plays a catalytic role in hydrogen activation. Spectra of NiA and NiB were also reported showing, for example, that the NiA site is inaccessible to solvent protons consistent with the suggestion that this state is inactive (Fan et al. 1991).

Hoffman and collaborators (Doan et al. 1994) also performed ENDOR experiments on the oxidized [3Fe-4S] cluster in D. gigas hydrogenase. The authors detected resonances from strongly coupled protons which were assigned to the β-CH_2 of the cysteines and exchangeable protons that are probably involved in three different hydrogen bonds to the sulfurs of the cluster. Based on these data a model for the binding of the cluster to the protein was developed.

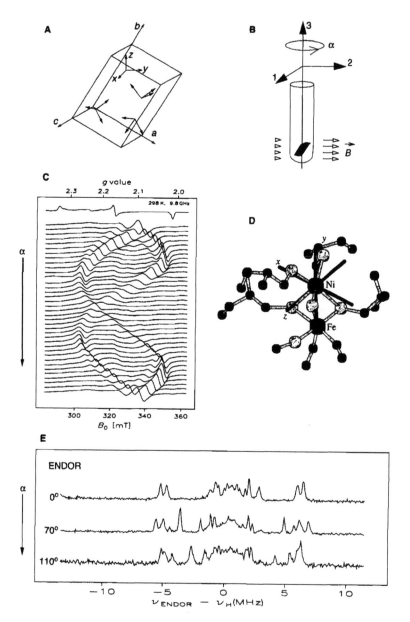

Figure 7.17 EPR/ENDOR experiments on [NiFe] hydrogenase single crystals from *D. vulgaris* Miyazaki F. (A) Unit cell (space group P2₁2₁2₁)with four magnetically distinguishable sites, X, Y, Z denote the intrinsic axes, *a, b, c* the crystallographic axes. (B) Crystal rotation in the external magnetic field *B* of the EPR spectrometer. (C) Orientation-dependent EPR spectra (rotation in an arbitrary plane) from 0 to 180° in five steps. (D) Orientation of the *g* tensor in the active centre (NiB) as obtained from analysis of EPR spectra. (E) Selected Pulsed ENDOR traces at $T = 10\,K$ obtained for NiB at specific rotation angles and magnetic field positions. The hyperfine splittings are elucidated according to eqn (p. 144).

Recent studies performed by Gessner *et al.* (1999) aimed at the elucidation of the electronic structure of the NiB ('ready') state in *A. vinosum* hydrogenase. The authors performed orientation-selected ('single crystal-type') ENDOR experiments and determined the hf tensors of three protons. Two large proton hfc's were assigned to the bridging cystein in the Ni-Fe cluster, the third smaller hfc might belong to a proton attached to a terminal cysteine but could not be assigned with confidence. This study showed that spin density is significantly delocalized from the Ni onto the surrounding cysteine ligands. This is in agreement with recent density functional calculations (see Section 15).

In this respect the possible presence of electron spin density at the iron in the Ni-Fe cluster is of importance. This question has been answered by ENDOR experiments performed on ^{57}Fe enriched [NiFe] hydrogenase (Huyett *et al.* 1997). In this study no significant ^{57}Fe hfc was detected in the NiA, NiB and NiC states ($A_{57_{Fe}} \lesssim 1$ MHz). This shows that in all relevant states of the enzyme the iron is in the low-spin Fe^{2+} state.

First ENDOR experiments on NiA and NiB of [NiFe] hydrogenase single crystals of *D. vulgaris* Miyazaki F have recently been performed by Brecht *et al.* (1998). They gave conclusive assignments of the measured proton hf tensors to the cysteines. Figure 7.17 illustrates the angular dependence of the ENDOR line splittings yielding the anisotropic hf couplings as an example. Similar experiments on NiC are currently in progress, which should show where in the complex the substrate hydrogen is bound. This information is very difficult to obtain from X-ray crystallography. It is, however, of central importance for an understanding of the mechanism of catalytic hydrogen conversion by these enzymes.

7.13. The application of X-ray absorption spectroscopy (XAS) to [NiFe] hydrogenases

Michael J. Maroney

The analysis of X-ray absorption spectra can be used to provide information regarding the structure of atoms, usually metal ions, in proteins (Scott 1985). The spectrum arises from the absorption of ionizing radiation near an absorption edge and requires an intense light source, such as synchrotron light, in order to obtain data from dilute (millimolar) samples. The edge is the abrupt absorption that occurs when a core electron is removed from the element of interest. Thus, a K edge corresponds to the loss of a 1s electron, an L edge from the loss of a 2p electron, etc. The technique can be used to monitor redox chemistry associated with a metal centre (shifts in edge energy), to determine the coordination number and geometry of the metal site, and to get metric details regarding metal-ligand distances. One advantage of XAS as a structural probe lies in the fact that it does not require crystalline samples; any enzyme intermediate that can be freeze trapped in a nearly pure (80 per cent) state can be examined. Thus, the method can be used to get structural information on states that are too difficult or unstable to crystallize, and to rapidly compare the structures of metal sites in a number of related enzymes. Further, the metal-ligand distances obtained are generally accurate to ca. ± 0.02 Å (with higher precision) and generally exceed the accuracy of these distances determined by single crystal X-ray diffraction of proteins. Since each element has characteristic and usually well-separated edge

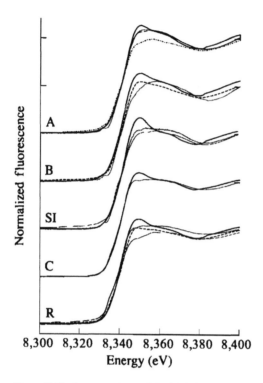

Figure 7.18 A comparison of the Ni K-edge X-ray absorption spectra for redox-poised samples of hydrogenases from different bacteria in the edge and XANES regions. Spectra are separated by redox level, with line types indicating the different bacterial sources (bold line, *T. roseopersicina*; light line, *D. gigas*; dotted line, *A. vinosum*; dashed line, *D. desulfuricans* ATCC27774; dashed-dot line, *E. coli*). Reprinted with permission from Gu, *et al.* (1996) and the American Chemical Society.

energies, the technique is element specific. However, if more than one atom of the same element is present in the sample, an average spectrum is obtained. Thus, in the case of [NiFe] hydrogenases such as those found in *D. gigas* or *A. vinosum*, information specifically pertaining to the Ni site may be obtained from the X-ray absorption, while the Fe XAS spectrum would represent an average of the twelve Fe centres present in these enzymes.

7.13.1. Edge-energy shifts as a probe of Ni redox chemistry

The edge energy of Ni metal is 8,333 eV. However, in $Ni^{(II)}$ coordination compounds, the energy of the edge shifts to about 8,340 eV, a reflection of the fact that the Ni centre has a partial positive charge in these complexes that makes it more difficult to ionize the metal ion further. In general, for first row transition metals a shift of ca. 2–3 eV in the K-edge energy is expected for a one-electron change in the oxidation state of the metal. Clearly, what is being measured is charge density, not oxidation state, and thus anything that changes the partial charge residing on the metal, such as changes in the ligand environment, will affect the observed edge energy. However, changes in one ligand (e.g. the loss or gain of a ligand, or changes from soft (polar-izable) donors to hard (non-polarizable) donors) lead to only a small shift in the edge energy, ca. 0.5 eV/ligand compared to a metal-centred redox process (Colpas *et al.* 1991). Thus, when large changes in the ligand environment can be ruled out (e.g. by XANES and EXAFS analyses), the edge-energy shift observed can be used to monitor metal-centred redox chemistry.

Examination of Ni K-edge spectra from redox-poised samples of [NiFe] hydroge-nases reveals that only a small shift (ca. 1 eV) in the edge energy occurs between the most oxidized (Forms A and B) and fully reduced states (Fig. 7.18) (Bagyinka *et al.* 1993; Gu *et al.* 1996). Coupled with the lack of large changes in the ligand environ-ment, this observation effectively rules out redox schemes employing Ni-centred redox involving both Ni(III) and Ni(I) centres. The data are consistent with delocal-ized one-electron redox chemistry involving formal Ni(III) and Ni(II) states of the active site. In addition, studies of dark-adapted Form C and photolyzed Form C do not reveal a change in the charge density of the Ni centre upon photolysis, at least in *T. roseopersicina* hydrogenase (Whitehead *et al.* 1993).

7.13.2. Coordination number and geometry from XANES analysis

Features in the XAS data lying below the edge in energy are frequently observed. This X-ray absorption near edge structure (XANES) arises from high-energy electronic transitions, such as 1s→3d transitions in the case of K-edge spectR. A 1s→3d transi-tion is forbidden by electronic absorption selection rules ($\Delta 1 = 2$), but can gain inten-sity in non-centrosymmetric complexes where p–d mixing can occur. Thus, measuring the intensity of the feature assigned to the process involving a 1s→3d transition gives

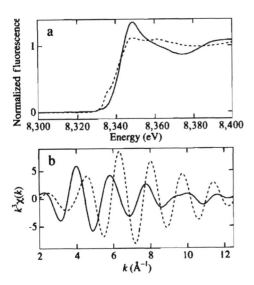

Figure 7.19 Ni K-edge X-ray absorption spectra for 'as-isolated' (solid line) and 'NADH- and H$_2$-reduced' (dashed line) samples of the hydrogenase from *R. eutropha* H16. (a) Edge region; (b) Fourier-filtered EXAFS (backtransform window = 1.1–2.6 Å). Reprinted with permission from Gu, *et al.* (1996) and the American Chemical Society.

information about the electronic symmetry of the metal site. For example, octahedral complexes would be expected to have weak or absent 1s→3d transitions, while tetrahedral metal complexes would have a more intense feature. In general, the effect of coordination number and geometry on the XANES spectrum is larger than the effect of the asymmetry due to differences in ligand donors, and thus the intensity can be used to determine the coordination number and geometry (Bagyinka *et al.* 1993). Nickel is unique among first-row transition metals in that it has two common high-symmetry coordination geometries: six-coordinate octahedral and four-coordinate square planar. Fortunately, these may be distinguished using a second XANES feature. In complexes lacking one or more axial ligands, a feature corresponding to a process involving a 1s→4p$_z$ electronic transition can be observed between the 1s→3d transition and the K edge. In the case of planar coordination environments, the 1s→4p$_z$ transition is usually observed as a resolved maximum, which is diagnostic for planar geometry. Five-coordinate pyramidal, but not trigonal bipyramidal, complexes exhibit a shoulder in this region.

Data regarding the 2p→3d transitions can be obtained from L-edge spectroscopy (Van Elp *et al.* 1995; Wang *et al.* 2000). This transition is allowed by electronic absorption selection rules ($\Delta 1 = 1$), and thus dominates the L-edge spectrum, often obscuring the L-edge transition itself. Transitions involving specific 2p and 3d orbitals can often be resolved, revealing a wealth of detail regarding the electronic structure of the metal. However, the soft X-rays involved (ca. 850 eV), lead to a number of technical difficulties that must be overcome in order to obtain spectR.

The Ni K-edge XANES spectra of [NiFe] hydrogenases never reveal a resolved peak corresponding to a 1s→4p$_z$ transition (Fig. 7.19), demonstrating that a planar Ni site is not observed in any of the states of the enzyme that have been examined (Gu *et al.* 1996). The intensity of the 1s→3d transition is appropriate for a five- or six-coordinate site, depending on the specific enzyme and its redox poise. In the case of the *A. vinosum* enzyme, the oxidized species appear to be five-coordinate and the reduced species are six-coordinate based on the intensity of the 1s→3d transition. This trend is also observed in more limited data sets from the *D. gigas* enzyme, and may indicate the loss or modification of the third (oxo/hydroxo) bridging atom upon reduction of the enzyme. However, this trend is not observed in data obtained from *T. roseopersicina* hydrogenase, where the intensity of the 1s→3d transition is most consistent with six-coordination in all redox levels (Bagyinka *et al.* 1993). Another exception is the soluble hydrogenase from *R. eutropha*, which reveals a large change in the Ni coordination environment between oxidized and reduced samples (Fig. 7.19) (Gu *et al.* 1996; Müller and Henkel 1997; Müller *et al.* 1997a,b). The conversion of dark-adapted Form C to its photoproduct is not accompanied by a change in the XANES spectrum, indicating that the coordination number/geometry and ligand environment is not affected by photolysis (Whitehead *et al.* 1993).

7.13.3. Metric details from EXAFS analysis

The extended X-ray absorption fine structure (EXAFS) portion of the spectrum extends above the edge in energy and arises from interferences between the photoelectron produced at the edge and photoelectrons backscattered by nearby atoms (e.g. ligand donor atoms) (Scott 1985). The frequency of EXAFS oscillations is directly

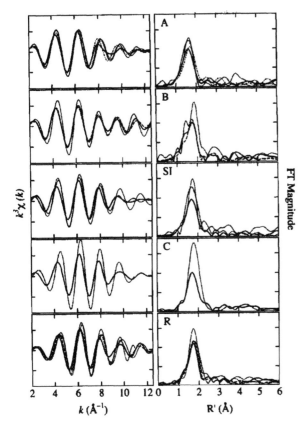

Figure 7.20 A comparison of the filtered EXAFS data (1.0–2.6 Å) for redox-poised samples of hydrogenases. Spectra are separated by redox level, with line types indicating different bacterial sources (see Figure 7.19 caption for linetype definitions). Reprinted with permission from Gu, *et al.* (1996) and the American Chemical Society.

related to the distance between the absorbing atom and the backscattering atom, and thus a frequency analysis of the EXAFS features provides distances between the metal and ligand atoms. Average distances obtained for groups of similar atoms with bond lengths that differ by less than about 0.05 Å. The intensity of the EXAFS damps out rapidly with distance from the absorbing atom and is related to the type of atom (atomic numbers differing by more than about two units of Z may be distinguished), the number of atoms in a given shell (a group of similar atoms at a similar distance), and the disorder in the distances to atoms in a shell. The disorder arises from both static (different but unresolved M-L distances) and dynamic (vibrations) components, and is handled by a Debye–Waller factor in the fitting procedure. Data are typically collected at low temperature in order to minimize the dynamic disorder. The disorder places a number of limitations on EXAFS analysis. Unless the motion of the absorbing atom and scattering atom are correlated, no EXAFS is expected regardless of proximity. Since both the number of scattering atoms and their disorder affect the intensity of the EXAFS, without knowledge of the disorder involved it is not possible to determine the number of atoms involved to a high accuracy from EXAFS analysis. Using information about the coordination number and geometry of the metal site from XANES analysis to constrain the EXAFS fits can alleviate this problem.

The EXAFS analysis of samples of [NiFe] hydrogenases generally reveals a spectrum that is dominated by Ni-S interactions at a distance of about 2.2 Å (Fig. 7.20) (Gu *et al.* 1996). The average Ni-S distance found does not change by more than ca. 0.03 Å as function of redox poise, again suggesting little change in the charge density residing on the Ni centre. Similarly, the Ni-S distances found for Form C and its photoproduct do not differ significantly (Whitehead *et al.* 1993). The number of S-scattering atoms is subject to the limitations described above but is generally found to be four for reduced samples of the enzyme. The number of O or N donors in the Ni coordination sphere is more variable. For samples of the *D. gigas* and *A. vinosum* enzymes in Form A, there is evidence of a short Ni-O(N) bond (ca. 1.9 Å) that is absent from the spectra of reduced enzymes. The EXAFS data on hydrogenases from *T. roseopersicina* (Gu *et al.* 1996) and *D. baculatum* (Eidsness *et al.* 1989) consistently fit with more O or N donors (ca. 2) at somewhat longer distances. And the *R. eutropha* soluble hydrogenase shows evidence of a large structural change upon reduction that is also reflected in the XANES region (Fig. 7.19) (Gu *et al.* 1996; Müller and Henkel 1997; Müller *et al.* 1997a,b). The spectrum of the oxidized *R. eutropha* soluble enzyme is the only [NiFe] hydrogenase sample examined so far where the EXAFS is dominated by scattering from O or N, and the Ni-S distance is significantly longer than 2.2 Å. Upon reduction, the spectrum of the *R. eutropha* enzyme changes to one that is more typical of other hydrogenases (four Ni-S at 2.2 Å).

Although hints of an Fe atom at a longer distance from the Ni site were suggested by some early EXAFS analysis, the Fe centre was not incorporated into EXAFS fits until it was revealed by crystallography. The reason for this is that for oxidized enzymes with an Ni-Fe distance of ca. 2.9 Å, the scattering due to Fe is a very small component of the overall EXAFS and inclusion of the Fe in the fits does not improve the goodness of the fits. However, in most reduced enzymes the Ni-Fe distance is shortened to 2.5–2.6 Å, and the inclusion of the Fe is necessary in order to get an optimal fit (Gu *et al.* 1996; Davidson *et al.* 2000).

7.14. Chemical and theoretical models of the active site

The large size of the hydrogenase molecule, and the presence of several metal centres of different nature, makes some types of detailed chemical studies very difficult. It is therefore useful to study the chemistry of smaller compounds for comparison. Models of the catalytic sites of hydrogenases fall into three categories:

1 Structural models, which are synthesized to imitate features of the proposed structure of the active site. These may be used to demonstrate the chemical conditions, which allow such structures to exist, to investigate their chemical properties and to give a better understanding of the spectroscopic characteristics of the native proteins. Examples of these include the mixed carbonyl/cyano complexes of iron, used to verify the infrared spectra to the hydrogenases (Fig 7.4) (Lai *et al.* 1998); and the nickel-thiolate complexes which have low redox potentials like the hydrogenases (Franolic *et al.* 1992).

2   Functional models, which have similar catalytic activities to those of hydroge-
    nases, regardless of structure.
3   Computational models, which calculate the optimized parameters for structures
    based on the crystallographic and spectroscopic information. These can be used
    to determine the likely patterns of electron density and protonation for the
    complexes.

Chemical model compounds

After the discovery of Ni in hydrogenases in the early 1980s, a variety of nickel model
compounds with various ligand systems have been described (examples are given in
Sections 8.6 and 8.7). More realistic approaches, however, were only possible after
the first crystal stuctures appeared (1995 for [NiFe] hydrogenases, and 1998/1999
for [Fe] hydrogenases). One of the first, elegant models was brought forward by
Darensbourg et al. (Darensbourg et al. 1997; Lai et al. 1998), who inspected a
compound in which an $Fe(CN)_2(CO)$ unit was bound to a cyclopentadienyl group
(Fig. 7.4). Not only its structure, but also its FTIR spectrum in hydrophobic media
matched those from the $Fe(CN)_2(CO)$ unit in [NiFe] hydrogenases remarkably well,
reinforcing the spectral interpretations of the enzyme.

A six-coordinate iron atom in the model complex $[Ph_4P]_2[Fe^{(II)}(PS3)(CO)(CN)]$
model compound was described by Hsu et al. (1997). Its spectral properties
(Mössbauer and FTIR) showed similarities with those of the low-spin Fe site in the
Ni-Fe active site of hydrogenases. Interestingly, the EPR spectrum of the oxidized
compound had g values and line widths very much like the one of active, oxidized [Fe]
hydrogenases (Hsu, H.F., Koch, S.A., Popescu, C.V., Münck, E. and Albracht, S.P.J. as
mentioned in Pierik et al. (1998a)), the structure of which was not known at the time.

Recently, the structure and spectral properties of a well-known diiron $Fe^{(I)}$ model
complex $(\mu\text{-}SCH_2CH_2CH_2S)Fe_2(CO)_6$ were reinvestigated by Darensbourg and
coworkers (Lyon et al. 1999). Also the $Fe_2(CO)_4(CN)_2$ derivative was prepared and
investigated. The $\nu(CO)$ and $\nu(CN)$ bands of the latter complex best fitted those
found for the reduced D. vulgaris [Fe] hydrogenase and hence the possibility of an
Fe^+-Fe^+ pair in the reduced enzyme was suggested by the authors.

The properties of sulfur-rich, mononuclear $Fe(CO)$ compounds as described by the
group of Sellmann (Sellmann et al. 1996) alerted the hydrogenase workers that the
protonation of a thiol ligand to Fe shifts the $\nu(CO)$ in such compounds by about
$40\,cm^{-1}$ to higher frequency, whereas the redox potential of the Fe ion increases by
500–600 mV.

Theoretical calculations

Over the last few years, theoreticians have used improved computational tools
(based on the Density Functional Theory, or DFT) to study the predicted electronic
properties of the bimetallic centres of hydrogenases. Several groups are presently
involved in this approach (Pavlov et al. 1998; Amara et al. 1999; Niu et al. 1999;
Stein et al. 2001; see Section 7.15). The interplay between theoretical expectations
and experimental results will undoubtedly have a beneficial impact on the hydrogenase research.

7.15. Density functional theory (DFT) – a promising tool for studying transition metal containing enzymes

Matthias Stein

7.15.1. Introduction

The field of computational quantum chemistry has received enormous recognition and impact since the award of the 1998 Nobel prize in chemistry to John Pople and Walter Kohn. Density functional theory (DFT) was around since the pioneering work of Thomas and Fermi in the 1920s and the sound theoretical foundations by Kohn and coworkers, but only recently was able to yield results of chemical accuracy when reliable functionals became available in the late 1980s and early 1990s. Because of its favourable scaling with molecular size and the accuracy of the results in particular for transition metal complexes, DFT has become an important player in the field of bioinorganic chemistry and hydrogenase research.

7.15.2. Hartree–Fock versus DFT

In traditional quantum chemistry, the electronic Schrödinger equation is solved:

$$\hat{H}_{el}\Psi = E\,\Psi,$$

where Ψ represents the wave function, $\hat{H}_{el}$ is the electronic Hamilton operator and E is the energy of the system. The total wave function is expanded as a linear combination of atomic orbitals ϕ_i.

$$\Psi = \Sigma_i c_i\,\phi_i.$$

Since the wave function is generally unkown, the equation must be solved iteratively until self-consistency is reached and the minimum in energy is found by varying the coefficients c_i of the atomic basis functions. In the simplest Hartree–Fock (HF) picture, electron correlation (e.g. the anticorrelated movement of electrons which sense each other) is neglected. This can be corrected by going to more sophisticated computational methods such as pertubation theory (Møller–Plesset 2nd order pertubation theory MP2), mixing in of different electronic configurations (configuration interaction CI) or coupled cluster theory (CC). The cost for increasing accuracy of the results is paid by increasing computational demands (both computing time and hardware requirements). While HF scales like N^4 with the size of the system (N being the number of basis functions and thus proportional to the number of atoms in the system), MP2 scales as N^5, CI and CC N^{6-7}. This limits the applicability of high-level correlated methods to systems of about ten to fifteen atoms.

The DFT approach (for an excellent introduction, see Parr and Yang 1989) is different and somewhat simpler. The electron density $\rho(r)$ has been recognized to be a feature that uniquely determines all properties of the electronic ground state (1st Hohenberg–Kohn theorem). Instead of minimizing E with respect to coefficients of the wave function as in HF, E is minimized with respect to the electron density ρ

(2nd Hohenberg–Kohn theorem). In addition to standard terms, the Hamiltonian in DFT contains the exchange-correlation functions $E_{XC}[\rho]$:

$$E_{XC}[\rho] = E_X[\rho] + E_C[\rho],$$

where E_X and E_C are the exchange and correlation energies, respectively. It can be shown, that E_X implicitly contains the static and E_C the dynamic correlation energies at least to a certain extent which are otherwise only considered by resorting to expensive post-HF methods. A further, even more valuable feature of DFT is its advantageous scaling with molecular size (being N^3) and thus enabling much larger systems to be investigated quantum mechanically. The recent success of DFT owes to energy functionals $E_{XC}[\rho]$ which yield results of chemical accuracy appearing only in the late 1980s and early 1990s. A major drawback of DFT is that there is no systematic way to improve the accuracy of the results. Varying basis sets and functionals does not necessarily lead to a systematic improvement of the results. The derivation of new energy functionals is still somewhat of an art. More and more functionals appear but the applicability of which to systems of interest must be validated.

7.15.3. DFT and transition metals

A major area of success of DFT is the investigation of transition metal compounds. Quantum mechanically, transition metals represent the most challenging task due to the large number of close lying electronic states. HF is doing very poorly for this type of systems and expensive post-HF techniques are needed to correctly describe atomic states, energies and bonding parameters. DFT calculations yield results superior to HF and sometimes comparable to the most expensive computational methods. The success of DFT for studying transition metal systems has been impressively demonstrated (for reviews, see e.g. Siegbahn 1996; Koch and Hertwig 1998; Bauschlicher *et al.* 1995).

7.15.4. The electronic structure of [NiFe] hydrogenase

With the advent of DFT methods, quantum mechanical calculations of systems as large as the active centre of [NiFe] hydrogenases have become feasible. Of course, not the complete protein can be treated as such but a cluster model has to be chosen small enough to be treatable on today's computers but still sufficiently large to correctly describe the electronic structure of the active site. The work by several groups (Pavlov *et al.* 1998, 1999; DeGioia *et al.* 1999a,b) aimed at a proposal of the mechanism of H_2 splitting by [NiFe] hydrogenase based on DFT calculations. In their work, the iron atom in the active center was the catalytically active metal. Niu *et al.* (1999) characterized the intermediate states in the reaction cycle by comparing calculated and experimental C≡O stretching frequencies. Amara *et al.* (1999) performed hybrid QM/MM calculations on [NiFe] hydrogenases and suggested a cysteine base-assisted H_2 cleavage.

The approach taken here is a more direct way to compare with available experimental findings. Single crystals of oxidized 'as-isolated' [NiFe] hydrogenase from *D. vulgaris* Miyazaki F were investigated by EPR (Gessner *et al.* 1996) and ENDOR

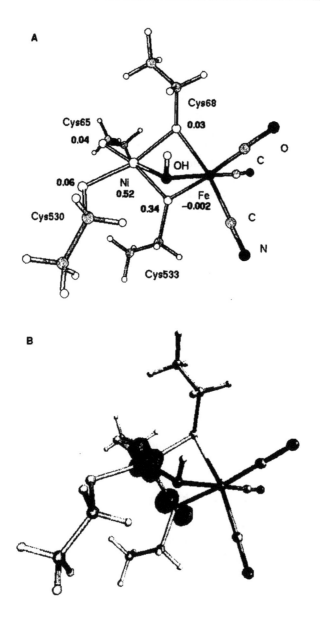

Figure 7.21 (A) BLYP/DZVP optimized cluster model of the active centre of *D. gigas* [NiFe] hydrogenase. The unassigned bridging ligand was assumed to be an OH^-. Numbers in bold font
are the calculated total Muliken atomic spin densities. The largest part (0.52 spins)
resides on the nickel atom, 0.34 is located on the sulfur atom of the bridging cysteine
Cys533. The iron is essentially low spin (-0.002). Spin densities at the remaining three
cysteines are an order of magnitude smaller than that of Cys533. (B) Contour plot at
$0.005\,e/a.u.^3$ of the unpaired spin density distribution in the active centre of *D. gigas*
[NiFe] hydrogenase. The unpaired spin density is localized along the Ni-SCys533 bond
and in good agreement with EPR and ENDOR studies.

(Brecht *et al.* 1998) spectroscopy which yield a maximum of information about the electronic structure, e.g. both magnitude and spatial orientation of unpaired spin density distribution and orbital occupancies. Further support comes from different other experimental data, i.e. ^{33}S hyperfine splitting due to one sulfur atom (Albracht *et al.* 1986b), very small to no ^{57}Fe-ENDOR effect in some paramagnetic states of the enzyme (Huyett *et al.* 1997), no ^{77}Se hyperfine splitting in oxidized [NiFeSe] hydrogenase from *M. voltae* (Sorgenfrei *et al.* 1997b), where a selenoCys residue replaces the terminal cysteine Cys530 present in the *D. gigas* enzyme.

DFT calculations of a large cluster model comprising forty-two atoms of the active centre of *D. gigas* were performed (Stein and Lubitz 1998). The Becke exchange and Lee–Yang–Parr correlation functional (BLYP) (Becke 1988) in conjunction with a special DFT-optimized basis set of sufficient quality (DZVP) (Godbout *et al.* 1992) was used. Complete geometry optimizations were carrried out imposing no constraints on the structure. The agreement in structural parameters between the computationally optimized structure and the most recent X-ray structural analysis (Volbeda *et al.* 1996) is about 0.1 Å in bond lengths.

DFT calculations remarkably well reproduce the experimental bonding parameters which indicate that the protein environment does not impose an energetically unfavourable conformation on the active centre. Amara *et al.* (1999) report an effect of the protein environment in their QM/MM calculations. This may, however, be due to their convergence on an $S = \frac{3}{2}$ spin state for the cluster *in vacuo* and on an $S = \frac{1}{2}$ spin state when the protein environment is considered.

In addition to structural parameters, results about the electronic structure from calculations should be comparable to experimental data. Although, X-ray data collection were done on crystals mainly consisting of Ni-A and calculations refer to Ni-B, for which more experimental data are available, no drastic geometrical differences are expected between Ni-A and Ni-B because the two are magnetically very similar. Indeed, the calculated spin density distribution is in good agreement with the experimental data mentioned above. Figure 7.21A shows a contour plot of the calculated unpaired spin density distribution. The g tensor was shown to have its g_z axis along the Ni-SCys533 bond (supposed to be associated with a d_{z^2} orbital). Two large isotropic hyperfine couplings were assigned to β-CH$_2$ protons of the bridging cysteine Cys533 by pulsed-ENDOR spectroscopy on protein single crystals (Brecht *et al.* 1998) orientation-selected cw-ENDOR studies (Gessner *et al.* 1999). The large isotropic hyperfine interactions of 12 and 11 MHz indicate significant unpaired spin at the sulfur in question. In addition the sulfur of Cys533 is presumably the one which gives rise to the ^{33}S hyperfine interaction (Albracht *et al.* 1986b).

Figure 7.21A also contains calculated total atomic spin densities. The majority of unpaired spin density resides at the nickel atom (0.52 spins) and explains the typical Ni EPR spectra. Samples enriched in ^{61}Ni exhibited hyperfine splittings that are significantly smaller than the one expected for a complete spin at a nickel atom. It must be added that it is very difficult to estimate the atomic spin density from the hyperfine splitting in EPR. The sulfur atom of the bridging cysteine Cys533 bears about 1/3 (0.32) of the unpaired spin giving rise to hyperfine splitting of ^{33}S in EPR and large isotropic hyperfine interactions of β-CH$_2$ protons adjacent to it. The calculated spin density at the iron atom is -0.002 (negative due to spin polarization from the Ni via the bridge to the Fe atom) and very close to zero. The strong ligand field caused by

CN⁻ and CO ligands keeps the Fe in a low-spin state (S = 0) and prevents the detection of hyperfine splitting of ^{57}Fe in EPR and ^{57}Fe-ENDOR (Huyett *et al.* 1997). Also, very good agreement with experimental data was obtained for model clusters of the Ni-A, Ni-C and Ni-L states.

7.15.5. Outlook

DFT calculations are able to treat systems as large as the active centre of [NiFe] hydrogenase and yield results in good agreement with experiments. Ni-B can best be described as an Ni(III)- μ-OH-Fe(II) system with 52 per cent of the spin density at the nickel atom, 34 per cent at the bridging sulfur of Cys533 and 0 per cent at the iron atom. Further work will help to clarify the differences between the individual paramagnetic states Ni-A, Ni-B, Ni-C and Ni-L. The accuracy of a direct calculation of spectroscopic observables, e.g. *g* and hf tensors, from DFT wave functions was demonstrated for Ni model complexes (Stein *et al.* 2001a). This was also done for the [NiFe] hydrogenase and encouraging results were obtained (Stein *et al.* 2001b). It is daring to postulate a reaction mechanism solely based on calculations. At every stage, theoretical calculations must be able to reproduce experimental data if available. After an atomistic description of the paramagnetic states of the active center has been achieved, a reaction mechanism may be proposed in which the Ni is the place of hydrogen splitting.

Chapter 8

The catalytic machinery

Richard Cammack

With contributions from *Christina Afting, Elisabeth Bouwman, Gerrit Buurman, Richard K. Henderson, Arnd Müller, Jan Reedijk* and *R.K. Thauer*

In Chapter 2 we saw that hydrogenases of the three basic types are made by organisms that have existed over billions of years. In Chapter 6, the structures of the proteins were laid out in three dimensions. In Chapter 7 we saw that the metal centres of the protein could exist in particular chemical states. We can now begin to understand how the hydrogenases catalyse their reactions with such extraordinary efficiency. Furthermore we ask, can similar catalysts be constructed artificially?

8.1. Significant features of the active sites of hydrogenases

There must be something special about the hydrogenase active site, that allows it to react easily with H_2. It is safe to assume that the nickel–iron centre in the [NiFe] hydrogenases, and the H cluster in the [Fe] hydrogenases, are the active sites at which hydrogen is produced and consumed. They are unique to hydrogenases, whereas the other iron–sulfur clusters are similar to those found in many other proteins which cannot produce hydrogen. Most iron–sulfur proteins, although they can achieve low redox potentials, do not react with or produce H_2. An exception is the active centre in nitrogenase, which produces hydrogen as it fixes N_2 to ammonia (Section 3.2 in Chapter 10).

Nickel occurs in many hydrogenase active sites, though not all. This element is used in relatively few other enzymes (Hausinger 1993; Cammack and van Vliet 1999; Maroney 1999). In order to use it, the bacterial cell requires complex and energetically costly systems to take up and store nickel (Chapter 3). What are the advantages of the [NiFe] centre over the [Fe] centre? We note that the nickel-containing hydrogenases tend to be less sensitive than the iron-only hydrogenases, to inhibition by carbon monoxide and oxygen. The methanogens, which grow in the most extreme anaerobic conditions, have been found to contain only [NiFe] hydrogenases. CO is a common metabolite in anaerobic environments. Strict aerobes, such as *Ralstonia eutropha*, also use [NiFe] hydrogenases, which may be related to the resistance to O_2.

Free coordination site. A significant characteristic of the structures of the nickel–iron site in [NiFe] hydrogenases and the H cluster in [Fe] hydrogenases, is a vacant, or potentially vacant, position. The two ions in the dinuclear centre each have a free coordination site. In the reduced NiFe hydrogenase, there is a position that can accommodate a third bridging ligand; in the oxidized state this site is occupied by oxygen or sulfur. In the [Fe] hydrogenases, the position is occupied by a bridging carbonyl and there is a free terminal coordination site. Such an open site is probably the key to efficient binding of H_2. It is difficult to create in synthetic homogeneous catalysts, because ligands and solvent have a strong tendency to bind to any vacant site.

Diatomic ligands. The diatomic ligands -CN and -CO are likely to be an essential feature. Both the iron-only and nickel-iron hydrogenases contain at least one iron ion with cyanide and carbonyl ligands, bridged by thiolates to a second metal ion (iron or nickel) which can be oxidized and reduced. Hydrogen-bonded cyanide has σ-donor properties, similar to those of a phosphine. The combination of sulfur, CO, and CN ligands tend to make the metal site 'softer', so that it behaves more like the zero-valent metal in the platinum electrode. The dinuclear sites in the hydrogenases are beautifully adapted for the binding of H_2 or hydride.

We do not know exactly where the hydrogen binds at the active site. We would not expect it to be detectable by X-ray diffraction, even at 0.1 nm resolution. EPR (Van der Zwaan *et al.* 1985), ENDOR (Fan *et al.* 1991b) and electron spin-echo envelope modulation (ESEEM) (Chapman *et al.* 1988) spectroscopy have detected hyperfine interactions with exchangeable hydrons in the NiC state of the [NiFe] hydrogenase, but have not so far located the hydron. It could bind to one or both metal ions, either as a hydride or H_2 complex. Transition-metal chemistry provides many examples of hydrides and H_2 complexes (see, for example, Bender *et al.* 1997). These are mostly with higher-mass elements such as osmium or ruthenium, but iron can form them too. In order to stabilize the compounds, carbonyl and phosphine ligands are commonly used (Section 6).

The spectroscopic evidence (Chapter 7), notably the very weak ^{57}Fe hyperfine coupling observed in all the EPR-detectable states of the [NiFe] hydrogenase, indicates that the iron stays in the diamagnetic Fe(II) state.

In summary, the active site has control over the following features, which are difficult to replicate in a chemical catalyst:

1   access to water hydrons, but not water itself;
2   electron transfer to specific donor and acceptor molecules;
3   free ligand site on each ion for binding of hydride and H_2;
4   an iron atom with unusual diatomic ligands which favours the formation of a hydride; and
5   protection of the active site from reaction with oxygen.

8.2. Connections to the active site

Compared to most enzymes, which often undergo considerable conformational changes on binding of the substrates, hydrogenases are rather rigid proteins. The substrates are small and mobile, and can penetrate to the active site. Probably the only parts of the enzyme that move significantly are amino acid side-chains and bound water molecules involved in transfer of hydrons to the active site (Fig. 8.1).

The structures of the metal-containing hydrogenases (Chapter 6) reflect, in an unexpectedly literal way, the different components of the reaction:

$$2H^+ + 2e^- \leftrightarrow H_2.$$

There are three pathways for hydrons, electrons and hydrogen molecules leading into the enzyme to the catalytic site.

A. [Ni-Fe] hydrogenase

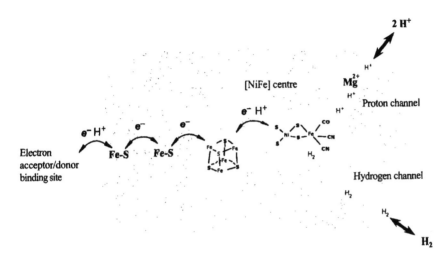

B. [Fe] hydrogenase

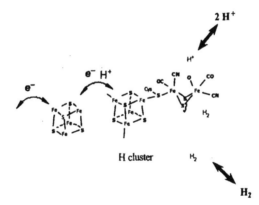

Figure 8.1 Diagrams of the catalytic machinery of [NiFe] and [Fe] hydrogenases.

The structures of the [NiFe] and [Fe] hydrogenases have been refined by evolution over billions of years. The relative positions of the metal atoms in the catalytic site, the electron-carrying iron–sulfur clusters, and the hydron and hydrogen channels (Fig. 6.11), all appear to be optimized to facilitate the reaction cycle. This provides clues to the movement of H_2, hydrons and electrons during the catalytic cycle. Cavity calculations, performed on a model of the [NiFe] hydrogenase from which all active site atoms are removed, showed that the internal hydrophobic channel network is directly connected to the one vacant nickel coordination site that is observed in the crystal structures of the oxidized enzymes. The electron-transfer pathway is directed toward the nickel site in the [NiFe] hydrogenases, and the Fe-[4Fe-4S] group in the [Fe] hydrogenases.

8.2.1. Hydron channels, and sites of hydron binding at the active site

The features of a putative hydron channel have been described by Volbeda et al. (See Section 3 in Chapter 6). Hydrons, like electrons, can tunnel from one site to another, but only over very limited distances (less than 0.1 nm). Therefore, the movement of hydrons must rely on the movement of transferring groups. These are probably the only moving parts in the machine. They comprise groups that can exchange hydrons (water, carboxyl, amino, amido and hydroxyl groups) in a chain from the protein surface. In D. gigas hydrogenase the chain incorporates water molecules coordinated to the magnesium ion. The motion of these groups is very rapid (on the nanosecond timescale) so that proton transfer is unlikely to restrict the rate of the reaction.

The next question is, where the protons go to in the active site during the catalytic cycle. For the base, there are too many possibilities to be certain. Groups near to the dinuclear cluster that can accept or exchange protons include the side chains of the amino acids arginine and histidine, thiolate ligands to the cluster, and peptide NH groups.

In addition there are water molecules. In D. gigas hydrogenase, no water molecules are seen in the active site by crystallography, but there are plenty of them in cavities leading through the protein to the surface, the nearest being 0.68 nm from the nickel. Hydrons can readily move between these sites by diffusion. So it is possible that multiple hydrogen-binding sites are involved in the reaction.

8.2.2. The electrical connection

Electron-transfer proteins have a mechanism that is quite different from the conduction of electrons through a metal electrode or wire. Whereas the metal uses a continuous conduction band for transferring electrons to the centre of catalysis, proteins employ a series of discrete electron-transferring centres, separated by distances of 1.0–1.5 nm. It has been shown that electrons can transfer rapidly over such distances from one centre to another, within proteins (Page et al. 1999). This is sometimes described as quantum-mechanical tunnelling, a process that depends on the overlap of wave functions for the two centres. Because electrons can tunnel out of proteins over these distances, a fairly thick insulating layer of protein is required, to prevent unwanted reduction of other cellular components. This is apparently the reason that the active sites of the hydrogenases are hidden away from the surface.

Electron acceptors and donors for hydrogenases, such as cytochromes, ferredoxins or NAD, bind to an electron-transfer site, which is usually close to the surface of the protein. This site binds the specific electron-transfer molecules, like a socket for an electrical plug. Other molecules that can accept or donate electrons to hydrogenase, such as methyl viologen, may bind at the same or different sites. In the structures of the hydrogenases, we can see the iron–sulfur clusters, neatly arranged in a chain, leading from the catalytic site to the electron-transfer site (see Fig. 6.8). The number and types of iron–sulfur clusters vary from one type of hydrogenase to another. In the [NiFe] hydrogenases, the clusters are in a different subunit from the [NiFe] centre, and there is considerable variation in this part of the molecule, sometimes involving several ancillary protein subunits.

The redox potentials of the electron-carrying groups have been the subject of much debate. In order to transfer electrons efficiently, one would expect that the electron-carrying groups along the chain should have progressively more positive midpoint redox potentials. In many hydrogenases this appears not to be the case. For example in some [NiFe] hydrogenases, such as those from *D. gigas* and *D. fructosovorans*, the middle iron–sulfur cluster is a [3Fe-4S] cluster that has a midpoint redox potential about 300 mV more positive than the [4Fe-4S] clusters on either side. This means that a reducing electron in this centre would be trapped in an energy minimum, and it is difficult to see how it could transfer to the next [4Fe-4S] cluster. Nevertheless, it appears to work. Electrons can transfer down the chain so rapidly that they do not hinder the hydrogenase reaction at all.

Would the hydrogenase be improved by having a more even distribution of redox potentials along the chain? Rousset *et al.* (1998) examined this possibility. By means of some difficult genetic engineering, Rousset mutated a proline to a cysteine in *D. fructosovorans* hydrogenase, and thus artificially converted the [3Fe-4S] cluster into the [4Fe-4S] type. However the enzyme with this smoother electron pathway did not perform any better than the native protein; moreover, it was more sensitive to oxygen damage. One way to interpret this result is that the reduced [3Fe-4S] cluster is good at conducting electrons by tunnelling, even if it is almost fully charged with an electron.

Electrostatic neutrality: Hydrons go with the electrons

In the hydrogenase active site, the effective dielectric constant is low, which means that there is a considerable energetic premium on introducing an electrostatic charge. In general, the addition of each electron is compensated by the introduction of a hydron. We know this, at least in *D. gigas* hydrogenase, because the pH dependence of the redox potentials has been determined for the nickel centre (Cammack *et al.* 1987). This is also true for the [4Fe-4S] clusters which, unusually, show pH-dependent redox potentials. The distal [4Fe-4S] cluster has a histidine ligand near the surface, which could acquire a hydron from the solution on reduction. The proximal cluster is distant from the surface but there is a possible channel for hydrons, leading to the NiFe centre, and thence to the surface (see Fig. 8.2). Thus an electron and a hydron can be transferred from the NiFe centre to the cluster during hydrogen oxidation. The [3Fe-4S] cluster does not transfer hydrons on reduction and this channel does not seem to extend right through the protein.

It is worth noting here that from the point of view of biological evolution, some of the subunits of hydrogenases are homologous, in their amino-acid sequences, to the proton-pumping NADH:ubiquinone reductases of present-day mitochondria (see Section 3.3 in Chapter 2). Molecular mechanisms that have evolved in one type of protein are often found in proteins with different functions. Proton channels are a feature of the enzymes that pump protons across membranes in order to produce a 'proton motive force' which is used to generate ATP (see, e.g., Puustinen and Wikstrom 1999; Luecke *et al.* 1999). It seems likely that hydrogenases were essential in some of the earliest organisms on Earth. Their pathways for proton transfers, required to maintain electrical neutrality at the active site, may have been the precursors of other enzymes that generate transmembrane electrochemical gradients.

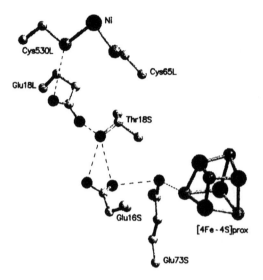

Figure 8.2 One of the possible pathways for hydron transfer between the [NiFe] centre and the proximal [4Fe-4S] cluster in *D. gigas* hydrogenase (A. Volbeda, unpublished).

8.2.3. How H_2 gets to the active site

Since the active site is deeply buried in the protein molecule, the reactants, H_2 and hydrons, have to diffuse to the catalytic centre. The first conjecture for hydrogenase was that H_2 diffused through the protein (Cammack 1995), just as oxygen is believed to diffuse into the oxygen-carrier molecule of the blood, hemoglobin. However, the heme in hemoglobin is more accessible to the surface than the hydrogenase active site. Moreover hemoglobin oxygenation does not have to be particularly fast; it has only to take up each oxygen molecule once in the lungs and release it once in the tissues. By contrast hydrogenase, as an enzyme, turns over its substrates very rapidly, so H_2 transfer might become a limiting factor. In retrospect, it is no surprise that hydrogenases have specific channels for transfer of H_2 to the active site.

Topological analyses of hydrogenase crystallographic models, along with X-ray diffraction studies of the diffusion of xenon within crystals and molecular dynamics calculations, first suggested that molecular hydrogen exchanges are mediated by large hydrophobic internal cavities interconnected by narrow channels (Fig. 6.8). One end of the channel network points to the active site nickel and several other ends lead out into the external medium (Montet *et al.* 1997; see Section 3 in Chapter 6). The major cavities and channels are conserved in each of the four structurally characterized NiFe hydrogenases, including the *Desulfomicrobium baculatum* enzyme which shows a much lower sequence identity to *D. gigas* hydrogenase than the others. Site-directed mutagenesis experiments to further investigate the role of the 'gas channels' are in progress.

The number of atoms of Xe or H_2 in the putative gas channels represents a higher solubility for H_2 in hydrogenase, than in water. The channels appear to concentrate the gas. It is interesting that one of the ways that has been discovered for storing H_2

is in carbon nanotubes, designed to be of a size such that H_2 gas molecules can line up inside (Dillon *et al.* 1997). These can achieve very high densities of H_2.

8.3. What happens at the active site

In the years since the discovery of nickel and iron in the catalytic centres, numerous different descriptions of the catalytic cycle of hydrogenase have been proposed. These were based on different oxidation states of the metal centres, and different sequences of transfer of electrons and hydrons. Although the reaction cycle has not been definitively resolved, the spectroscopic evidence places constraints on possible models that should be considered.

8.3.1. Hydride formation

Hydrogenases belong to a select class of oxidoreductases that can convert two hydrogen reducing equivalents to electron equivalents. The active sites of such enzymes all have the property that they can accept one or two electrons at a time. From the hydrogen isotope exchange experiments, we know that the metal-containing hydrogenases operate by heterolytic cleavage of H_2 into a hydride and a hydron. To bind a hydride, we need a suitable electron-deficient metal centre, M, and to hold a hydron we need a base, B, thus

$$H_2 + B + M \rightleftharpoons MH + BH^+$$

(Note that, by convention in organometallic chemistry, the hydride is written MH and not MH^-). The hydride represents two reducing equivalents, equivalent to two electrons, or two hydrogen atoms. Now, it is very difficult to withdraw the electrons one at a time from a hydride or H_2 molecule, because it would be energetically unfavourable to create hydrogen in the intermediate state $H^\bullet$, a hydrogen atom. Therefore, one or both of the metal ions in the active site *must accept two electrons simultaneously* from the hydride MH. The electrons are then transferred, *one at a time*, to the iron–sulfur clusters, which are normally one-electron acceptors. That is the trick that the hydrogenase active centre is specially designed to do.

The reaction mechanism can be likened to the cycle of an internal combustion engine, going through a series of steps (fuel and air induction, ignition, exhaust). When it is running, we only see a blurred view of the engine. It can be stopped and turned into various positions. However, some of the crucial steps (such as ignition of the fuel) are transient and it is not possible to stop the engine just then. The states we can observe are energy minima. The situation with the hydrogenase active site is analogous. As described in Chapters 5 and 7, hydrogenases can be isolated or frozen in various states of oxidation and reduction. Examination of the structures and spectroscopic properties of these states can provide clues to the course of the enzyme-catalysed reaction. When we gradually decrease the redox potential, we are applying an increasing 'electron pressure' which forces the centres to lower redox states, rather like turning over the engine. For the hydrogenase it is more complicated than looking at a single engine, because we are dealing with a solution containing many molecules. What we observe by spectroscopy, etc., is the macroscopic state of the system, which is the average of many molecules, each in a particular microscopic state. The states that can be frozen are stable ones; the mechanism is the way in which the enzyme converts between these states.

We should also remember that not all of the states that we see when freezing the enzyme (Section 7.4) are necessarily part of the mechanism. The most stable enzyme molecule is a dead one, so we must be aware that some of the spectroscopic signals represent damaged molecules. In the [NiFe] hydrogenases, the NiA and NiB states probably are not involved in the catalytic cycle, because they react slowly, if at all, with H_2. In the mechanism shown in Fig. 8.3, it is assumed that the relevant active states are NiSR, NiA and NiR.

In the NiFe centre, EPR spectroscopy has shown that the nickel is redox active, being reduced from Ni(III) to Ni(II) before formation of the hydride. A fraction of the electron density is distributed the thiolate ligands, as predicted by DFT calculations (Section 7.15). The next steps in the reaction cycle involve the transfer of the two electrons from the hydride to the iron–sulfur clusters. We propose that the nickel is oxidized to Ni(III); then one electron from the hydride reduces the nickel, while the other is transferred to the proximal iron–sulfur cluster. During this process the nickel can be transiently reduced by both electrons, Ni(III) → Ni(I). The Ni(I) state, though paramagnetic, would be very short-lived and probably undetectable by EPR spectroscopy. In the Fe hydrogenases, we can argue by analogy, that the iron atom that is

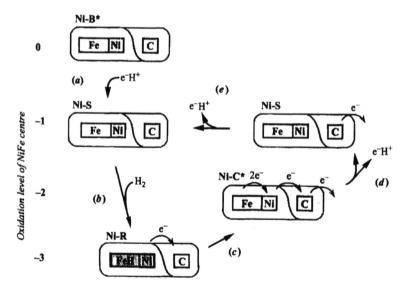

Figure 8.3 Outline reaction cycle of NiFe hydrogenase. The 'minimal hydrogenase' is depicted, consisting of the [NiFe] centre in the large subunit, and the proximal [4Fe-4S] cluster (C) in the small subunit. The reaction is written in the direction of the oxidation of H_2. Electrons are transferred out through the other iron–sulfur clusters to an acceptor protein (not shown). The equivalent states of the NiFe centre B, SR, R and C are indicated. Reduced centres are shaded. Electron transfers are accompanied by transfers of hydrons (not shown).

reduced in the H cluster, is the one that is linked to the [4Fe-4S] cluster. Like nickel, this would be able to accept two electrons transiently.

A reaction cycle that is consistent with most of the available evidence for the NiFe hydrogenases is illustrated in Fig. 8.3. The reaction is drawn in the direction of hydrogen uptake, starting from the oxidized ready state, NiB.

- Activation: the Ni ion is reduced by an external donor.
- The metal ions are in the Ni(II) and Fe(II) oxidation state (reaction of H_2 with Ni(III) or Fe(III) is unlikely on chemical grounds). H_2 reacts with the centre, forming a bound H_2 or hydride on one of the metal ions, here assumed to be Fe(II).
- The Ni(II) is oxidized to Ni(III), by transfer of an electron to the proximal [4Fe-4S] cluster. To maintain charge neutrality, a hydron released from the H_2 is transferred to a base adjacent to the [4Fe-4S] cluster, leaving a hydride on the iron.
- The proximal cluster donates an electron to the external acceptor. Two electrons are transferred from the hydride, at first reducing the nickel from Ni(III) to Ni(I). An electron is transferred from the nickel to the proximal cluster.
- The second electron is transferred from the cluster to the acceptor protein, leaving the enzyme in the NiSR state. Meanwhile two hydrons are transferred to the solution.
- The production of H_2 comprises the same steps in the opposite direction, with electrons coming from a donor protein.

A similar reaction can be written for the [Fe] hydrogenases with a Fe-[4Fe-4S] complex replacing the nickel. Note that the nickel atom in the NiFe cluster, and the Fe-[4Fe-4S] sites are nearest to the electron carrier [4Fe-4S] clusters, indicating that electron transfer occurs through these atoms. The other atom in each of the centres is an iron atom with -CN and -CO ligands, and it seems likely that this is a binding site for hydride (Fig. 8.1).

This mechanism easily accounts for the other reactions catalysed by hydrogenase. Exchange of the hydron and/or hydride with another hydron from the water, and reversal of step 1 would explain the ^1H – ^2H exchange reactions of hydrogenase (Chapter 5).

8.4. The metal-free hydrogenase from methanogenic archaea

Gerrit Buurman, Christina Afting and Rudolf K. Thauer

So far, the hydrogenases described have all used transition-metal ion clusters to react with H_2, and transfer electrons. However, it appears that there is another mechanism by which enzymes can catalyse reactions with H_2, without transition metal ions. This exception is the H_2-forming methylenetetrahydromethanopterin dehydrogenase from methanogenic archaea which contains neither nickel nor iron (Thauer *et al.* 1996). This unusual metal-free hydrogenase, abbreviated Hmd, catalyses the reversible reaction of N^5,N^{10}-methenyltetrahydromethanopterin (methenyl-H_4MPT$^+$) with H_2 to N^5,N^{10}-methylenetetrahydromethanopterin (methylene-H_4MPT) and a proton ($\Delta G^{o\prime}$ = −5.5 kJ/mol). This reaction is involved in the pathway of methane formation from CO_2 and H_2 (Afting *et al.* 1998; Thauer 1998). Tetrahydromethanopterin (H_4MPT) is an analogue of tetrahydrofolate (H_4F). However, the redox potential of the methenyl-H_4MPT^+/methylene-H_4MPT couple ($E^{o'} = -390$ mV) is more negative than that of the methenyl-H_4F^+/methylene-H_4F couple ($E^{o'} = -300$ mV); it is close to that of the H_2 electrode at pH 7.0 ($E^{o'} = -414$ mV).

Hmd is composed of only one type of subunit with an apparent molecular mass of approximately 40 kDa and has a specific activity of above 1,000 U/mg (Zirngibl et al. 1992). Evidence for the absence of nickel and iron in the metal-free hydrogenase is based on several findings (Thauer et al. 1996): (i) the purified active enzyme contains nickel and iron only in substoichiometric amounts (<0.1 mol/mol) that do not correlate wit activity; (ii) the primary structure of the enzyme shows no binding motifs for nickel or iron and is not sequence similar to Ni/Fe and Fe-only hydrogenases; (iii) the specific activity of the enzyme increases rather than decreases in cells growing under conditions of nickel limitation (Afting et al. 2000); and (iv) the activity is not inhibited by CO, NO or acetylene which inhibit Ni/Fe and Fe-only hydrogenases most likely by binding to the transition metal in the active site.

With respect to the catalytic mechanism the following findings are of importance (Thauer et al. 1996): (i) in contrast to NiFe and Fe-only hydrogenases, the enzyme per se neither catalyses an exchange between H_2 and the protons of water nor the conversion of $para$-H_2 to $ortho$-H_2; however, in the presence of methenyl-H_4MPT^+ the enzyme does catalyse both exchange reactions; (ii) the enzyme catalyses a direct hydride transfer from H_2 into the pro-R position of methylene-H_4MPT, the rate of incorporation being almost identical for H_2 and D_2; and most importantly (iii) the enzyme catalyses a direct exchange of the pro-R hydrogen of methylene-H_4MPT with protons of water.

The results indicate that in methenyl-H_4MPT^+ reduction with H_2, the methenyl group is activated by the metal-free hydrogenase in a way that it can directly react with H_2. They further indicate that in methylene-H_4MPT dehydrogenation, the pro-R hydrogen of the methylene group is activated such that it can directly react with a proton. Thus in case of the metal-free hydrogenase the hydrogen donor/acceptor rather than H_2 is activated. How this could be achieved is depicted in Fig 8.4. The proposed catalytic mechanism is similar to the one assumed for the reversible reaction of carbocations with H_2 under superacidic conditions (Berkessel and Thauer 1995). At present only this mechanism explains the reversible exchange of the pro-R hydrogen of methylene-H_4MPT with protons of water catalysed by the metal-free hydrogenase (Thauer et al. 1996).

In Fig. 8.4 it is assumed that methenyl-H_4MPT^+ undergoes a conformational change upon binding to the enzyme. The conjugational stabilization of the planar formamidinium ion is thus eliminated and a cationic centre at C14a is created which in its properties corresponds to a carbocation normally generated only under superacidic conditions. The activated carbocation is able to reversibly react with H_2, a hydride being incorporated into the pro-R position of the methylene group and a proton being released into the solvent. Methylene-H_4MPT is generated in a constraint conformation which relaxes upon dissociation from the enzyme as deduced from the stereochemical course of the reaction and the reversed conformation of methylene-H_4MPT free in solution (Geierstanger et al. 1998; Bartoschek et al. 2001). The mechanism is supported by ab $initio$ molecular orbital calculations (Cioslowski and Boche 1997; Scott et al. 1998; Teles et al. 1998).

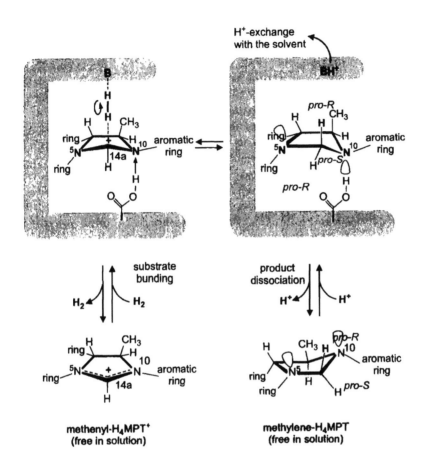

Figure 8.4 Proposed mechanism for the reversible reaction of N^5, N^{10}-methenyltetrahy-dromethanopterin (methenyl-H_4MPT$^+$) with H_2 to N^5, N^{10}-methylenetetrahy-dromethanopterin (methylene-H_4MPT) and a proton catalysed by the metal-free hydrogenase from methanogenic archaea. For complete structures of methenyl-H_4MPT$^+$ and methylene-H_4MPT, see reference Thauer *et al* (1996). Methylene-H_4MPT free in solution is in a conformation in which the methylene pro-R C-H bond is syn-clinal to the lone electron pair of N^{10}. Upon binding to the enzyme the imidazolidine ring of methylene-H_4MPT is forced into an activated conformation in which the pro-R C-H bond is antiperiplanar to the lone electron pair of N^{10} resulting in the prefor-mation of the leaving hydride (Bartoschek *et al*. 2001).

Very recently evidence was provided that Hmd contains a low-molecular-mass, thermolabile cofactor that is tightly bound to the enzyme but could be released upon enzyme denaturation in urea or guanidinium chloride (Buurman *et al*. 2000). No indi-cations were found that the cofactor contains a redox-active transition metal. Further studies are needed to determine the structure of the cofactor and its putative role in the catalytic mechanism.

8.5. Can we mimic nature, or even improve on it?

8.5.1. Hydrogenase biomimetics as substitutes for platinum

A worldwide transition from fossil energy to renewable energy sources such as solar energy seems inescapable. Research on hydrogenases and biomimetic model compounds is highly relevant. One of the major long-term goals of the research on hydrogenases is to synthesize a stable and cheap catalyst for interconversion of electricity and H_2. Such a catalyst would be essential for processes such as the production of H_2 by solar energy (electrolysis of water) and the conversion of H_2 into electricity (fuel cells). Present-day systems still make use of platinum as a catalyst, which prevents such systems from becoming economically viable for large-scale applications (Appleby 1999). Although there have been improvements, using platinum in the form of small particles, the supplies of this precious metal would not be sufficient for a worldwide hydrogen economy (Berger 1999). An alternative catalyst is needed for the reversible and rapid activation of molecular hydrogen, preferably at normal temperature and pH. However, the catalyst would also have to be more durable. The discovery of the structure and mechanism of the hydrogenases is a considerable step forward, which should help to guide further spectroscopic, mechanistic and biomimetic studies on hydrogenases.

It has already been shown that a very efficient catalyst can be produced by adsorbing the enzyme onto a carbon electrode (see Section 9 in Chapter 5). It produces H_2 at high rates. This can be considered as the biological equivalent of platinum. Though stable, by biological standards, it cannot compare with metallic catalysts. Nature tends to use such catalysts as disposable, and regenerates them as necessary. We have learned much about the intricate processes involved in assembling the hydrogenase molecule (Chapter 4), and it is clear that, in order to construct and maintain these catalysts, all the resources of a living cell are required. One possible approach would be to use bacterial cells as 'cell factories'. The catalytic properties of the enzymes could be refined either by targeted mutagenesis of parts of the structure, or by random mutation with evolutionary selection, or, most probably a combination of these approaches.

The alternative approach is to use the knowledge gained about the enzymes, to construct chemical catalysts. Here the field is in its infancy, since the construction of structural models could only start when the structure of the proteins began to emerge, in 1995. Previously, many heterogeneous catalysts had been known, for example for hydrogenation processes. But most of these used heavier, and scarcer metals.

With these possible applications in mind, we should review the significant characteristics of hydrogenase as a catalyst. Compared with most chemical catalysts, hydrogenases are large molecules. The protein has been selected by evolution, from an almost infinite number of possible structures. The whole protein is part of the machinery. Therefore even minor tampering with the protein, for example by site-directed mutagenesis, is likely to lead to unexpected changes in the properties of the enzyme.

Hydrogenases, like most enzymes, are extraordinary in their specificity. They catalyse a specific reaction with a particular set of substrates, and produce particular products. The reaction can go backwards and forwards many times without producing unwanted by-products. They do not, for example, reduce CO_2 to CO, which would poison most of them (there are other enzymes that do catalyse this reaction, and interestingly, they also contain nickel and iron–sulfur clusters). Chemical catalysts, by contrast, are plagued by unwanted side reactions. This is because in a chemical catalyst such as platinum, the reactions take place at the surface. In an enzyme the catalytic centre is protected from molecules in the solution by a layer of protein, and access to the catalytic site is controlled.

The side reaction that most hydrogenases find difficult to avoid is the reduction of oxygen. In the presence of oxygen, low-potential catalysts tend to generate oxygen radicals such as superoxide and hydroxyl radical, which are damaging to cells as well as the enzyme itself. However, most hydrogenases from organisms that are exposed to oxygen have some sort of protective mechanism, usually to switch them off in the presence of oxidants. The hydrogenases of hydrogen-oxidizing bacteria such as *R. eutropha*, and the sulfhydrogenase of *Pyrococcus furiosus*, have a specially protected active site that allows them to reduce compounds such as NAD, even in the presence of oxygen (see Chapter 7). Under these conditions, platinum would catalyse the wasteful reaction between oxygen and hydrogen. Thus oxygen is less of a problem with hydrogenase than it is with chemical catalysts.

The mechanism of action, and organization of the catalytic sites, in hydrogenases are different from a solid catalyst such as platinum. For a start, the reaction of H_2 with hydrogenase involves heterolytic cleavage into a hydron and a hydride. This contrasts with the reaction of H_2 at the surface of a metal such as platinum, which is usually considered to involve the homolytic cleavage into two hydrogen atoms. Moreover in the enzyme, the catalyst is a cluster of metal ions (with oxidation states +2 or +3) rather than the metal (oxidation state 0).

8.6. Synthetic model compounds – how chemists mimic nature

Arnd Müller

From the point of view of an inorganic chemist, the metal sites in metalloproteins can be defined as classic coordination compounds, also called complex compounds, with exceptionally large ligands (the proteins), because their chemistry cannot be separated from that of the corresponding metals. Most of the structural, physical, spectroscopic, and chemical properties of coordination compounds are determined by the composition of the immediate centre consisting of the metal ion(s) and all directly bound, surrounding atoms, that are referred to as ligands. Complex compounds, which resemble the biological originals in important properties, are then termed 'model compounds'. These are usually low-molecular-mass compounds, in which the large protein is replaced by smaller biomimetic ligands and that can be characterized and manipulated more readily than the metalloproteins.

An ideal model compound should correspond to the biological metal centre in terms of structure, composition as well as coordination and oxidation states of the individual metal ions, and also possess comparable spectroscopic and chemical properties. However, real model compounds rarely meet all these requirements at once. Usually, only special aspects of a metal centre are modelled, such as the structure, magnetic or electronic properties (spectroscopy), or the reactivity (function), and the corresponding model compounds are accordingly called 'structural', 'spectroscopic' or 'functional' models, respectively. Synthetic model compounds yield only information on the intrinsic properties of a metal centre, as the influence of the surrounding protein is not included. They provide information that is necessary for the understanding of basic principles and is difficult to derive from investigations on the complex biological systems. The concept of synthetic model compounds is not restricted to the replicative modelling of structural, spectroscopic and functional properties of already known metal centres, but can also be utilized to check the validity of hypotheses, that are based on spectroscopic studies. Important parameters necessary for the interpretation of spectroscopic data of metalloenzymes can only be obtained and calibrated by measurements on suitable model compounds. Synergic studies, that include the progressive improvement of the model compounds, are required in order to develop significant theories. The following paragraphs should demonstrate the principles outlined above with some examples for model compounds relevant to hydrogenases, but by no means attempt to give a comprehensive review of all the excellent and extensive work done in this field so far.

A textbook example for synthetic model compounds is provided by the thermodynamically stable iron–sulfur cluster compounds, that were designed to resemble the biological metal centres found in the iron–sulfur proteins (e.g. Holm 1977). These proteins range from electron transport proteins of low molecular mass (ferredoxins) to highly complex enzymes comprising several subunits, such as the hydrogenases, in which they are believed to form electron-transfer pathways through the protein matrix. Synthetic compounds such as the 'cubane'-type clusters $[Fe_4S_4(SR)_4]^{2-}$ in Fig. 8.5 (1) are readily formed in solution by the reaction of Fe^{3+} ions with hydrosulfide anions (HS^-) or elemental sulfur in the presence of thiolates with small hydrocarbon residues (RS^-, with R = methyl, ethyl, benzene, etc.), which are the biomimetic equivalent of cysteine (i.e. they have the same functional group, called thiolate) (Herskovitz *et al.* 1972). It is believed that similar spontaneous self-assembly reactions might also be involved in the *in vivo* synthesis of iron–sulfur centres, and it was even speculated that related reactions might have taken place at the origin of life (Hall *et al.* 1971). In fact, the cluster cores of [4Fe-4S] or [2Fe-2S] centres are stable enough to be removed intact from a protein under denaturing conditions, or (re)-inserted into an apo-protein.

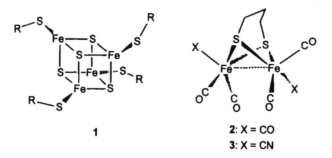

1

2: X = CO
3: X = CN

Figure 8.5 (**1–3**) Structures of chemical models for the active sites of hydrogenases. For references, see text.

The ability to synthesize reasonable amounts of model compounds, not only identical to the biological originals, but also suitably modified, has helped substantially with the investigation and the understanding of structure and function of biological iron–sulfur centres during the last thirty years. Of special importance are the high-precision structures, that were obtained by single-crystal X-ray diffraction on model compounds and are routinely used for the structure solution of metalloproteins, when only data sets of less than atomic resolution are available (see Chapter 6). For small molecules such as model compounds, X-ray diffraction usually provides important structural parameters such as bond distances and angles as accurate as 0.1 pm and 0.1°, respectively, while the accuracy determined in protein crystallography is rarely within 10 pm or several degrees for bond distances or angles, respectively. Another important piece of information was learned from electrochemical studies on model compounds: the [4Fe-4S] cluster can be reversibly oxidized twice, which explained the extremely wide range of electrochemical potentials covered by ferredoxins. Model compounds were also employed to probe the complex electronic and magnetic properties of iron–sulfur clusters.

Although reported procedures for synthesis of inorganic coordination complexes typically sound straightforward, they are mostly the result of extensive studies undertaken to overcome the unique problems encountered in transition-metal chemistry. First, complex compounds differ from the majority of other molecules in that they normally are kinetically unstable, i.e. they exist in equilibrium with other species of comparable thermodynamic stability in solution. Hence, the preparative task is not just restricted to synthesizing the desired compound, but also to find the appropriate conditions to stabilize and isolate it as the main product. It should preferably be in crystalline form, because X-ray diffraction is usually the only method to determine its structure unequivocally. A second problem characteristic of nickel-thiolate chemistry is that nickel ions in oxidation states higher than +2 have under most circumstances enough oxidizing potential to convert thiolate molecules to the corresponding disulfides. Moreover, nickel ions can catalyse reactions of the ligands involving the breakage or formation of covalent bonds, respectively, which can lead for example in the unexpected appearance of sulfide ions in the isolated product (e.g. Krüger et al. 1989; Müller and Henkel 1996), or can lead to the formation of sulfoxy groups in the presence of oxygen (Kumar et al. 1989; Farmer et al. 1992). Although many of the model compounds reported do not seem to be related to the biological centre at all, they help us to understand what is chemically possible, and what chemical obstacles the process of biological centre assembly has to overcome.

8.6.1. Models of the hydrogenase active sites

The structure of the dimetallic [2Fe] moiety of the unprecedented 'H cluster' found at the active site of 'iron-only' hydrogenases has only been known for a short time (Chapter 6, and Peters et al. 1998; Nicolet et al. 1999), but has already led to the construction of many new model compounds (Darensbourg et al. 2000). The chemistry of iron complexes is well investigated, and there are a large number of known complexes with mixed carbonyl and thiolate ligands which provided a good starting point. The complex compound $[(\mu\text{-SCH}_2CH_2CH_2S)Fe_2(CO)_6]$ in Fig. 8.5(2) has been known for over twenty years as a member of the comprehensively studied family of [(μ-SR)$_2$Fe$_2$(CO)$_{6-x}$L$_x$] compounds (e.g. Seyferth *et al.* 1982). These stable complexes are thermodynamic sinks in low-valent iron-thiolate carbonyl chemistry. By using well-established chemical reactions, [(μ-SCH$_2$CH$_2$CH$_2$S)Fe$_2$(CO)$_6$] in Fig. 8.5(2) could be readily converted to the complex [(μ-SCH$_2$CH$_2$CH$_2$S)Fe$_2$(CO)$_4$(CN)$_2$]$^{2-}$ in Fig. 8.5(3), which was shown to be a good structural and spectroscopic (IR) model for the dimetallic [2Fe] moiety (Lyon *et al.* 1999; Schmidt *et al.* 1999). This compound could only be oxidized irreversibly; the logical next step is to connect it via a sulfur bridge to a [4Fe-4S] cluster. The significance of this second moiety of the 'H-cluster' may be to serve as an oxidation level buffer in the enzyme.

The history of model compounds for the heterodimetallic nickel–iron centre of [NiFe] hydrogenase is an excellent example of the synergistic process in bioinorganic research (Darensbourg *et al.* 2000). In fact, the extended field of nickel-thiolate chemistry known today principally originates from the interest triggered by the discovery of nickel–sulfur bonds in the [NiFe] hydrogenases from *D. gigas* and *Methanothermobacter marburgensis* by EXAFS spectroscopy (Scott *et al.* 1984; Lindahl *et al.* 1984). Hence, the first synthetic goals were mononuclear nickel compounds with complete sulfur ligation, which preferably should show EPR signals similar to the hydrogenase. Attempts to use spontaneous self-assembly reactions, similar to the ones that work so well for iron–sulfur complexes, did not produce the desired model compounds, but did reveal some important general principles of nickel-thiolate chemistry. The nickel compounds formed using thiolate ligands with saturated hydrocarbon residues, contained nickel in the divalent oxidation state (Ni^{2+}), whose electronic configuration (d^8) favours square-planar coordinated complexes; these have no unpaired electrons and are therefore diamagnetic ('EPR silent'). Moreover the square-planar NiS$_4$ units were usually found to form polynuclear complexes via common edges, either resulting in zigzag chains or closed ring structures (for examples, see e.g. Krebs and Henkel 1991). However, none of the compounds synthesized was found to possess any of the chemical or spectroscopic characteristics of the biological metal centre. The same is true for mononuclear complexes with tetrahedral symmetry, which were formed with *thio*-phenolate or ring-substituted ligands – Fig. 8.5(4) (e.g. Müller and Henkel 1995). Although mononuclear complexes with square-planar symmetry, which are formed with difunctional ligands of the 1,2-dithiolate type in Fig 8.5(5), could be reversibly oxidized in cyclic voltammetric experiments (Fox *et al.* 1991, Krüger *et al.* 1991), none of the oxidation products could be isolated (Köckerling and Henkel 2000).

Stable mononuclear complexes of nickel in formal oxidation states of +1 or +3 were found with – at least partially – a coordination of ligands, that are classified as 'harder' (i.e. more ionic in bonding character) than the rather 'soft' (i.e. covalent in bonding character) thiolate sulfur function, such as nitrogen or oxygen donor atoms. These complexes usually possess anisotropic EPR signals, and can have coordination numbers in the range of four to six – Fig. 8.5(6, 7, 8) (Krüger and Holm 1989, 1990; Krüger *et al.* 1991; Köckerling and Henkel 1993). Penta-coordinate nickel complexes with a mixed nitrogen (oxygen) and sulfur coordination have attracted much interest, because of the results of XAS measurements on the [NiFe] hydrogenase from *Thiocapsa roseopersicina* and other bacterial sources (see Section 13 in Chapter 7; Gu *et al.* 1996; Müller *et al.* 1997a,b). In order to obtain penta-coordinate complexes, however, one has to employ multifunctional chelating ligands,

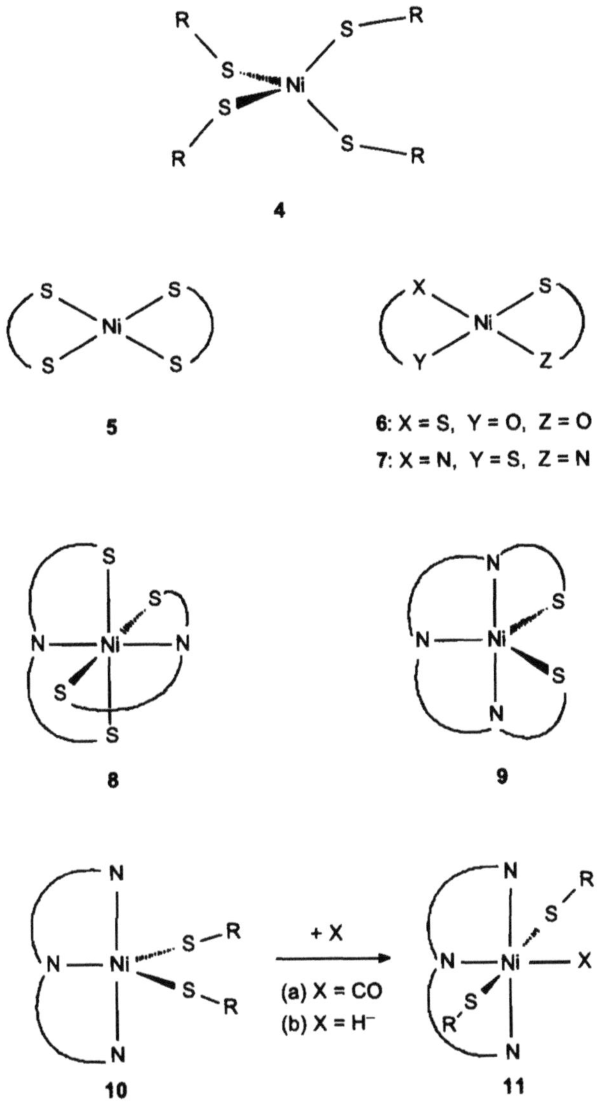

Figure 8.5 (4–11) Structures of chemical models for the active sites of hydrogenases. For references, see text.

that provide a set of donor functions in a suitable arrangement and thus determine the structure of the formed complex in a similar way to the protein. An example is the specially designed pentadentate ligand in complex 9 (Shoner *et al.* 1994). Some remarkable knowledge on the reactivity of such nickel compounds was gained from the penta-coordinate complexes with trigonal bipyramidal geometry, which are formed with a trifunctional nitrogen donor ligand and two monofunctional thiolates – Fig. 8.5(10) (Baidya *et al.* 1992; Marganian *et al.* 1995). These compounds could not only be reversibly transferred through nickel oxidation states of +1, +2 and +3, but were also shown to bind a carbon monoxide molecule or a hydride ion, respectively, in the Ni^{1+} state. The penta- and hexa-coordinate compounds in Fig. 8.5(10, 11) showed EPR spectra that closely resembled the characteristic hydrogenase Ni-EPR signals. This series of model compounds was the first not restricted to modelling only a single aspect (structural, spectroscopic, functional) of the biological nickel centre, as they modelled most of the properties known at this date reasonably well.

However, despite the great progress made in the field of nickel-thiolate chemistry, the surprising discovery of the dinuclear nature of the active nickel site in the [NiFe] hydrogenase by protein crystallography (Chapter 6) found model chemistry rather ill prepared. Although at that time numerous dinuclear nickel complexes with two thiolate bridges were known, only two contained nickel ions coordinated solely by thiolate ligands, in a geometry other than square planar (Halcrow and Christou 1994; Halcrow 1995; Fontecilla-Camps 1996). The homodinuclear complexes $[Ni_2(S^tC_4H_9)_6]^{2-}$ in Fig. 8.5(12) and $[Ni_2(S-2,4,5-{}^iPr_3C_6H_2)_5]^-$ in Fig. 8.5(13) each contain two nickel ions with a distorted tetrahedral thiolate coordination, linked by two or three bridging thiolate ligands, respectively (Müller and Henkel 1995; Silver and Millar 1992). Heterodimetallic compounds containing thiolato-bridged nickel–iron units were only known with nickel ions chelated by ligands of the N_2S_2-type and additional chloride ligands on the iron ions (Mills *et al.* 1991; Colpas *et al.* 1992), but S-metalation reactions using similar nickel chelates as precursors readily produced dinuclear nickel–iron complexes such as [Ni(N,N′-bis-2-mercaptoethyl-1,

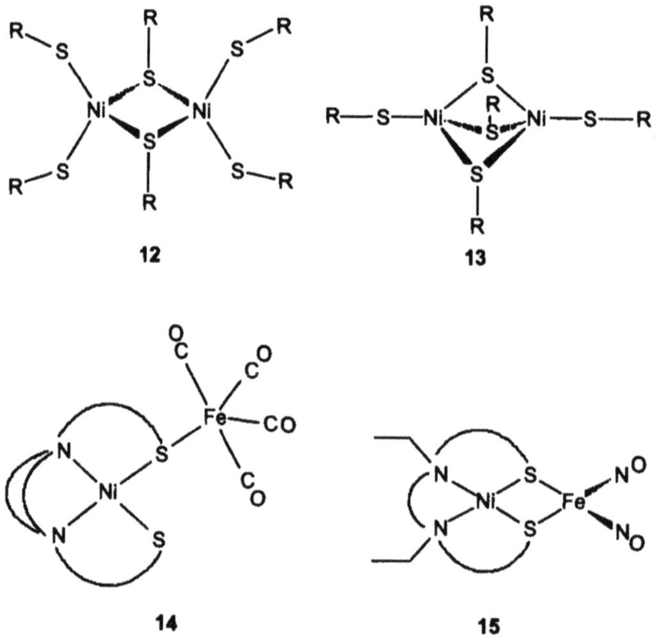

12   **13**

14   **15**

Figure 8.5 (12–15) Structures of chemical models for the active sites of hydrogenases. For references, see text.

5-diazacyclooctan)Fe(CO)$_4$] in Fig. 8.5(14) or [Ni(N,N'-diethyl-3,7-diazanonane-1,9-dithiolate)Fe(NO)$_2$] in Fig. 8.5(15) (Lai *et al.* 1996; Osterloh *et al.* 1997). These contain iron-carbonyl or iron-nitrosyl fragments that are bridged to a nickel moiety by one or two thiolate functions, respectively. An impressive model for the iron part of the heterodimetallic centre was found with the organometallic compound [(η^5-C$_5$H$_5$)Fe(CN)$_2$(CO)]$^-$ in Figs 7.4 and 8.5(16), in which the pyramidal Fe(CN)$_2$(CO) fragment is bound to a cyclopentadienyl ligand, that mimics the six-electron donating ability of the Ni(μ-SCys)$_2$(μ-O) donor face of the enzymatic centre (Darensbourg *et al.* 1997). Subsequently, a series of related compounds [(η^5-C$_5$R$_5$)Fe(CO)$_{3-x}$(CN)$_x$]$^{n-}$ (R = H, Me; x = 0–2; n = 0–1) (Lai *et al.* 1998) was used to probe the effects of electronic pressure and hydrogen bonding, respectively, on the vibrational modes of the triply bonded diatomic molecules, that are used to distinguish between the different enzymatic states in FTIR spectroscopy (Sections 2.1 and 8 in Chapter 7).

Although hitherto no structural analogue of the complete heterodimetallic nickel–iron centre of [NiFe] hydrogenase has been reported, some compounds with similarities to the biological centre were recently synthesized using specially designed

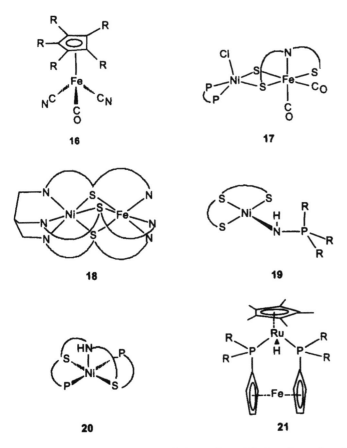

Figure 8.5 (16–21) Structures of chemical models for the active sites of hydrogenases. For references, see text.

chelate ligands in Fig. 8.5(17, 18) (Davies *et al.* 1999; Steinfeld and Kersting 2000). The necessity for the restriction induced by these tailored ligands is not surprising, as all the evidence points to the extraordinary character of the enzymatic centre, for whose construction even Nature has to resort to complicated procedures (Chapter 4). However, studies towards functional models for the hydrogenase suggest that the complete heterodimetallic centre might not be necessary. Mononuclear nickel complexes such as [Ni(NHP(C_3H_7)$_3$)(bis(2-sulfanylphenyl)sulfide)] in Fig. 8.5(19) catalyse the hydrogenase-specific H_2/D_2 exchange reaction (Sellmann *et al.* 2000), and compounds such as [Ni(bis(5-(diphenylphosphino)-3-dithiapentanyl)amine)]$^{1+}$ in Fig. 8.5(20) were found to produce H_2 gas from protons without even having thiolate ligands (James *et al.* 1996). The catalytic activity of the hydrogenase in hydrogen consumption coupled to methyl viologen reduction was modelled by the organometallic ruthenium–iron compound in Fig. 8.5(21) (Hembre *et al.* 1996), which might indicate an active function of the iron part in the reaction, other than the mere fine-tuning of reactivity properties.

In conclusion, model compound studies have contributed valuable information for our understanding of the structure and reactivity of the enzymatic centre by probing the chemical possibilities. But apart from the help in the complete understanding of the catalytic principle, they also point to potential alternatives for functional catalysts, which are needed for the cheap production of hydrogen on a large scale to meet the increased demand expected after its introduction as a fuel in the future (Chapters 9 and 10).

8.7. A dinuclear iron(II) compound mimicking the active site of [Fe] hydrogenases

Elisabeth Bouwman, Richard K. Henderson and Jan Reedijk

We have reported the first dinuclear iron(II) compound which may be regarded as a promising structural model for [Fe]-only hydrogenases (Fig. 8.6A). The synthesis and crystallographic characterization of this mixed-spin dinuclear iron(II) complex, containing sulfur bridges and terminal carbon monoxide ligands, have been described (Kaasjager *et al.* 1998). The compound has been synthesized by refluxing a mixture of [Fe(II)(dsdm)]$_2$ (H_2dsdm = N,N'-dimethyl-N,N'-bis(2-mercapto-ethyl)ethylenediamine) with K[HFe(CO)$_4$] in the presence of 2-bis(mercaptoethyl)sulfide (H_2bmes). The structure shows two octahedrally coordinated iron(II) atoms, each having different ligand environments. Fe1 is coordinated to the ligand dsdm, and the two thiolate sulfurs from bmes, resulting in an FeN$_2$S$_4$ chromophore. Fe2 is bound to the ligand bmes, to two terminally coordinating carbon monoxide groups and one of the thiolate sulfurs from dsdm, resulting in an FeC$_2$S$_4$ chromophore. The two octahedrons are face sharing through three asymmetric μ_2-thiolato bridges. The dinuclear molecule is electrically neutral, and charge considerations lead to the conclusion that both irons are divalent.

The IR spectrum of the compound in the solid state reveals strong absorptions for the carbonyl ligands at 2,011 cm^{-1} and 1,957 cm^{-1}, well within the range as is observed for the [Fe]-only hydrogenases.

The spin states of the iron(II) atoms have been verified with Mössbauer spectroscopy. A Mössbauer spectrum recorded at room temperature, shows two doublets

A

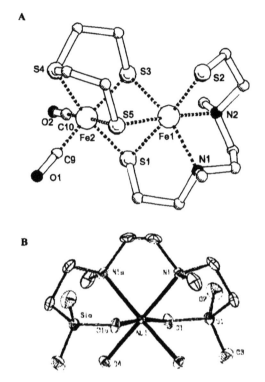

B

Figure 8.6 (A, B) Dinuclear models for the active sites of hydrogenases. For references, see text.

of equal intensity but with different isomer shifts and quadrupole splittings, confirming the presence of two distinct Fe(II) atoms.

In conclusion, the presented dinuclear iron structure is the first example of a biomimetic iron compound, which can be regarded as a first generation model for the class of [Fe]-only hydrogenases. The complex incorporates both relevant carbon monoxide ligands, as well as three bridging thiolato ligands, which could be possibly present in the active site of these enzymes.

8.7.1. A nickel disulfonato complex obtained by oxidation of a mononuclear nickel dithiolate complex

The oxidation with H_2O_2 of the mononuclear nickel dithiolate complex [Ni(dsdm)] (H_2dsdm = N,N'-dimethyl-N,N'-bis(2-mercaptoethyl)ethylenediamine) led to the high-yield formation of a disulfonato nickel complex [Ni(dsodm)($H_2O)_2$] ·$2H_2O$ (dsodm = N,N'-dimethyl-N,N'-bis(2-sulfonatoethyl)ethylenediamine) (Fig. 8.6B). This is the first example of oxidation of a nickel thiolate to a nickel sulfonate compound (Henderson *et al.* 1997). The coordination around the nickel is octahedral, with the nickel sitting on a two-fold axis. The tetradentate ligand occupies four coordination sites with the two sulfonate oxygens in trans-positions. Two water molecules cis to each other complete the octahedral coordination set. The nickel to amine nitrogen distances are 2.143(2) Å. The nickel to oxygen bonds in [Ni(dsodm)(H$_2$O)$_2$] ·2H$_2$O are 2.044(1) and 2.084(1) Å for the sulfoxides and the water molecules, respectively.

The use of six equivalents of dihydrogen peroxide leads to a clean conversion of the dithiolate complex to the disulfonate compound. Earlier studies on oxidation of nickel thiolates showed that oxidations with dioxygen stop at monosulfinates. Our observation and the characterization of the first chelating bis-sulfonato nickel complex formed from the direct oxidation of a mononuclear nickel dithiolate, may also provide new insight into the chemistry of sulfur-rich nickel-containing enzymes in the presence of oxygen.

Chapter 9

Hydrogenase in biotechnology
Hydrogen consumption

Richard Cammack

With contributions from *Reinhard Bachofen, Christof Holliger, Michael Hoppert, Kornél L. Kovács, Birgit Krause, Lynne Macaskie, F. Mayer, Iryna Mikheenko, K. Mlejnek, Paul Péringer, Thomás Ruiz-Argüeso, Jean-Paul Schwitzguébel, Mylène Talabardon, Esther van Praag and Ping Yong*

As described in Chapter 2, microbial systems have a natural propensity to use H_2 and hydrogenase for many purposes. Some of these processes have, or could have in the future, great potential in practical applications. The progress on some of these areas is reviewed in this chapter.

The production of methane from waste exploits the abundance of natural H_2-producing and H_2-consuming organisms. These processes are being developed by selection of the optimum types of organisms and culture conditions.

The green revolution in agriculture has been at the expense of nitrogen fertilizers, most of them produced with the energy of fossil fuels. In order to progress to a sustainable agriculture it may be necessary to move toward a greater reliance on biological N_2 fixation. Only prokaryotes can fix N_2, and most of the biological N_2 fixation in agriculture is done by symbiotic N_2-fixing bacteria such as *Rhizobium* in root nodules of leguminous plants. As previously noted nitrogenase itself produces excess H_2 gas during its reaction at the expense of ATP. Some, but not all, strains of rhizobia have hydrogen-uptake (Hup) hydrogenases to recover some of the chemical energy Knowledge of hydrogenase catalysis and genetics can be used to improve productivity of agricultural N_2 fixation, for example in selecting or constructing strains of the symbiotic N_2-fixing organisms to increase the yields of leguminous plants.

The pharmaceutical and fine chemical industry might use pure hydrogenase or partially purified enzyme preparations in bioconversion applications such as regio and stereoselective hydrogenation of target compounds (van Berkel-Arts *et al.* 1986). Enzymes are able to catalyse such stereospecific syntheses with ease. However, the cofactors for the NAD-dependent oxidoreductases are expensive. The pyridine nucleotide-dependent hydrogenases such as those from *Ralstonia eutropha* and hyperthermophilic archaea (Rakhely *et al.* 1999) make it possible to exploit H_2 as a low-cost reductant. The use of inverted micelles in hydrophobic solvents, in which H_2 is soluble, has advantages in that the enzymes appear to be stabilized.

9.1. Utilization of hydrogen metabolism in biotechnological applications

Kornél L. Kovács

Hydrogen evolution and consumption by intact bacterial cells are frequently observed in Nature. This is how hydrogenases were discovered originally seventy years ago

(Stephenson and Stickland 1931). In microbial ecosystems the role of these micro-organisms is creation and maintenance of an anaerobic, reductive environment as well as supplying the universal reducing agent, molecular hydrogen (Kovács *et al.* 1996). Gaseous hydrogen is usually not released from the natural ecosystems unless there is an excess of reductive power which needs to be disposed of in order to ensure the optimal metabolic and growth equilibrium in the population (Sasikala *et al.* 1993). H_2 generated *in vivo* by hydrogen-forming bacteria is utilized by hydrogen-consuming members of the microbiological community. Hydrogen is transferred to the recipient microorganism very effectively by interspecies hydrogen transfer (Belaich *et al.* 1990). The molecular details of this process are not fully understood, but its significance in safeguarding the optimum performance of the entire ecosystem and the delicate regulatory mechanisms should be appreciated. In the mixed population bacterial systems presented here, the advantages of interspecies hydrogen transfer are exploited.

9.1.1. Bioremediation: Generation of reductant

Environmental contamination usually consists of a mixture of pollutants and their partially degraded derivatives. Such an ill-defined chemical mixture will eventually lead to the formation of an ecosystem of microbes. The individual member species cannot survive in the toxic and hostile environment. Effective bioremediation tech-nologies should therefore invoke a mixture of microorganisms forming synergistic consortia. Any realistic bioremediation concept is based on the recognition that it is the concerted action of various species, which may bring about the desired clean-up effect.

A significant drawback for remediation microbiology is that there is very limited basic microbiological knowledge on cooperative effects and interactions among microorganisms. This lack of information derives from the first and most important prerequisite in a classical microbiology experiment: the purity of the strains. The requirement of working with isolated individual strains is fully justified by the rules of fundamental microbiology, yet this strategy automatically excludes any observa-tion of the potential beneficial effects arising from interactions among various strains. Therefore no such effects can be taken into account when biodegradation of a com-plex hazardous chemical mixture is considered and the simplistic view of assuming the mere summation of individual microbial contributions in a microbial consortia is inherently included in most approaches. One consequence is the widespread belief, that given certain minimum nutrient conditions, the microbial activity is not rate lim-iting in biodegradation systems. It follows from this principle that microorganisms, specially adopted to decompose a certain type of contamination, will be most abundant at the site of contamination. The more incompatible a given chemical is to common forms of life, the stronger is the selection pressure to give preference to the growth of microorganisms that are capable of metabolizing that compound (Blackburn and Hafker 1993; Lowe *et al.* 1993).

From an applied environmental biotechnology point of view this lack of fundamen-tal microbiology knowledge results in poorly understood system characteristics and inferior or unpredictable performance. A solution for this dilemma, employed in sev-eral laboratories, including ours, is the assembly of controlled mixed cultures, in which pure cultures of bacteria are deliberately mixed in order to enhance bioconversion/ biodegradation yields (Perei *et al.* 1995).

An increasing amount of evidence supports the significance of bacterial interactions as an important additional dimension in applied microbiology, e.g., in the utilization of bacterial hydrogen metabolism for environmental biotechnology applications. These advantages are demonstrated in the case of biogas production and denitrification. Other examples show the importance of selective pressure in developing laboratory scale microbiological answers to pressing bioremediation needs (Kovács *et al.* 1996).

9.1.2. Biogas production from wastes

Biological methane production is carried out in three stages, performed by separate groups of microorganisms (Fig. 9.1). Complex organic materials are first hydrolysed and fermented by facultative and anaerobic microorganisms into fatty acids. The fatty acids are then oxidized to produce H_2 and organic acids (primarily acetate and propionate), processes termed dehydrogenation and acetogenesis, respectively. The last stage is methanogenesis (see Chapter 2). Some methanogens can combine the hydrogen directly with carbon dioxide to form methane, others split acetate into carbon dioxide and methane (Lusk 1998).

Among the significant recent advances in understanding the ecology of anaerobic biodegradation of organic wastes is the recognition of the close syntropic relationship among the three distinct microbe populations and the importance of H_2 in process control. The regulatory roles of hydrogen levels and interspecies hydrogen transfer optimize the concerted action of the entire population. The concentration of either acetate or hydrogen, or both together, can be reduced sufficiently to provide a favourable free-energy change for propionate oxidation.

During anaerobic biodegradation, the H_2 concentration is reduced to a much lower level than that of acetate. The acetate concentration in an anaerobic digester tends to

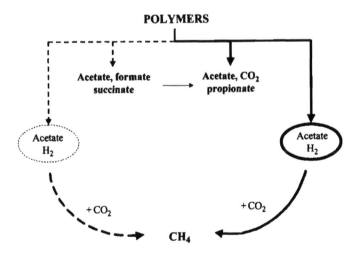

Figure 9.1 Scheme of microbiological metabolic pathways for biogas production.

range between 10^{-4} and 10^{-1} M, whereas the H_2 ranges between 10^{-8} and 10^{-5} M, or about four orders of magnitude less (Kovács and Polyák 1991). In addition, the hydrogen partial pressure can change rapidly, perhaps varying by an order of magnitude or more within a few minutes. This is related to its rapid turnover rate. The energy available to the acetate-using methanogens is independent of hydrogen partial pressure, whereas that of the hydrogen-producing and hydrogen-consuming species is very much a function of it.

The low concentrations of free hydrogen within the digester have implications for the rate of H_2 turnover in the system. In a typical system the turnover rate of the hydrogen pool is 6.8×10^6/day or about 80/s. The interspecies H_2 transport is therefore a very rapid, sensitive and rate limiting step.

For processes on this microscale, diffusion is the main mechanism for hydrogen transport between species. The calculations clearly indicate that the optimal distance between interacting hydrogen producing and consuming species is about 10 μm under practical conditions. This distance is equal to about 10 bacterial widths. Clearly, the bacteria must indeed be close together, otherwise interspecies hydrogen transfer will be the rate limiting step in the overall process (Kovács and Polyák 1991). This is the case in an ideal, completely mixed or continuously stirred tank reactor where the individual bacterial species are dispersed uniformly throughout the system, and the distribution of reactants, intermediates, products of biotransformation and bacterial species is homogeneous throughout the reactor.

New process designs, such as the biofilm reactors, encourage different species to live in proximity to one another and the result is a much higher rate of conversion per unit volume of reactor, although not yet optimal. Biofilms also permit a diversity of environments to develop in proximity to one another so that suitable conditions for the degradation of each substrate and intermediate can exist somewhere within the biofilm.

We have shown that, under these circumstances, addition of hydrogen producers to the system and thereby shifting the population balance brings about advantageous effects for the entire methanogenic cascade. The decomposition rate of the organic substrate, which was animal manure in our first experiments, increases and both the acetogenic and methanogenic activities are remarkably amplified. In laboratory experiments some 2.6-fold intensification of biogas productivity has been routinely observed and the same results were obtained in a 100-litre digester scale-up experiment (Fig. 9.2).

The intensification of biogas formation from solid municipal waste in an anaerobic digester of waste water sludge has been demonstrated in a field experiment. Municipal solid waste is usually difficult to break down and the biodegradation process is particularly slow in landfill type of 'bioreactors'. Nevertheless, an economic biogas-producing system has been installed in the municipal waste landfill site in Hungary (Fig. 9.3). The facility receives about $200,000 \, m^3$ of wastes annually, mostly solid household garbage. The biogas has been collected through an underground pipeline system, compressed, and fed into a natural gas-based heating centre that serves a residential area. Biogas production reached $700,000 \, m^3$ by 1990. Since the system is simple and inexpensive and the biogas collection and utilization is solved in an ingenious and fortuitous way, the installation operates smoothly and efficiently. Inoculation with the pure culture of a specially selected hydrogen-producing bacterium resulted in a >30 per cent overall increase of biogas yield.

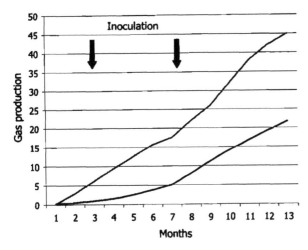

Figure 9.2 Cumulative gas production curves showing intensification of biogas production from pig manure in a 100-litre digester.

The environmental conditions in a waste water sludge treatment system are better controlled. It is probably for this reason that an overall 80 per cent increase of biogas productivity has been reached in a 2,500 m^3 mesophilic digester upon inoculation with the hydrogen-producing microorganism.

Proper management of the bacterial population is expected to facilitate the startup of the fermentation. In order to reduce the costs of this treatment, supplemented bacteria are grown in diluted industrial waste water. Spraying the microbes onto the surface of the garbage layer before covering the layer with dirt accelerates the formation of biofilms, the highly effective microbial centres of fermentative degradation (Sanchez *et al.* 1994). In addition to an increased biogas production, the faster decomposition rate is expected to shorten the long lifetime of the landfill site and the land will be ready for development within ten to fifteen years instead of the usual twenty to thirty years needed for conventional treatment.

9.1.3. Denitrification for removal of nitrate from water

Nitrate contamination of natural waters is gradually increasing worldwide, particularly among the industrialized nations. Several methods have been considered for removal of nitrate from water (Knowles 1982). Currently ion exchange is the only effective technique used in full-scale treatment facilities to remove nitrate from drinking water. Continuous regeneration of the ion exchange resin is possible and plants to clean groundwater containing 20–30 mg/l of nitrate-nitrogen can produce lower than 2 mg/l nitrate levels. Several ion exchange systems are in operation worldwide with mixed results as far as system performance and economy is considered (van der Hoek *et al.* 1988a,b).

Biological denitrification has been studied for the purification of both waste water and drinking water (Soares *et al.* 1991). The technique, as applied today, consists of

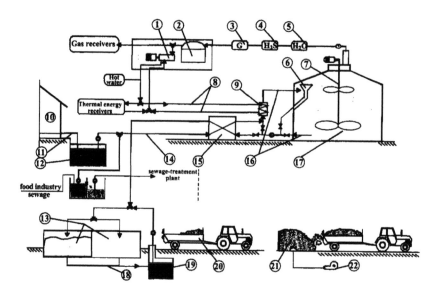

Figure 9.3 Block diagram of the designed facility for large-scale production according to the EUREKA 1438 project. 1 – 380 V generator, 2 – biogas tank, 3 – gasmeter, 4 – desulfurization, 5 – dehydrator, 6 – overflow, 7 – mixer, 8 – water pipeline, 9 – heat exchanger : water/manure, 10 – piggery, 11 – manure transporting pipeline, 12 – preliminary tank, 13 – manure chambers, 14 – feeding pipelines, 15 – heat exchanger: manure/manure, 16 – mixing pipeline, 17 – fermentation chamber, 18 – drainage, 19 – reflux tank, 20 – dung spreader + tractor, 21 – composting plate, 22 – fan.

the use of denitrifying microorganisms in a filter bed to reduce the nitrate to N_2 (Fig. 9.4). The biological filter is followed by a conventional filter to remove the carry-over bacterial growth. The heterotrophic denitrifying organisms require an organic energy source. An organic substance therefore must be provided for treating drinking water because groundwater supplies are essentially free of organic material. In spite of being able to truly eliminate nitrate contamination, biological denitrification has not been enthusiastically received by water utilities yet, primarily because of the potential to introduce contaminations of microbial origin into the purified water (Kurt *et al.* 1987).

The system developed in our laboratories includes two novel features, that, to our knowledge, have not been applied in denitrification technologies before (Kovács and Polyák 1991). We intend to exploit the benefits of interspecies hydrogen transfer and that of our novel immobilization technology. Interspecies hydrogen transfer is essential to supply the necessary hydrogen for nitrate reduction by denitrifying microorganisms (Hochstein 1988). To accomplish this task we employ well-defined mixtures of bacteria. One species in this artificial microbiological ecosystem is responsible for hydrogen production from added organic substrate. Certain industrial waste waters, sugars or cellulose can serve as organic substrate (Fig. 9.5). Because of the close spatial proximity, hydrogen is effectively transferred from the helper microorganisms to the denitrifying bacteria immobilized together with the hydrogen producer species (Fig. 9.6).

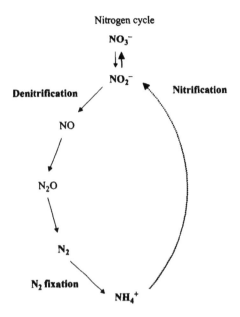

Nitrogen cycle

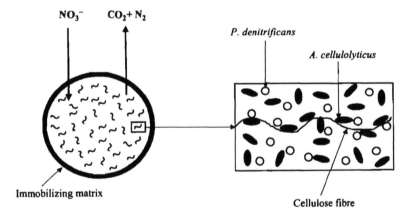

Figure 9.4 Nitrogen cycle scheme. The steps are catalysed by enzymes in various chemolithotrophic bacteria.

Figure 9.5 Diagram of denitrification process.

Interspecies hydrogen transfer brings about synergistic effects for the denitrifying bacteria since it is a very efficient way to administer hydrogen for their function. From the point of view of operational safety, generation and utilization of hydrogen *in situ* is evidently superior to bubbling the system with explosive H_2 gas. Immobilization of the participating bacteria in beads of high physical resistance provides better performance because of dramatically increased bacterial population density and high flow rate. Moreover, with the immobilization techniques (Martins dos Santos *et al.* 1997),

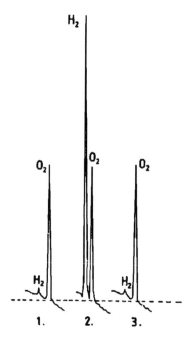

Figure 9.6 Gas chromatographic traces from a defined mixed culture system for water denitrification, showing efficient interspecies H_2 transfer. 1 – Denitrifying bacterium. 2 – H_2 producer. 3 – H_2 producer + denitrifying bacterium.

axenic fermentation conditions can be maintained which allows a strict control of the microbiological ecology within the immobilized system.

The process in the simplest form contains an ion exchange column producing potable water while another ion exchange column is regenerated through a biological denitrification reaction (van der Hoek *et al.* 1988b). During ion exchange resin regeneration a brine solution containing nitrate in high concentration is generated. A suitably engineered bacterial population is able to convert nitrate to N_2. In this way the regenerant can be used again after it has been subjected to denitrification and thus salt requirement and brine production are minimized. In laboratory experiments a reduction of 90 per cent in brine volume has been routinely achieved.

Biological regeneration of the brine solution from the nitrate-loaded anion exchange resin can be achieved in the bioreactor containing immobilized microorganisms. In the system developed in our laboratory, a continuous flow circuit is used, containing the ion exchange column which has to be regenerated and a bioreactor accommodating the biological material. The regenerant is recirculated in the system through the ion exchange column and the denitrification bioreactor, which converts nitrate into N_2 (Fig. 9.7).

The best H_2-producer helper bacterium strains convert the organic material found in dilute waste waters from the food processing industry (sugars and proteins). They are selected because in this way, a treatment of polluting waste waters can be linked to the elimination of polluting nitrate contamination. The ability to bring about denitrification

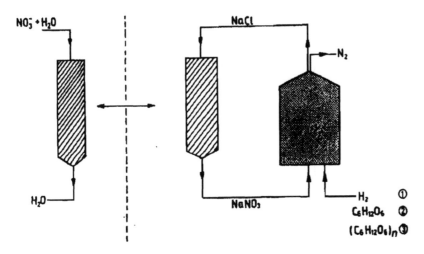

Figure 9.7 Diagram of a denitrification plant.

is characteristic of a wide variety of common bacteria, including the genera *Pseudomonas*, *Achromobacter* and *Bacillus*. In our hands such a biological denitrification system can handle 1,000–2,000 mg nitrate/l (100 mmol $NaNO_3$/l) at a conversion rate of 10 mg nitrate/g bead/h. Nitrite was not detectable in the effluent of our experimental bioreactor even after several weeks of continuous operation.

9.2. Nitrogen fixation

Thomás Ruiz-Argüeso

Hydrogen production by legume nodules has been estimated to be associated with energy losses of 40–60 per cent of the energy available for nitrogen fixation, and constitutes the main source of inefficiency of the legume-rhizobial symbiosis (Fig. 9.8). Hydrogen production results from the concomitant, unavoidable reduction of protons to H_2, catalysed by nitrogenase. Rhizobial strains that express uptake hydrogenase activity are expected to carry out a more efficient fixation of N_2 by re-utilizing the H_2 produced by nitrogenase (Albrecht *et al.* 1979). Several mechanisms by which the presence of a hydrogenase might increase the overall efficiency of nitrogen fixation have been postulated, including: provision of additional source of energy and reductant to the N_2-fixing system, protection of nitrogenase from O_2 damage, and prevention of H_2 inhibition of nitrogenase-catalysed N_2 reduction. Convincing evidence from *in vitro* experiments has shown that the hydrogenase system has a potential to improve N_2 fixation (see Arp 1992; Evans *et al.* 1987; Maier and Triplett 1996; Van Soom *et al.* 1993, for reviews). However, the actual contribution of H_2 recycling ability to increase productivity of nodulated legumes is still a conflicting issue.

Positive effects of the *hup* system on legume productivity have been reported for the *Bradyrhizobium japonicum*-soybean system and the bean (*Phaseolus vulgaris*) system. However, negative results have also been obtained in other experiments

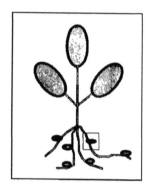

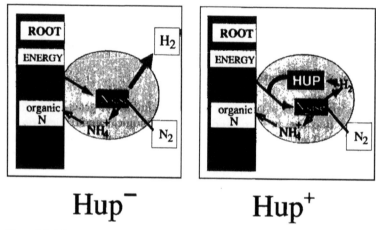

Figure 9.8 Nitrogenase and energy consumption in nodules, with and without the H_2-uptake *hup* genes.

or systems. The discrepancies may be attributed to a number of factors, including variations in experimental conditions; use of rhizobial strains that do not efficiently recycle H_2; inefficient coupling of H_2 oxidation to ATP generation in the host legume; harvesting of plants before maturity; use of insufficient numbers of replicates for statistical significance, particularly in field experiments; and the use of inadequate Hup$^-$ control strains (Arp 1992; Evans *et al.* 1988). The need to compare Hup$^+$ and Hup$^-$ strains that are isogenic except for the H_2 oxidation characteristic has repeatedly been emphasized (Evans *et al.* 1988). However, since we know now that the H_2 oxidation capacity is coded by a multigenic system, Hup$^-$ mutants deleted of the entire *hup* cluster should be more appropriate controls than Hup$^-$ mutants affected in a single hydrogenase gene. Alternatively, the effect of Hup on legume productivity can be examined by comparing Hup$^-$ wild-type strains of root nodule bacteria and genetically engineered, derivative Hup$^+$ strains that have received the entire *hup* cluster. The construction of these strains is a first step in a more general biotechnological objective aimed to extend the Hup phenotype to rhizobia that nodulate important legume crops and normally lack the *hup* system.

A number of surveys of legume root nodules have been carried out to assess directly the frequency of strains with the capacity to recycle H_2 (reviewed by Arp 1992). The ability to oxidize H_2 is not a predominant phenotype among root nodule bacteria. The frequency of Hup$^+$ strains is very low among the fast-growing group of rhizobia, and even in the slow-growing group, where the frequency of the Hup phenotype is higher, the presence of Hup$^+$ strains in species such as B. japonicum appears to be scarce in agricultural soils. Strains capable of efficiently recycling H_2 have not been found in rhizobia nodulating important legume crops such as alfalfa, clovers or chickpeas.

Progress in identification and characterization of genetic determinants for hydrogenase synthesis and in the regulation of their expression in rhizobia makes transfer of hup genes to heterologous backgrounds feasible. Several strategies have been used to transfer the hup system from Hup$^+$ strains of B. japonicum and R. leguminosarum bv. viciae – the two best-known systems – into Hup$^-$ recipient strains of different rhizobia. Transfer of the hup system in a plasmid or cosmid basis frequently faces problems of instability of the Hup character in root nodules in the absence of selective pressure for maintenance, but maintenance of the Hup phenotype can be improved by stabilizing the cosmids with par genes. When native Hup$^-$-containing symbiotic plasmids from R. leguminosarum have been used, the presence of non-hup DNA information imposes an extra metabolic load to the recipient cell which can mask the effect of the hup system on the symbiotic behaviour. Integration of the plasmid-borne hup cluster into the chromosome appears to be the strategy of choice, although its large size (over 15 kb) restricts its usefulness for in vitro recombination. The most promising strategy for integration of the hup system into the chromosome of recipient strains appears to be the use of engineered mini-transposons. Genes to be integrated are cloned between the inverted repeats of Tn5 adjacent to a marker antibiotic resistance gene and, since the rest of the transposition system is located outside these repeats, the procedure results in a stable integration. Using adequate suicide vectors one can eventually transfer any DNA fragment into the chromosome. This system allowed us to stably integrate the entire 18 kb hup cluster from R. leguminosarum bv. viciae strain UPM791 into the chromosome of Hup$^-$ strains of B. japonicum, Mesorhizobium ciceri, M. loti, R. leguminosarum bv. viciae and bv. trifolii, R. etli and Sinorhizobium meliloti (Báscones et al. 2000). Hydrogenase expression analysis of these newly generated Hup$^+$ strains in symbiosis with their corresponding legume hosts showed a broad range of expression levels depending on the bacterial species and even on the strains (Fig. 9.9). From the Hup$^-$ mini-transposon-containing vectors, it is straightforward to remove the antibiotic resistance gene in order to construct environmentally friendly Hup$^+$ strains to be used in field trials or for commercial release.

Stable integration of the hup system into the chromosome still does not ensure acquisition of the H_2-recycling capacity by the recipient Hup$^-$ strains. Efficient expression of hydrogenase genes, and hydrogenase activity in symbiosis with the corresponding host legumes, are also required. At least three factors have been identified that affect the expression of hydrogenase activity in legume nodules:

The bacterial genetic background drastically affects the levels of heterologous expression of the R. leguminosarum hup system (Báscones et al. 2000). High levels of hydrogenase expression were detected in R. etli, B. japonicum and M. loti and certain strains of R. leguminosarum bv. viciae in symbiosis with P. vulgaris, Glycine

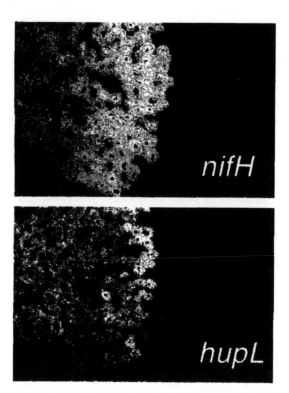

Figure 9.9 Dark field micrographs obtained from *in situ* hybridization experiments, demonstrating the colocalization of *hupS* and *nifH* mRNAs (white areas) in pea nodules containing *R. leguminosarum* bv. viciae.

max, Lotus corniculatus and *Pisum sativum*, respectively. In contrast, very low levels of activity were observed in *M. cicer* and *S. meliloti* in symbiosis with *Cicer arietinum* and *Medicago sativa*, respectively. The molecular basis for this variation in Hup expression is not known at present, although differences in regulation of the *hup* genes by NifA and Fnr, the two major activators of *R. leguminosarum hup* system (Brito *et al.* 1997), might account for the observed differences.

Nickel availability to the host plants severely limits the expression of the *R. leguminosarum* hydrogenase genes in the *P. sativum* symbiosis (Brito *et al.* 1994) and probably in other symbioses such as *M. loti–L. corniculatus* (Brito *et al.* 2000). This limitation occurs at the level of processing of the enzyme subunits (Brito *et al.* 1994). It is not clear, however, whether Ni limitation is due to the bacterial or the plant component of the symbiosis. Recent results of nickel transport experiments with intact pea symbiosomes indicate that the peribacteroid membrane is not a specific barrier for Ni transport into the bacteroid (Báscones *et al.* unpublished).

Some legume hosts have been shown to be more permissive than others for expression of hydrogenase. Host-plant mediated control of the Hup phenotype has been reported for the symbionts of pea, soybean, cowpea and common bean (Maier and Triplett 1996). The molecular basis for this effect is not yet understood. In the case of *R. leguminosarum* bv. viciae, it has been demonstrated that strain UPM791 induces hydrogenase activity in pea bacteroids but is unable to express this activity in lentils (López *et al.* 1983). Preliminary data suggest that the blocking of Hup expression is exerted at the level of transcription of *hup* structural genes (Toffanin *et al.*, unpublished).

In summary, although the tools for introducing the *hup* system in any strain of rhizobia in a stable form are now available, more research is needed to achieve an efficient expression of the *hup* system in the different legume hosts and to evaluate their effect on legume productivity before succeeding in obtaining marketable, improved inoculant strains.

9.3. A novel anaerobic thermophilic fermentation process for acetic acid production from milk permeate

Mylène Talabardon, Jean-Paul Schwitzguébel and Paul Péringer

In Switzerland, the cheese industry produces large amounts of lactose in the form of milk permeate. Ultrafiltration of milk is used for concentrating milk in several cheese-producing plants (Feta cheese) as well as in manufacturing special milk products. This cheese-making technology produces, instead of whey, a deproteinated permeate which needs further processing. The permeate contains about 5 per cent lactose, 1 per cent salts, and 0.1–0.8 per cent lactic acid; it is practically free of N-compounds and thus not comparable with whey which contains up to 0.8 per cent protein. Because of its lack of protein, it is unsuitable for food or animal feeding. It has a high biological oxygen demand of ~50 g/l and is a major disposal problem. Its direct discharge to sewers is costly and overloads sewage treatment plants. This lactose source, however, could serve as an excellent feedstock for the production of acetic acid and calcium magnesium acetate (CMA), a non-corrosive road deicer.

Anaerobic fermentations to produce acetic acid from whey lactose have been studied under mesophilic conditions. However, a thermophilic fermentation process, which generally has a higher production rate, is more resistant to contamination, is easier to maintain anaerobic conditions required for homoacetogens, and thus might be more advantageous for industrial production of acetic acid from milk permeate. Unfortunately, no known thermophilic bacteria can produce acetate directly from lactose, although most thermophilic homoacetogens can readily convert glucose to acetate with a product yield as high as 100 per cent.

In this work, several potential pathways for lactose fermentation to acetic acid under anaerobic thermophilic conditions (~60°C) were studied (Fig. 9.10). Although none of the known thermophilic homoacetogens could ferment lactose, one strain could use galactose and two strains could use lactate. *Moorella thermoautotrophica* and *M. thermoacetica* homofermentatively converted lactate at thermophilic temperature (50–65°C) and at pH between 5.8–7.7. Approximately 0.95 g of acetic acid was formed from each gram of lactic acid consumed. *M. thermoautotrophica* grew at an optimal pH of 6.35–6.85 and an optimal temperature of 58°C. There was no known thermophilic homolactic bacterium. In this study, twelve thermophilic fermentative bacteria were screened for their abilities to produce acetate and lactate from lactose: one heterolactic bacterium, *Clostridium thermolacticum*, was found to produce large

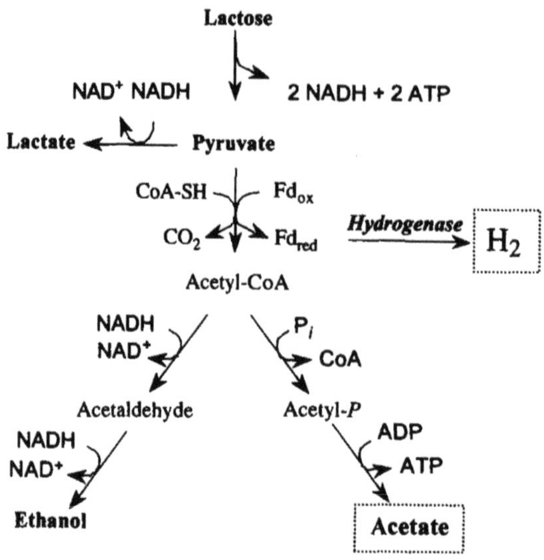

Figure 9.10 Pathways of heterolactic fermentation of lactose. Fd_{ox}, oxidized ferredoxin, Fd_{red}, reduced ferredoxin.

amounts of lactate and acetate, but it also produced significant amounts of ethanol, CO_2 and H_2. However, there was a metabolic shift from heterolactic to homolactic pathway during the fermentation, depending on the growth conditions: at pH 6.4, acetate, ethanol, CO_2 and H_2 were growth associated, while lactate was produced with a maximal yield in the stationary phase. It is thus possible to produce 4 mol of lactate from each mole of lactose using this bacterium under thermophilic conditions. Therefore, 6 mol of acetic acid can be produced from lactose using lactate and hydrogen as intermediary products in a co-cultured fermentation with the homoacetogen, *M. thermoautotrophica* able to assimilate both hydrogen and lactate. The interactions between the heterolactic bacterium and the homoacetogen, which were affected by growth conditions, suggest an efficient hydrogen transfer between both species. As a consequence, homoacetic fermentation of lactose was observed in the co-culture. The production, transfer and utilization of H_2 or other reducing equivalents are expected to be critical factors in the dynamic equilibrium between different metabolic pathways, and hydrogenases could play a key role in such balance. The optimal conditions (pH, temperature, and medium composition) for acetate production in the co-cultured fermentation are being evaluated using both free-cell fermentation and an immobilized-cell bioreactor. Higher acetate yield (>80 per cent), production rate (>1 g/l·h) and final acetate concentration (>40 g/l) can be produced in fed-batch fermentations with immobilized cells in a fibrous-bed bioreactor. This co-cultured fermentation process could be used to produce acetate from milk permeate and could become an attractive method for conversion of community waste to a community resource, with the double environmental benefit of reducing milk permeate disposal problems and pollution from salts used as road deicer in winter.

9.4. Electron production by hydrogenase for bio-remediation of Se-oxoanions

Esther van Praag and Reinhold Bachofen

Selenium is an element found in a solid form in the Earth's crust. Erosion can solubilize it, and so it will eventually be transported into the hydrosphere. Selenium is generally found as oxidized molecules and, when it is allowed to accumulate in high concentrations, it becomes toxic for living beings (1). Various bacteria have however been found to resist and survive in presence of the toxic Se-oxoanions. Photosynthetic bacteria, such as *Rhodospirillum rubrum*, were even shown to be able to detoxify the oxo-anion SeO_3^{2-}, by reducing it into a metallic form and eventually methylate it (Moore and Kaplan 1992; McCarty *et al.* 1993) (Fig. 9.11). Although little is known about the biochemical pathway, it is believed that the process involved is enzymatic and not simply due to a chemical reduction by the reduced electron carriers. In anaerobic conditions, hydrogenase is thought to play an important role by producing molecular hydrogen and electrons. The electrons produced this way could be accepted by the oxo-anions of Se.

R. rubrum was shown to possess hydrogenase and its activity was followed for 72 h. The bacteria were grown synchronously under anoxic conditions, with N_2 or CO_2 gas, at 30°C and in absence or presence of SeO_3^{2-}. The production of dimethyl selenide and dimethyl diselenide was detected by chemiluminescence detection of samples of the gas phase (Fig. 9.12). After a 24 h exposure to complete darkness, the bacteria were transferred to the experimental conditions and changes in the activity of hydrogenase with time were studied in non-growing cells, cultivated in white light.

Interestingly, the cells, synchronized by this treatment, exhibited a cycle of hydrogenase activity under all 'growing' conditions, with periods of either 16 h or 22–24 h. During the 16 h cycle, observed when the bacteria grow in the presence of gaseous N_2, the activity of hydrogenase increased over 8–10 h and then decreased. This was repeated over four to five cycles during a 72 h incubation period. The activity of hydrogenase in *R. rubrum* was not affected by the presence of SeO_3^{2-}, but EDTA was found to induce a strong increase of hydrogenase, both in the absence and in the presence of SeO_3^{2-}. Regular cycles observed in hydrogenase activity in intervals of

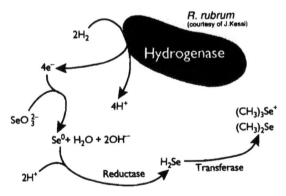

Figure 9.11 Enzymatic reduction pathway of Se by *R. rubrum* (micrograph by courtesy of J. Kessi).

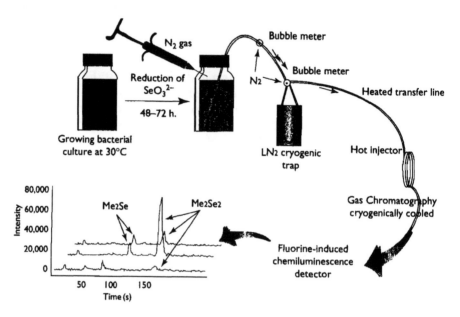

Figure 9.12 Apparatus for measurement of reductive metabolism of selenite.

16–24 h suggest that *R. rubrum* possesses mechanisms of endogenous circadian rhythm. Such fluctuations of hydrogenase activity were also observed in other prokaryotes (Kondo *et al.* 1993), especially in the cyanobacterium *Synechcoccus* sp. (Huang *et al.* 1990).

9.5. Hydrogenase in the reduction of halogenated pollutants

Christof Holliger and Birgit Krause

The solvent tetrachloroethene (PCE) is one of the most abundant halogenated pollutants in soil, groundwater and sediments, and, as far as is known, cannot be degraded aerobically. In the absence of oxygen, PCE is biologically reduced to less chlorinated ethenes and ethene itself. *Dehalobacter restrictus*, a Gram negative anaerobe, was isolated from an anaerobic fixed-bed column supplemented with lactate and PCE. It uses PCE as a respiratory acceptor, reducing it to *cis*-1,2-dichloroethene (Holliger and Schumacher 1994). The reaction is coupled to the oxidation of H_2:

$$2H_2 + C_2Cl_4 \rightarrow C_2H_2Cl_2 + 2Cl^- + 2H^+.$$

The reduction is mediated by a PCE reductase and hydrogenase, both of which are associated with the cytoplasmic membrane. The hydrogenase faces the periplasm of the cell and thus releases the protons into the periplasm (Schumacher and Holliger 1996). Two electrons are transferred across the membrane to the PCE reductase, via cytochrome *b* and menaquinone (Fig. 9.13). Since dihydrogen is the only electron donor used by *Dehalobacter restrictus*, the hydrogenase is a crucial enzyme of this

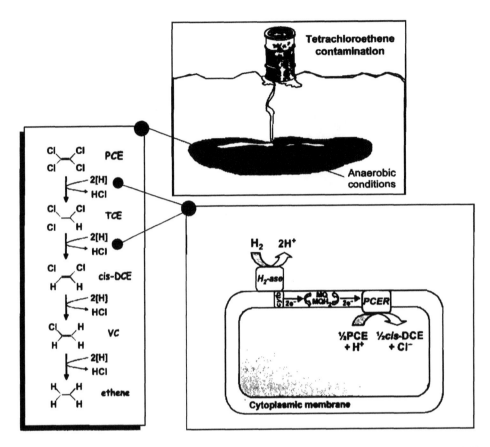

Figure 9.13 Reduction of PCE under anaerobic conditions: dehalorespiration by *D. restrictus* with H_2 and tetrachloroethene.

bacterium. The hydrogenase of *D. restrictus* is located on the outside of the cytoplasmic membrane and is membrane associated. It oxidizes dihydrogen producing the protons for energy conservation and transferring the electrons via menaquinone to the terminal reductase. The best artificial electron acceptor was methylene blue, followed by benzyl viologen and methyl viologen. The hydrogenase has an activity optimum at pH 8.1, is inhibited by $CuCl_2$ and is very oxygen sensitive. Since iron was the only trace metal needed for growth, it was proposed that this is an [Fe] hydrogenase.

9.6. Microbial recovery of platinum group metals

Lynne Macaskie, Ping Yong and Iryna Mikheenko

The widespread use of platinum, palladium and other metals in automotive catalytic converters has been driven by environmental considerations and the increasing costs of the metals. This has not been matched by the development of clean reprocessing technologies for the catalysts themselves. The spent catalyst metals are oxidized to their cations via leaching into concentrated acids.

Figure 9.14 Hydrogen-dependent deposition of metallic palladium onto cells of *D. desulfuricans*. (A) Cells before exposure to Pd^{2+}; (B) cells exposed to soluble Pd^{2+} and H_2. Black deposits of palladium are seen on the outer surface.

Sulfate-reducing bacterial cells, which contain hydrogenase, have been successfully used to re-reduce the metal cations to the base metals (Lloyd *et al.* 1998). In the presence of H_2 and palladium ions, the cells accumulate Pd metal at the cell surface (Fig. 9.14). The soluble Pd^{2+} was obtained from spent automotive catalysts by standard leaching technology and the bacteria were able to recover Pd from the leachate. The cells do not have to be viable and the process can operate at acidic pH values (pH 2.5). The process is clean and economically viable. For use the bacteria are immobilized as a layer (biofilm) onto the surface of a Pd-Ag alloy membrane which has the property of transporting hydrogen to the cells as atomic hydrogen. The hydrogen is provided from the back-side electrochemically, and the front-side with the attached bacteria is suspended in a flow containing Pd^{2+}. This allows the process to be operated continuously. The metals are precipitated in particulate form, which can be re-used as a catalyst.

9.7. Hydrogenases in water-in-oil microemulsions

Michael Hoppert, K. Mlejnek, F. Mayer

Usually, activities of enzymes (hydrogenases included) are investigated in solutions with water as the solvent. However, enhancement of enzyme activity is sometimes described for non-aqueous or 'water-limiting' surroundings, particular for hydrophobic (or oily) substrates. Ternary phase systems such as water-in-oil microemulsions are useful tools for investigations in this field. Microemulsions are prepared by dispersion of small amounts of water and surfactant in organic solvents. In these systems, small droplets of water (1–50 nm in diameter) are surrounded by a monolayer of surfactant molecules (Fig. 9.15). The water pool inside the so-called 'reverse micelle' represents a combination of properties of aqueous and non-aqueous environments. Enzymes entrapped inside reverse micelles depend in their catalytic activity on the size of the micelle, i.e. the water content of the system (at constant surfactant concentrations).

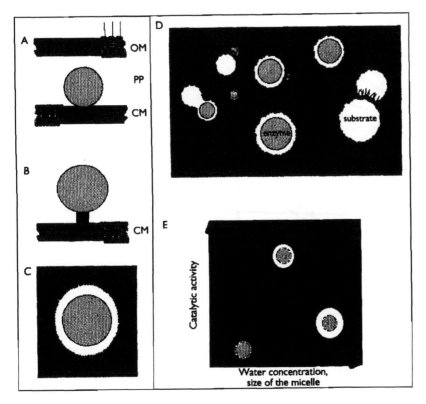

Figure 9.15 Enzymes in aqueous (light-coloured) and hydrophobic (shaded) phases. (A) A protein in the periplasm (PP) of a cell (OM = outer membrane, CM = cytoplasmic membrane); (B) membrane-bound protein in a lipid bilayer; (C) hydrophilic protein in an inverted micelle; (D) interaction between enzyme and substrates in aqueous micelles; (E) graph of catalytic activity as a function of micelle concentration.

For several hydrogenases, especially the membrane-bound hydrogenase from *R. eutropha* (HoxP) and the F_{420}-reducing hydrogenase from *Methanobacter marburgensis*, unusual activities in microemulsions have been observed. The temperature optima of hydrogen production in microemulsion were about 15° higher. Even at 95°C the hydrogen produced by HoxP was about 75 per cent of the maximal production rate. Furthermore, when aliquots of the hydrogenases were inactivated by prolonged incubation in aqueous buffer at elevated temperatures and subsequently dispersed in microemulsion, enzyme activity was partially recovered. This behaviour was not restricted to hydrogenases: the temperature optima of dehydrogenases (alcohol dehydrogenase, lactate dehydrogenase) was raised from 20–30°C in aqueous solutions to 35–50°C in microemulsions, and turnover rates were increased by a factor of 5. By addition of compatible solutes (betaine, glycerol, trehalose) to the aqueous phase of the microemulsion, the thermal stability of the entrapped enzymes was further improved (Fig. 9.16A,B). Turnover rates of hydrogenases and dehydrogenases were increased by a factor of 2 (i.e. a factor of 10, as compared to the aqueous buffer

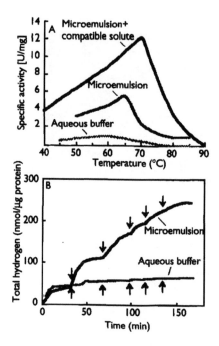

Figure 9.16 (A) Activity of hydrogenase as a function of temperature under various conditions. (B) Production of hydrogen with successive additions of reductant, at 70°C.

solution). Kinetic investigations showed that increased turnover was based on increased maximal reaction rate (HoxP) or enhanced stability (F_{420}-reducing hydrogenase) of the enzymes, K_m values remained unaffected. Enhanced activity or stability of enzymes is considered of major value for use in biotechnological applications. For instance, microemulsions may provide a more favourable surrounding for artificial light-dependent hydrogen-producing systems than aqueous systems.

Chapter 10

Producing hydrogen as a fuel

K. Krishna Rao and Richard Cammack

With contributions from *Catherine E. Gregoire Padró, Bärbel Hüsing, Kornél L. Kovács, Peter Lindblad, Paula Tamagnini, Paola M. Pedroni* and *Paulette M. Vignais*

Over the last thirty years, many approaches for H_2 production by whole-cell and cell-free systems have been explored, and a number of pilot-scale feasibility studies have been performed. These have been reviewed in recent books (Zaborsky 1998; Miyake *et al.* 2000). In this chapter we review the ways in which our knowledge of hydrogenases in Nature can guide our future research on hydrogen energy, focusing on biotechnological and biomimetic approaches.

10.1. Introduction

10.1.1. The hydrogen economy

Periodic crises in the supply and price of fossil fuels have drawn attention to the fact that renewable energy sources are the only long-term solution to the energy requirements of the world's population. Molecular hydrogen is a future energy source/carrier that is being actively investigated as an alternative to fossil fuels. It reacts with oxygen, forming only water; hence, it is a clean renewable energy source. It has a high calorific value, and can be transported for domestic consumption through conventional pipelines. Contrary to a widely held belief (the 'Hindenburg disaster syndrome'), hydrogen gas is safer to handle than domestic natural gas.

When fossil fuels are no longer abundant, or their use is curtailed because of concerns over changes in the atmosphere, the way in which we use energy will be fundamentally changed. For example, the present methods of generating electricity are a compromise between the efficiency of large power stations and the losses in transmission over long distances. It is more efficient to transmit H_2 gas through pipelines, than electricity through power lines. Electricity could be produced locally, even domestically, from H_2 and air, in fuel cells. The risks of using H_2 are offset by the use of lower electric voltages. The switch to a hydrogen economy could be a gradual transition. Hydrogen can be mixed with methane in domestic gas supplies with minimal change to the equipment. But the greatest benefits for H_2 will come when exploiting the thermodynamic advantage of fuel cells, converting chemical energy directly to electricity.

A great deal of research is being applied to the use of hydrogen as a fuel in transportation, for cars and aeroplanes. The goal here is the promise of near-zero emissions. H_2 is also being considered as an alternative to batteries for electronic equipment.

For transport, the difficulties are concerned with finding a compact storage for the low-density fuel; and the expense of the catalysts. Various approaches are being used for storage. One is to store it as a higher-density liquid such as methanol, and reform it to H_2 as required. This is somewhat analogous to the biological approach (Chapter 1), though it leads to the release of CO_2. Other options are to compress the H_2, store it as liquid hydrogen at very low temperatures, or combine it with metals to form hydrides from which the gas can be released at will. Another method, which again resembles the biological solution (Section 8.2.3) is to store the H_2 in carbon nanotubes, which offer high-density and lightweight storage.

At present most of the H_2 is produced industrially by conversion of fossil fuels, either directly or indirectly. This leads inevitably to the net production of the greenhouse gas CO_2. New methods will have to involve recycling of organic matter, or direct production of H_2 from water using energy sources such as sunlight. This can be achieved either directly in photochemical fuel cells, or by using photovoltaic cells, which use solar radiation to electric current for electrolysis of water into H_2 and O_2. A great deal of effort has gone into the development of silicon solar cells (photovoltaics) for production of energy from sunlight, which have become less expensive and improved in efficiency. The costs of production of H_2 from the electricity produced are decreasing steadily, but they still involve noble-metal catalysts.

10.1.2. Biological hydrogen production

The most challenging option is the production of hydrogen by photosynthetic microorganisms. This biological approach is environmentally friendly, as it uses renewable materials, produces no toxic waste products, CO_2 or NO_x, and avoids the need for precious metals. It is not a new idea. It was known that algae and cyanobacteria could evolve small amounts of H_2 under certain growth conditions (Gaffron and Rubin 1942). Microbes produce H_2 for two principal reasons. The first is to dispose of excess reducing equivalents during fermentative metabolism either carried out in a dark anaerobic process or associated with the anoxic photosynthetic activity. An example is the production of hydrogen during the phosphoroclastic reaction of pyruvate and other 2-oxoacids (Fig. 10.1). This reaction often uses an [Fe] hydrogenase, although bidirectional [NiFe] hydrogenases are also used. A cell may produce H_2 in some stages of growth, then consume it in others (Belaich 1991). Second, H_2 is a by-product of the action of nitrogenase (Fig. 10.2), the enzyme that is needed for the fixation of atmospheric N_2. This is an extremely complex reaction; the enzyme is slow, and produces at least one H_2 for every N_2 fixed. The H_2 released by nitrogenase is often recovered within the nitrogen-fixing cells, by means of the membrane-bound uptake [NiFe] hydrogenase, and so less than proportional net H_2 production is observed. The membrane-bound respiratory chain, using H_2 as a fuel, produces ATP, which partly compensates for the energy consumed in the nitrogenase reaction.

Photoheterotrophic bacteria conduct a simple type of photosynthesis, in which there is only one photosystem, and reduced inorganic or organic compounds instead of water act as electron donors. Photosynthesis by these organisms can be used to produce hydrogen, with various forms of organic compounds as electron donors. They can be used, in principle, to produce H_2 from dairy and agricultural waste matter using light energy.

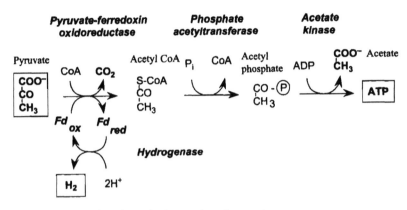

Figure 10.1 The phosphoroclastic reaction of pyruvate.

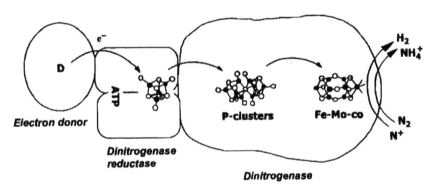

Figure 10.2 Cartoon of the electron transfer pathway in nitrogenase.

Artificial cell-free systems have been investigated, to test models of photosynthetic production of H_2. Benemann *et al.* (1973) demonstrated that it was possible to produce H_2 and O_2 by combining chloroplasts from green plants and bacterial hydrogenase, with ferredoxin as the intermediate electron carrier:

$$2 H_2O \rightarrow 2 H_2 + O_2.$$

In this system, oxygen is produced by photosystem II, as in green plants and cyanobacteria. The photosynthetic electron transfer, via photosystem I, is linked by low-potential electron carriers to hydrogenase, which produces H_2 (Fig. 10.3). Benemann and Weare (1974) then went on to investigate H_2 evolution by N_2-fixing cyanobacterial cultures as a whole-cell source of hydrogen energy.

An alternative approach, using semiconductors as light-driven electron donors, has been demonstrated in model systems (Grätzel 1982; Nikhandrov *et al.* 1988). These are more stable than the photosystems, but show lower photochemical conversion efficiencies owing to short-circuiting of reducing equivalents. The presently used

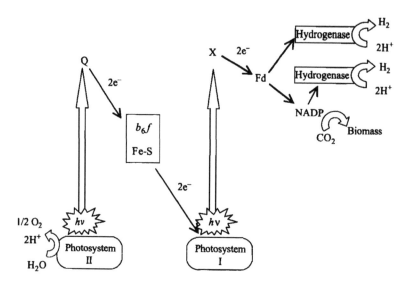

Figure 10.3 'Z-scheme' of oxygenic photosynthesis in green algae and cyanobacteria, showing links to hydrogenase. Q (plastoquinone) and X (an iron–sulfur cluster) are electron acceptors from photosystems II and I, respectively. The two hydrogenases shown are the NADP-dependent bidirectional hydrogenase and a ferredoxin-dependent enzyme.

semiconductors such as titanium dioxide (TiO_2) use only the blue/ultraviolet end of the whole energy spectrum.

10.2. Technologies to produce hydrogen by biological systems

Hydrogen can be produced from water using a variety of energy sources such as nuclear, solar, electrical, biological and chemical. Since both water and solar radiation are abundant, ubiquitous and 'free', photolysis of water (either direct or via additional components) is a preferred route to generate hydrogen from water. Other sources for hydrogen that are currently used, such as the reforming of hydrocarbons, are not neutral in terms of net CO_2 production.

The use of solar energy to split water to hydrogen and oxygen is an attractive means to harvest the Sun's energy falling on the Earth, and convert it to storable chemical energy. Photochemical, photoelectrochemical and photobiological technologies are being investigated for the development of sustainable, economically viable hydrogen production systems. Photosynthetic microorganisms (phototrophs) containing hydrogenases, as well as isolated hydrogenases, are used in these studies.

A major drawback in obtaining sustainable hydrogen production is the instability of the enzyme during continued operation. One approach is to use living cells, which have the ability to repair, maintain and reproduce themselves. The alternative would be to create a stable and inexpensive synthetic catalyst, which would mimic the properties of the natural enzyme, which is the rapid and reversible activation of H_2 at water temperature (below 100°C) and near-neutral pH. Such a catalyst would be most welcome in processes like the production of hydrogen by photobiological/photochemical methods and in the generation of hydrogen for fuel cells. In short, the 'cheap' synthetic hydrogenase would replace the expensive platinum as a hydrogenation catalyst. The development of such catalysts is at an early stage, as described in Chapter 8 (Section 8.6).

In principle, the conversion of energy to H_2 production by biological systems could be quite efficient. The light reactions of photosynthesis convert light to chemical energy with very high efficiency, and hydrogenases, as already described in this book, are highly efficient enzymes. However, the potential has not been realized in the small-scale studies done so far, where conversion efficiency of light energy is typically less than 1 per cent. It is generally considered that efficiencies of 10 per cent or more would be needed to support our energy needs. It has become apparent that the development of viable large-scale systems would need a very significant commitment to applied and fundamental research.

The principal problems that were encountered were:

1   The oxygen-evolving photosynthetic systems are exposed to the damaging effects of light, and without the repair machinery present in the plant, soon lose activity. Plants have mechanisms to minimize the damage, but these reduce the conversion efficiency.
2   The hydrogenase and nitrogenase, which produce H_2, are inhibited or irreversibly damaged by the oxygen produced.
3   The product is a mixture of hydrogen and oxygen in solution, which have to be transferred to the gas phase (requiring a large surface area) and separated from each other before storage.
4   Sunlight as an energy source, although free, is intermittent and diffuse.
5   Hydrogen is the lowest-density gas with a very low boiling point, and storing it requires either large volumes, very low temperatures, or high pressures.
6   Wild-type microorganisms such as hydrogen-oxidizing and methanogenic bacteria would be competitors, growing like weeds, in systems for H_2 production, and have to be excluded.

Over the last thirty years, investigators around the world have explored methods of overcoming these obstacles. The yields can be increased by genetic manipulation of the enzymes and the machinery for producing them. For example, the yield of H_2 production by nitrogenase can be enhanced by mutations, and expression of the enzyme can be increased by modifying the way in which it is repressed by fixed nitrogen. Further development is needed, to improve the resistance of hydrogenase to oxygen and carbon monoxide.

A fundamental difficulty is that the main purpose of living organisms is the efficient production of their own cell material. H_2 is a by-product of their metabolism. Ultimately, in order to progress towards large-scale H_2 production, it may be necessary to engineer, by molecular biology techniques, a new type of organism or consortium of organisms, in which the photosynthetic production of H_2 is the major metabolic activity. This is an ambitious aim, but there are historical precedents in the breeding of animals and agricultural plants for improved agricultural productivity. The new breed of H_2-producing organisms would have to be grown in an environment from which other microorganisms are excluded.

New directed evolution techniques make it possible to modify the genetic makeup of an organism in a wide variety of ways. They make it possible to speed up the natural process of evolution from millions of years to a few months. Many possibilities for improving the metabolism and enzymology of organism can be explored. What is required is an effective way of applying the right selection pressure to ensure that the most effective systems emerge.

If solar energy were to supply the energy needs of the world's population, it would require large areas to be dedicated to it. These could be in sunny areas of low population such as deserts, or the oceans. It should be remembered that salt water is no barrier to the growth of many of these organisms. However, technological advances on this scale are obviously long term.

We now present some reports on the state of research at present, and some forecasts of the way in which biological H_2 production might develop in the future.

10.3. Hydrogen production by cyanobacteria and algae

K. Krishna Rao, Paula Tamagnini and Peter Lindblad

Cyanobacteria (prokaryotes) and unicellular green algae (eukaryotes) have been considered as promising phototrophs for the development of environment-friendly photobiological H_2 production systems. They are ubiquitous, and relatively inexpensive to grow. For growth, they need only water, air and simple minerals, and light as energy source.

10.3.1. Photosystems

Biophotolysis of water involves the use of the photosynthetic apparatus to generate hydrogen from water and sunlight. Solar energy (photons) captured by the antenna pigment–protein complexes excites the reaction centre chlorophylls to produce strong reducing potentials (Fig. 10.3). Photosystem II of cyanobacteria and plants takes electrons from water via the manganese-containing oxygen-evolving complex, and donates electrons to quinone. The electrons are passed through the cytochrome b_6f complex, to Photosystem I, which provides a strong reductant for ferredoxin. Ferredoxin:NADP reductase provides NADPH, for fixation of CO_2. It could also be used to reduce the bidirectional hydrogenase. Photosynthetic electron transport also creates a proton gradient (proton motive force) across the photosynthetic membrane, which promotes the synthesis of ATP. ATP and reduced ferredoxin are the important molecules involved in H_2 photoproduction from nitrogenase. (Hall and Rao, 1999).

10.3.2. Enzymes for photosynthetic hydrogen production

Cyanobacteria and green algae are capable of H_2 photoproduction from water, although the enzymology of H_2 production is different in the two groups. H_2 production in cyanobacteria is observed principally in diazotrophic strains under nitrogen-fixing conditions, and is catalysed by nitrogenase. Green algae are eukaryotes and not capable of N_2 fixation; H_2 evolution is catalysed by hydrogenase. The hydrogenase of *S. obliquus* has recently been characterized, and shown to have a unique structure with a particularly short polypeptide sequence (Florin *et al.* 2001). Nitrogenases and the algal hydrogenases are all extremely O_2 sensitive and so are deactivated by photosynthetically generated O_2 (Rao and Hall 1996; Hansel and Lindblad 1998; Appel and Schulz 1998). Research efforts have concentrated on enhancing these activities, and on strategies for avoiding interference by O_2.

Nitrogenase

Nitrogenase, the enzyme that catalyses N_2 fixation, is essential for the maintenance of the nitrogen cycle on Earth, since fixed nitrogen is often limiting for the growth of living organisms (Postgate 1987; Burris 1991; Gallon 1992). The ability to fix N_2 is restricted to prokaryotic organisms. However, a representative number of eukaryotes, notably green plants, can establish symbiosis with N_2-fixing bacteria. A broad range of microorganisms, including both archaea and eubacteria, has been found to have the capacity to fix N_2. This diversity of organisms contrasts with the remarkable conservation of nitrogenase itself (Flores and Herrero 1994; Haselkorn and Buikema 1992; Smith 1999).

The most common form of nitrogenase is the molybdenum-containing enzyme. It consists of two proteins: the dinitrogenase (MoFe protein or protein I) and the dinitrogenase reductase (Fe protein or protein II). The dinitrogenase is an $\alpha_2\beta_2$ heterotetramer of about 220–240 kDa, which binds four 4Fe4S clusters, organized into two 'P clusters', and two FeMo cofactors (Flores and Herrero 1994). It is generally accepted that the FeMo-cofactors constitute the active site at which N_2 is reduced to ammonia. The α and the β subunits are encoded by the genes nifD and nifK, respectively. The Fe protein, encoded by the gene nifH, is a homodimer of about 60–70 kDa and has the specific role of mediating the transfer of electrons from the external electron donors (a ferredoxin or a flavodoxin) to the P clusters of dinitrogenase. Together, the subunits bind one intersubunit 4Fe4S cluster. In addition to the three structural genes mentioned above, many other genes are involved in the nitrogen-fixation process and its regulation (for a recent review, see Böhme 1998).

The reduction of nitrogen to ammonium, catalysed by nitrogenase, is a highly endergonic reaction requiring metabolic energy in the form of ATP. First, the dinitrogenase reductase is reduced by a ferredoxin or a flavodoxin and binds ATP, which lowers its potential. At this lower potential (around $-400\,mV$), the dinitrogenase reductase transfers electrons to dinitrogenase. The transfer is accompanied by the hydrolysis of ATP to ADP + P_i. This reaction is also accompanied by an obligatory reduction of protons to H_2. At infinite pressure of N_2, 75 per cent of the electrons would be allocated for N_2 reduction and 25 per cent for H^+ reduction. As two ATP are required for each electron transferred from dinitrogenase reductase to dinitrogenase, the reaction requires a minimum of sixteen ATP until the dinitrogenase has accumulated enough electrons to reduce N_2 to NH_3. The overall reaction can be written as follows:

$$N_2 + 8e^- + 10H^+ + 16ATP \rightarrow 2NH_4^+ + H_2 + 16ADP + 16P_i$$

although it should be noted that this represents an ideal efficiency under optimal conditions. In practice, nitrogenase tends to produce more H_2 than this equation would suggest.

The 'alternative' nitrogenases contain vanadium or iron instead of molybdenum. They are encoded by the *vnf* and *anf* gene clusters, respectively. Structurally they are similar to the molybdenum-containing nitrogenases, but have catalytic clusters containing vanadium (FeVa cofactor) or iron only (FeFe cofactor) respectively. Nitrogenase is usually detected in cells by its ability to reduce acetylene (ethyne) to ethylene (ethene), which can readily be detected by gas chromatography. The alternative nitrogenases are less active in this reaction, and also tend to continue the reduction as far as ethane. They are repressed by molybdenum and for this reason they were isolated and characterized much later than the molybdenum enzyme. Significantly for bio-hydrogen production, they allocate a higher proportion of electrons to the reduction of H^+ to H_2 when compared to the conventional Mo-enzyme complex. For the Fe nitrogenase a minimum ratio $H_2 : N_2$ of $7.5 : 1$ was estimated (Schneider *et al.* 1997).

In the heterocystous cyanobacterium *Anabaena variabilis*, four different nitrogenases have been identified and characterized. Two are Mo-dependent enzymes. While one (the so-called conventional nitrogenase, encoded by the nif1 gene cluster) functions only in the heterocysts under both aerobic or anaerobic growth conditions, the other (encoded by the nif2 gene cluster) functions strictly under anaerobic conditions in both the vegetative cells and the heterocysts. Furthermore the differences between the two nif clusters suggest that the conventional nitrogenase is developmentally regulated, while the other is regulated by environmental factors. The occurrence of a V-containing nitrogenase was first reported by Kentemich *et al.* (1988) and, subsequently, confirmed by Thiel (1993). This enzyme is encoded by the vnfDGK gene cluster, that is transcribed in the absence of molybdenum, and in which vnfDG are fused in a single ORF. Physiological evidence indicates that the fourth nitrogenase is an Fe enzyme similar to the one encoded by the anf gene cluster in *Azotobacter vinelandii* (Bishop and Premakumar 1992; may also occur in *A. variabilis*, although the gene characterization is still lacking, only in the Anabaena strain isolated from the fern Azolla, Kentemich *et al.* 1991).

The biosynthesis of nitrogenase is under strict regulation and it is suppressed by many factors, including the availability of other sources of fixed nitrogen, and the presence of oxygen. The optimal yields of nitrogenase are obtained by working under argon, when nitrogenase has no substrate to work on, but this is unlikely to be feasible on the large scale. It should be possible to engineer the regulatory mechanisms to increase the yield of the enzyme. Modified enzymes would be able to change the ratio of $H_2 : N_2$ produced. The disadvantages of nitrogenases are their low turnover numbers, high demand for ATP and oxygen sensitivity.

Uptake hydrogenase

All nitrogen-fixing unicellular and filamentous cyanobacteria examined so far have been found to possess an uptake hydrogenase. The enzyme seems to be membrane bound and, in some filamentous strains, is particularly expressed in the N_2-fixing heterocysts with no or minor activity in the photosynthetic vegetative cells. It is expressed under the same conditions as nitrogenase, with the evident function of catalysing the uptake of H_2 produced by nitrogenase. As a consequence release of H_2 is not normally observed. The uptake hydrogenase, like the membrane-bound hydrogenase of hydrogen bacteria such as *Ralstonia eutropha*, is linked to the respiratory chain and in cyanobacteria also to photosystem I, ultimately to cytochrome oxidase. The ATP generated is used for biosynthetic reactions, including further N_2 fixation.

Syntheses of uptake hydrogenases of cyanobacteria, as well as uptake hydrogenases of other bacteria, have been shown to be dependent on the availability of Ni^{2+} in the growth medium. The enzymes are insensitive to O_2 in whole-cell preparations, but become sensitive after cell extraction (Houchins and Burris 1981b).

Bidirectional hydrogenase

Cyanobacteria also contain a reversible/bidirectional hydrogenase, which is soluble and has the capacity to both take up and produce H_2 (Bothe *et al.* 1991; Flores and Herrero 1994; Appel and Schulz 1998; Hansel and Lindblad 1998). The latter enzyme was called 'bidirectional hydrogenase' until the respective structural genes were sequenced and characterized by Schmitz *et al.* (1995). It has been the subject of some interest as another possible means to produce H_2 photosynthetically.

In *Anabaena* sp. strain PCC7120, the uptake hydrogenase is less sensitive to carbon monoxide inhibition than the bidirectional enzyme. Both enzymes have low K_m for H_2 but only the uptake hydrogenase activity was elicited by addition of H_2 to the gas phase.

In contrast with the uptake hydrogenase, the soluble, or loosely membrane-associated bidirectional hydrogenase was believed to be a constitutive enzyme (Kentemich *et al.* 1989, 1991; Serebriakova *et al.* 1994) widely distributed among N_2-fixing and non-N_2-fixing cyanobacteria. The activity of this enzyme, in heterocystous strains, increases considerably under anaerobic or microaerobic conditions, whereas in the unicellular non-N_2-fixing *Gloeocapsa alpicola* the partial pressure of oxygen does not seem to have any significant influence (Serebryakova *et al.* 1998). The hydrogenase activity in vegetative cells of *A. variabilis* has been suggested to be under a form of 'redox control' whereby the enzyme is only activated upon removal of light and oxygen. This mechanism may involve a thioredoxin (Spiller *et al.* 1983).

Schmitz *et al.* (1995) sequenced a set of structural genes (hox genes) encoding a bidirectional hydrogenase in the filamentous heterocystous *A. variabilis*, and proposed that the bidirectional enzyme is a heterotetrameric enzyme consisting of a hydrogenase part (encoded by hoxYH) and a diaphorase part. Nucleotide sequence comparisons showed that there is a high degree of homology between the hox genes of cyanobacteria and genes encoding the NAD^+-reducing hydrogenase from the chemolitotrophic H_2-metabolizing bacterium *R. eutropha* as well as methyl viologen-reducing hydrogenases from species of the methanogens (Chapter 2). The bidirectional hydrogenase of *A. variabilis* has been partially purified and characterized (Serebryakova *et al.* 1996).

10.3.3. Oxygen sensitivity and how to avoid it

The nitrogenases are deactivated immediately on exposure to air or O_2; the process is almost irreversible. All diazotrophs must protect the enzymatic complex from the deleterious effects of O_2. Cyanobacteria perform oxygenic photosynthesis, and must, therefore, protect their N_2-fixing machinery not only from atmospheric O_2 but also from the intracellularly generated O_2. Diverse strategies have evolved for separating the production of H_2 and O_2, either spatially or temporally. Some filamentous nitrogen-fixing

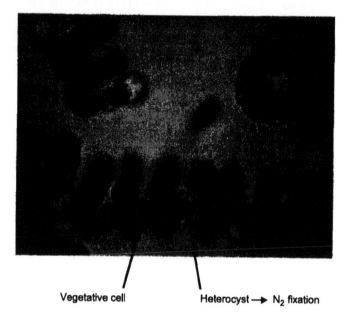

Vegetative cell   Heterocyst ➝ N₂ fixation

Figure 10.4 Cyanospira rippkae (courtesy of Prof. H. Bothe).

cyanobacteria have specially adapted cells, called heterocysts, which contain nitrogenase and do not evolve O_2 (Fig. 10.4).

The non-heterocystous cyanobacteria have attracted attention because of the
apparent paradox of being able both to fix N_2 and to evolve O_2. Temporal separation between photosynthetic O_2 evolution and N_2 fixation seems to be the most common strategy adopted. A very restricted number of strains are able to perform N_2
fixation under aerobic conditions, and there is no evidence that the two processes,
oxygenic photosynthesis and N_2 fixation, may occur simultaneously within a single
cell (Bergman *et al.* 1997). However, not all strains fix N_2 exclusively during the dark
phase of light/darkness cycle. N_2 fixation requires ATP and reductant, and the fermentation of stored carbohydrates may not be sufficient to cover the energy demand
of nitrogenase activity. In natural populations of *Oscillatoria limosa*, the nitrogenase
activity coincides with the transitions from dark to light and light to dark with a maximum at sunrise. The last event could be explained by light energy driven N_2 fixation
in an initially low O_2 environment. The situation in the marine filamentous non-
heterocystous *Trichodesmium* is far more complex. A spatial separation between the
N_2 fixation and the photosynthesis probably occurs, without any obvious cellular differentiation. In contrast with the permanent changes occurring during heterocyst differentiation, those occurring in non-heterocystous cyanobacteria can be reversed.

When filamentous cyanobacteria such as *Nostoc* or *Anabaena* are grown under limited combined N_2, 5–10 per cent of their vegetative cells differentiate to form heterocysts.
Heterocyst formation does not take place randomly within the filament. Most probably
it is a dynamic selection in response to nitrogen deprivation but neither the existence of
predetermination nor a combination of both events can be ruled out (Wolk 1996).

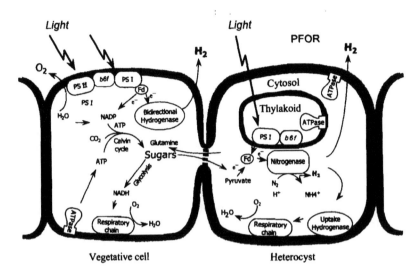

Figure 10.5 Scheme of electron transfer in a filamentous heterocystous cyanobacterium.

The heterocyst provides a virtually anaerobic environment suitable for the functioning of nitrogenase since: it lacks photosystem II activity, and therefore does not produce O_2 as a by-product of photosynthesis. The glycolipids on the thick outer envelope of the heterocysts restrict the diffusion of O_2 into the cells, which have a higher rate of respiratory O_2 consumption (Fay 1992; Wolk *et al.* 1996). The connection between the vegetative cell and the heterocyst is narrow and occurs via microplasmodesmata. Heterocysts import carbohydrates and in return export glutamine to the vegetative cells (Fig. 10.5). ATP formation in heterocysts is not completely understood. ATP can be produced either by cyclic photophosphorylation (PSI) in the light, or by oxidative phosphorylation; the latter process consumes oxygen and uses pyridine nucleotides or hydrogen as electron sources. It has been suggested that NADPH provides electrons for nitrogenase via a heterocyst-specific ferredoxin (FdxH) and ferredoxin:$NADP^+$ oxidoreductase (Böhme 1998).

An interesting feature of heterocyst differentiation in cyanobacteria is the occurrence of developmentally regulated genome rearrangements (Carrasco *et al.* 1995; Carrasco and Golden 1995; Golden *et al.* 1985, 1987, 1988; Matveyev *et al.* 1994). These rearrangements occur late during heterocyst differentiation at about the same time as the nitrogen-fixation genes are transcribed. Inserted elements in the genes for nitrogenase, uptake hydrogenase and ferredoxin are excised, during heterocyst differentiation, by site-specific recombinases.

10.3.4. Cyanobacterial hydrogen production: Present status and future potential

Filamentous cyanobacteria have been used in bioreactors for the photobiological conversion of water to hydrogen. However, the conversion efficiencies achieved are low because of the competing processes of hydrogen production and consumption within the cells. Fixation of N_2 is accompanied by H_2 production which is an inevitable side reaction of nitrogenase, but often the H_2 is quickly reabsorbed by a unidirectional uptake hydrogenase. Cyanobacteria contain, in addition, a bidirectional (reversible) enzyme, which under some growth conditions oxidizes the H_2, and in others evolves it.

Investigations are continuing on ways of improving the efficiency of photobiological hydrogen production by cyanobacteria. Screening of natural strains and mutants with high nitrogenase activity, and improved aerobic stability is being pursued vigorously. In laboratory-scale experiments the H_2-evolving activity of nitrogenase is enhanced by substituting molybdenum in the standard culture medium with vanadium. Improvements of the conversion efficiencies have been achieved through the optimization of the conditions for H_2 evolution by nitrogenase, through the production of mutants deficient in H_2 uptake activity. *A. variabilis* PK84, a mutant of *A. variabilis* ATCC29413, impaired in uptake hydrogenase activity, was used by Borodin *et al.* (2000) for the continuous production of H_2 in air and CO_2, in an outdoor photobioreactor (Fig. 10.6). Alternatively, increased H_2 evolution has been obtained by the bidirectional enzyme. Symbiotic cells are of fundamental interest since they function as 'bioreactors': high metabolism, transfer of metabolite(s) from symbiont to host but almost no growth.

Combining all the physiological and immunological works with the recent molecular data, notably with the presence/absence of the rearrangement within the gene

Figure 10.6 Helical photobioreactor for testing the effects of growth parameters during continuous H_2 production. The test reactor was constructed from 50 m of PVC tubing, wound round a wire former, and contained 3 l of a culture of *A. variabilis*. An airlift flushed the culture with air + 2 per cent CO_2, and circulated it through a degasser and computer-controlled monitoring equipment (Borodin *et al.* 2000).

hupL in the filamentous strains, it is impossible to establish one single pattern of cellular/subcellular localization and regulation of the uptake hydrogenase, even for closely related strains.

10.4. H$_2$ production by eukaryotic algae

Eukaryotic green microalgae such as *Chlamydomonas reinhardtii* and *S. obliquus* have been considered as sources of H$_2$ (Ghirardi *et al.* 2000). They have a more complex cellular organization than the cyanobacteria, and can live under a wide range of conditions. They do not have nitrogenase, but they contain inducible hydrogenases of the [NiFe] types and [Fe] types, and can produce hydrogen using hydrogenase under certain growth conditions. Dark hydrogen production, like the hydrogen-producing protozoa and anaerobic bacteria such as clostridia, occurs under anoxic conditions and is probably by the action of pyruvate:ferredoxin oxidoreductase, ferredoxin and hydrogenase. There is also evidence that hydrogenase can be coupled to photosynthetic electron transport, where it provides an overspill mechanism for excess reducing equivalents (Appel *et al.* 2000).

Melis *et al.* (2000) have described a novel approach for sustained photobiological production of H$_2$ gas via the reversible hydrogenase pathway in the green alga *C. reinhardtii*. Inhibition of the reversible hydrogenase by photosynthetic O$_2$ evolution was prevented by a two-stage procedure in which the photosynthetic phase was alternated with H$_2$ production. A transition from one stage to the other was by sulfur deprivation of the culture, which reversibly inactivated photosystem II (PSII) and O$_2$ evolution. Under these conditions, oxidative respiration by the cells in the light depleted O$_2$ and caused anaerobiosis in the culture, which was necessary and sufficient for the induction of the reversible hydrogenase. H$_2$ gas production was supported by photosynthetic electron transport through Photosystem I, from endogenous substrate to the ferredoxin and reversible hydrogenase (compare Fig. 10.5).

10.5. Hydrogen production by photoheterotrophic and heterotrophic bacteria

Paola M. Pedroni, Paulette M. Vignais, Kornél L. Kovács

10.5.1. Photoheterotrophic bacteria

Research at Enitechnologie (Italy) has been particularly addressed to develop technologies that combine the bioproduction of gas to the disposal of waste compounds exploiting sunlight as an energy source. This strategy is expected to reduce costs of process realization and management, and represents the basic concept of both *in vivo* and *in vitro* hydrogen bioproduction technologies developed (Fig. 10.7). In particular, the *in vivo* system is based on the cultivation of the photosynthetic bacterium *Rhodobacter sphaeroides*, which is able to evolve hydrogen utilizing organic wastes as growth substrates and harvesting sunlight as an energy source. Suitable photobioreactors have been designed to optimize bacterial cultivation with regard to solar irradiation.

Less investigated as potential methods for hydrogen bioproduction, *in vitro* systems make exclusively use of the enzymatic component responsible for hydrogen evolution

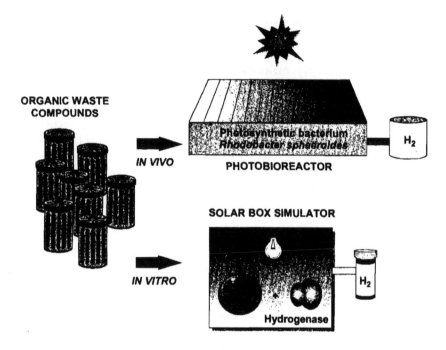

ORGANIC WASTE COMPOUNDS

IN VIVO

PHOTOBIOREACTOR

Photosynthetic bacterium *Rhodobacter sphaeroides*

H₂

SOLAR BOX SIMULATOR

IN VITRO

Hydrogenase

H₂

Figure 10.7 In vivo and in vitro hydrogen production technologies developed to combine the bioproduction of gas to the disposal of waste compounds exploiting the sunlight as an energy source (courtesy of Enitecnologie, Milano).

instead of the whole bacterial cell. In particular, in an *in vitro* system the gas is photocatalytically produced by coupling the hydrogenase enzyme with the inorganic semiconductor titanium dioxide (TiO_2). The reaction takes places in a solar simulator apparatus, which is used as a source of light. Organic waste compounds will be tested as renewable source of electrons.

The goal of research activities concerning *in vivo* hydrogen bioproduction is to improve the hydrogen evolution performance of *R. sphaeroides* wild-type strain by genetic engineering techniques. In this photosynthetic bacterium, at least two different enzymes are involved in hydrogen metabolism: the nitrogenase complex, responsible for hydrogen production, and a membrane-bound uptake hydrogenase which mediates hydrogen consumption (Fig. 10.8A). Since the physiological function of the latter enzyme is to recycle the hydrogen evolved by the nitrogenase, its activity reduces the net amount of this gas photoproduced by the bacterium. To isolate a hydrogen-overproducing strain starting from the wild type, an Hup⁻ mutant has been constructed in which the hydrogen recycling was eliminated by abolishing the uptake hydrogenase activity. This was achieved by interposon mutagenesis of the gene encoding the catalytic large subunit of the uptake hydrogenase heterodimer, since this disruption was considered as the most decisive at the functional level. Interposon mutagenesis technique led to the replacement of the functional gene copy (*hupL*) on the bacterial genome, with a copy of the same gene inactivated by the insertion in its reading frame of heterologous DNA. The hydrogen photoproduction performance of

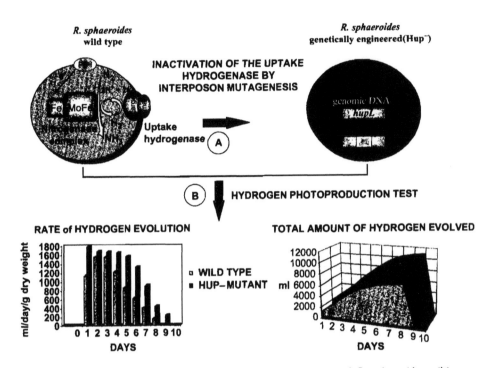

Figure 10.8 Improvement of the hydrogen production capacity of *R. sphaeroides* wild type. Construction of the Hup⁻ mutant strain by interposon mutagenesis (A) and evaluation test of its hydrogen photoproduction performace in comparison to the wild type (B) (courtesy of Enitecnologie, Milano).

the Hup⁻ mutant strain was then assayed in comparison to that of the wild type using an indoor 1 L-scale photobioreactor as an initial evaluation test (Fig. 10.8B). The hydrogen photoproduction rate of the mutant strain, measured as milliliters of gas evolved per day per gram of dry weight, was significantly higher than that of the wild type during all the fermentation period, thereby determining an increase of about 50 per cent in terms of total amount of hydrogen photoproduced.

In another approach, inactivation of hydrogen uptake activity in the mutant deficient in pleiotropic HypF accessory gene function, which is necessary to the assembly of active [NiFe] hydrogenases (Chapter 4), resulted in dramatic increase in the hydrogen evolution capacity of *Thiocapsa roseopersicina* under nitrogen-fixing conditions. Remarkably the *hypF* minus mutant strains showed a more than sixty-fold increase in H$_2$ evolution under nitrogen-fixing conditions when compared to the wild-type controls (Fig. 10.9). Firstly, this observation indicated that a substantial amount – probably all – hydrogen uptake activity has been inactivated in the *hypF*⁻ cells. Secondly, *hypF* mutation clearly offers an important avenue for the development of efficient biological hydrogen production systems using this organism. Growing under phototrophic and nitrogen fixing conditions, *T. roseopersicina* can evolve hydrogen at a practically significant level (B. Fodor, G. Rákhely, Á. T. Kovács and K. L. Kovács, University of Szeged, Hungary, in press).

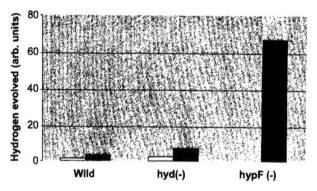

Figure 10.9 Relative hydrogen production of the wild type and the *hypF* defective (M539) *T. roseopersicina* strains *in vivo* under nitrogenase repressed (white columns) and derepressed (black columns) conditions. Samples were measured after cultivation for three days. The amount of H_2 evolved by the wild-type strain under non-nitrogen-fixing condition was chosen as 1.

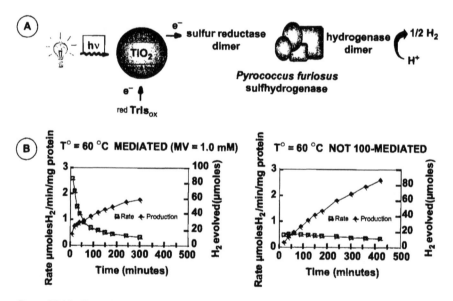

Figure 10.10 Optimization of the *in vitro* hydrogen bioproduction technology. Schematic representation of the electron transfer pathway among reaction components in un-mediated condition (A) and performance of the simplified system at 60°C compared to that of the previously developed version also including MV (B) (courtesy of Enitecnologie, Milano).

Figure 10.10A shows the reaction scheme of the system outlining the electron transfer pathway among the different components, namely the electron donor Tris, the semiconductor TiO_2 and the sulfhydrogenase of the hyperthermophilic archaeon *P. furiosus*, a bifunctional enzyme catalysing either proton or sulfur species reduction.

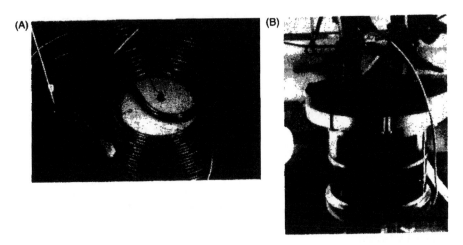

Figure 10.11 (A) Spiral-coiled plane photobioreactor filled with a culture of the photosynthetic bacterium *R. capsulatus*. A semiconductor temperature sensor, placed in contact with the culture in the center of the bioreactor is shown. (B) Degassing chamber (4.6 l) equipped with two level sensors.

This enzyme has been selected to enhance the stability of the biological catalyst, one of the most relevant problems connected with *in vitro* hydrogen photoproduction systems. Thermotolerant enzymes, in fact, are endowed with biochemical and structural features, such as high catalytic efficiency and resistance to chemical and denaturing agents, particularly suitable for practical purposes. Under light irradiation, the TiO_2 particle generates electron/hole couples. The hole is a relatively high-energy oxidative site and can accept electrons from Tris, the donor compound. As a consequence, an excess of negative charge is accumulated on the TiO_2 particle and electrons are then directly transferred from the light-excited semiconductor to the *P. furiosus* sulfhydrogenase which in turn reduces protons to hydrogen. This un-mediated reaction scheme represents a simplified and improved version of the *in vitro* system we previously set up. This latter included an additional component, the artificial redox mediator methyl viologen (MV) which mediated electron transfer from the irradiated TiO_2 particle to the biological catalyst. Its presence affected the life-time of the reaction, undergoing an irreversible degradation and therefore represented a limit to the total amount of hydrogen photoproduced. Besides optimizing the system with regard to reaction components, we also improved its performance as a function of temperature. In fact, lowering the reaction temperature from 80 to 60°C enhances the thermochemical stability of the reaction components and therefore had a positive effect on the life-time of the hydrogen photoproduction.

Figure 10.10B shows the results obtained at 60°C comparing the performance of the simplified un-mediated version to that of the mediated system in terms of rate and amount of hydrogen evolved. In the absence of the the artificial redox mediator MV, the initial rate of hydrogen production is lower, but its trend is more constant and the reaction life-time can be extended from 5 to 8 h, giving rise to a significantly higher amount of H_2 evolved.

A 10 l photobioreactor has been developed at CEA/GRENOBLE (France), which consists of transparent polyvinyl chloride (PVC) tubing, 30 m long, 15 mm internal diameter, volume 5.3 l), spiral-coiled so as to form a plane light captor of 1 m², was used to study the degradation of lactate by the photosynthetic bacterium *R. capsulatus*, strain B10 (Delachapelle *et al.* 1991) (Fig. 10.10). The bacterial culture was continuously circulated in the reactor so as to maintain a homogeneous suspension to optimize illumination of the cells, avoid bacterial self-shading, allow regeneration of ATP by photophosphorylation and make a well-mixed reactor for optimal nutrient transfers and for degassing of the medium. The bacteria were cultivated anaerobically under photoheterotrophic and N-limited conditions to allow the synthesis of the nitrogenase enzyme, the catalyst, which produces H_2.

To run it under automated conditions, the bioreactor was equipped with two temperature sensors, two pH electrodes, a water level detector, a manometer and a computer-controlled electric valve. Control of key parameters (pH, temperature, dilution rates) has allowed to define the culture conditions producing maximal amounts of molecular hydrogen.

The production of H_2 accompanying lactate degradation was maximal in diluted nitrogen-limited continuous cultures. It was observed at a dilution rate of $0.04 \, h^{-1}$ with 5 mM glutamate in the influent medium, the optical density of the culture being 2.1 at 660 nm. Under these conditions an average H_2 production of $85 \, ml \, h^{-1} l^{-1}$ ($0.85 \, l \, h^{-1}$ for the whole bioreactor) was observed over a 200 h period and up to 90 per cent of the added lactate (initial concentration 30 mM) was degraded. The concentration of degradation products (formate, acetate) remained below 2 mM. At higher bacterial concentrations, the limitation of light energy resulted in a decrease in nitrogenase activity and therefore in a drop in H_2 production.

10.5.2. Heterotrophic microorganisms

Fermentative bacterial hydrogen production has been studied extensively as H_2 is a major product of anaerobic fermentation and nitrogen fixation. The stoichiometric fermentation of glucose can, in principle, yield 12 mol of H_2/mol glucose. However, if all chemical energy stored in the sugar molecule is converted to H_2, the microorganisms cannot generate ATP. This is not preferred because generation of ATP, the general chemical energy carrier and storage molecule in any living cell, is the ultimate driving force to carry our metabolic conversion of glucose. In order to maintain energetic balance, it has been suggested that the conversion of only about one-third of the glucose energy to H_2 is feasible for microorganisms carrying out anaerobic dark fermentation. This is in line with observations in growing bacterial cultures, which produce only 1–2 mol of H_2/mol glucose. This compares poorly with other potentially useful products of intermediate metabolism, e.g., about 80–90 per cent of the chemical energy of glucose can be converted to either ethanol or methane. However, this line of argument only applies to vigorously growing bacterial cultures needing plenty of energy in the form of ATP. Under real life conditions, and under conditions existing in large fermenters, cell growth is very often limited by the availability of nutrients or by unfavourable environmental stresses (Keasling *et al.* 1998).

Heterotrophic fermentation leading to H_2 production can be carried out by a wide variety of mesophilic and thermophilic microorganisms under anoxic conditions.

Under mesophilic conditions (e.g. in *Escherichia coli* or *Enterobacter aerogenes*) maximum conversion yields of 4 mol of H_2/mol glucose have been reported (Kengen *et al.* 1996). Twice this yield would be sufficient for economic H_2 fermentation (Keasling *et al.* 1998). This is valid if the fermentation costs are similar to those currently used in ethanol or methanol fermentations. The remaining chemical energy of glucose appears in the form of various by-products, such as acetate and lactate. These compounds are difficult to convert to H_2 under regular dark anaerobic fermentation conditions. However, phototrophs can utilize acetate and/or lactate and produce H_2 (Section 10.2.1). In general, factors determining optimal H_2 yield (e.g. pH, temperature, light intensity and quality, cell density and nutritional status) differ from optimal growth conditions (Sasikala *et al.* 1993). This further emphasizes the need to separate cultivation from H_2 production.

Alternatively, the biological system also favours H_2 production at elevated temperature, because in hyperthermophiles the affinity for H_2 decreases and the thermodynamic equilibrium of H_2 formation from acetate is favoured (Adams 1990). For heterotrophic hyperthermophiles, such as the species *Pyrococcus furiosus* (Pedroni *et al.* 1995) and *Thermococcus litoralis* (Rakhely *et al.* 1999), energy for growth is derived from the fermentative metabolism of peptides and sugars. In these cases, excess reductant, formed during the redox metabolic processes, is disposed of by a tetrameric hydrogenase producing H_2. An additional feature of interest is that these H_2-producing NiFe hydrogenases contain nucleotide cofactors, in addition to the regular NiFe and Fe clusters (see Section 2.3.3). These cofactors are known to be very sensitive to heat and fall apart at hyperthermophilic temperatures (above 80°C) very quickly. One of the puzzles surrounding hyperthermophilic hydrogenases is how they manage to stabilize their nucleotide cofactors.

It has been recognized that dark, anaerobic fermentative H_2 production, similarly to H_2 photoproduction by photosynthetic bacteria, only takes place to a limited degree in Nature. In mesophilic dark fermentation, the reason is thermodynamic preference of other metabolic products by the bacteria. However, metabolic engineering, coupled with the beneficial effect of high temperature, may bring about dramatic changes as demonstrated recently (Woodward *et al.* 2000). In an *in vitro* attempt, mixing enzymes of the so-called pentose phosphate cycle and hydrogenase from *P. furiosus*, 11.6 mol H_2 were obtained from 1 mol of glucose-6-phosphate. This experiment proved that 97 per cent of the maximum stoichiometric yield is indeed attainable, although sustained H_2 production in such artificial systems can only be maintained for a short period because of the inactivation of the participating enzymes. At any rate, hydrogen fermentations rank highest in their potential for both the likely economics of biological hydrogen production systems and environmental sustainability. Both in photoheterotrophic and dark heterotrophic fermentations there is plenty of room for improvement through genetic modification of metabolic balances. Besides the production of hydrogen in high yield, the use of hyperthermophiles exploits the advantage of an automatic selection pressure for hyperthermophilic hydrogen producers, preventing other processes which compete for the substrate(s) or consume the evolved H_2. Furthermore, some hyperthermophiles excrete a variety of enzymes enabling them to utilize biopolymers, such as starch and cellulose, thus facilitating the use of cheap biomass.

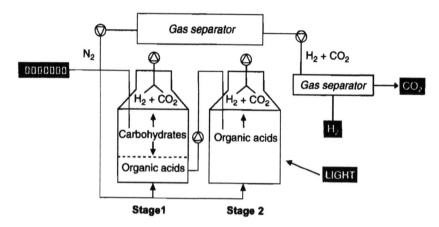

Figure 10.12 Scheme of the two stage biohydrogen fermentation system.

An interesting combination of photoheterotrophic and hyperthermophilic hydrogen fermentation was tested in a European Union 5th Framework Project, coordinated by the Wageningen University, The Netherlands. The main objective of this project was the development of an integrated process in which biomass is utilized for the biohydrogen fermentation. The biological fermentation process takes place in two steps (Claasen *et al.* 1999). In the first stage, the biomass is fermented to acetate, CO_2 and H_2 in a thermophilic dark fermentation. In a separate photobiological anaerobic fermenter (Fig. 10.12), acetate is converted to H_2 and CO_2. The ambitious aim is to get as close to the theoretical maximum of production 12 moles of H_2/mol glucose equivalent as possible. At present, the major environmental disadvantage of the utilization of hydrogen as a fuel is the vast amount of fossil carbon dioxide, which is released into the environment during the production of hydrogen from natural gas. In the proposed bioprocess, the ratio of carbon dioxide released to hydrogen produced will be the same. Moreover, biohydrogen fermentation of sugar or sugar containing agricultural waste can still be considered as 'green' technology and not a potential contribution to global warming because the released CO_2 originates from photosynthetic CO_2 fixation; therefore net CO_2 production does not take place.

10.6. The road to the hydrogen future: R&D in the US Hydrogen Program

Catherine E. Gregoire Padró NREL

The US Department of Energy (DOE) conducts R&D for the development of safe, cost-effective hydrogen energy technologies that support and foster the transition to a hydrogen economy. Of particular interest is the innovative research supported by the DOE's Hydrogen Program, focused primarily on exploration of long-term, high-risk concepts that have the potential to address large-scale energy needs.

Hydrogen can be produced directly from sunlight and water by biological organisms and using semiconductor-based systems similar to photovoltaics, or indirectly, via thermal processing of biomass. These production technologies have the potential to produce essentially unlimited quantities of hydrogen in a sustainable manner.

Storage of hydrogen is an important area for research, particularly when considering transportation as a major user, and the need for efficient energy storage for intermittent renewable power systems. Although compressed gas and liquid hydrogen storage systems have been used in vehicle demonstrations worldwide, the issues of safety, capacity and energy consumption have resulted in a broadening of the storage possibilities to include metal hydrides and carbon nanostructures. Stationary storage systems that are high efficiency with quick response times will be important for incorporating large amounts of intermittent photovoltaic and wind into the grid as base load power.

In addition to the extensive fuel cell development programs in other offices within DOE, the Hydrogen Program conducts fuel cell research focused on development of inexpensive, membrane electrode assemblies and the development of reversible fuel cells for stationary applications. The Program also supports research in the development of hydrogen/methane blends and hydrogen-fuelled internal combustion engines and generator sets.

A large hurdle to expanded use of hydrogen is public perception. Widespread hydrogen use represents an extraordinary educational challenge, as well as the absolute requirement that safety be intrinsic to all processes and systems. The development of reliable, low-cost hydrogen sensors is an important aspect of the Program, as is the development of codes and standards for the safe use of hydrogen.

The use of solar energy to split water into oxygen and hydrogen is an attractive means to directly convert solar energy to chemical energy. Biological, chemical and electrochemical systems are being investigated within DOE as long-term (more than ten years), high-risk, high-payoff technologies for the sustainable production of hydrogen.

10.6.1. Biological systems

In Nature, algae absorb light and utilize water and CO_2 to produce cell mass and oxygen. A complex model referred to as the 'Z-scheme' has been identified to describe the charge separation and electron transfer steps associated with this process that ultimately drives photosynthesis. A number of enzymatic side pathways that can also accept electrons have been identified. Of interest is a class of enzymes known as hydrogenases that can combine protons and electrons obtained from the water oxidation process to release molecular hydrogen. These algal hydrogenases are quickly deactivated by oxygen. Researchers have identified mutant algal strains that evolve hydrogen at a rate that is four times that of the wild type, and are three to four times more oxygen tolerant (Ghiradi et al. 1997).

Photosynthetic organisms also contain light harvesting, chlorophyll–protein complexes that effectively concentrate light and funnel energy for photosynthesis. These antenna complexes also dissipate excess incident sunlight as a protective mechanism. The amount of chlorophyll antennae in each cell is directly related to the amount of 'shading' experienced by subsequent layers of microorganisms in a mass culture. In a recent set of experiments, researchers have observed that green alga grown under high light intensities exhibit lower pigment content and a highly truncated chlorophyll antennae size. These cells showed photosynthetic productivity (on a per chlorophyll basis) that was six to seven times greater than the normally pigmented cells, a phenomenon that could lead to significant improvements in the efficiency of hydrogen production on a surface-area basis.

These technical challenges are being addressed by a team of scientists from Oak Ridge National Laboratory (ORNL), the University of California Berkeley, and the National Renewable Energy Laboratory (NREL). Various reactor designs are under development for photobiological hydrogen production processes (single-stage vs two-stage, single organism vs dual organism). At the University of Hawaii's Natural Energy Institute (HNEI), a new, potentially low cost, outdoor tubular photobioreactor is under development to test a sustainable system for the production of hydrogen.

In addition to the photosynthetic production of hydrogen from water, the Program supports the development of systems to convert CO (found in synthesis gas) to hydrogen via the so-called water-gas shift reaction ($CO + H_2O = CO_2 + H_2$). This reaction is essential to the widely used commercial steam methane reforming process for the production of hydrogen. In the industrial process in use today, high-temperature (450°C) and low-temperature (230°C) shift reactors are required to increase the overall hydrogen production efficiency and to reduce the CO content to acceptable levels. In this project, microorganisms isolated from Nature are used to reduce the level of CO to below detectable levels (0.1 ppm) at temperatures of around 25–50°C in a single reactor. This process, under development at NREL, has significant potential to improve the economics of hydrogen production when combined with the thermal processing of biomass or other carbon-containing feeds.

10.6.2. Photochemical systems

Among the technologies that have been investigated, photocatalytic water splitting systems using relatively inexpensive, durable and nontoxic semiconductor photocatalysts show promise. Supported catalysts such as $Pt-RuO_2/TiO_2$ have sufficient band gaps for water splitting, although the current rate of hydrogen production from these systems is too low for commercial processes. Modifications to the system are required to address issues such as the narrow range of solar wavelengths absorbed by TiO_2, the efficiency of subsequent catalytic steps for formation of hydrogen and oxygen, and the need for high surface areas. Binding of catalyst complexes that absorb light in the visible range to the TiO_2 should improve the absorption characteristics. Aerogels of TiO_2 as a semiconductor support for the photocatalysts have potential for addressing reaction efficiency and surface area issues. The University of Oklahoma is investigating these systems.

The Florida Solar Energy Centre (FSEC), in conjunction with the University of Geneva, is investigating tandem/dual bed photosystems using sol/gel-deposited WO_3 films as the oxygen-evolving photocatalyst, rather than TiO_2. In this configuration, the dispersion containing the wider band gap photocatalyst must have minimal light scattering losses so that the lower band gap photocatalyst behind it can also be illuminated.

10.6.3. Photoelectrochemical systems

Multijunction cell technology developed by the photovoltaic industry is being used to develop photo-electrochemical light harvesting systems that generate sufficient voltage to split water and are stable in a water/electrolyte environment. The cascade structure of these devices results in greater utilization of the solar spectrum, resulting in the highest theoretical efficiency for any photoconversion device. In order to develop cost-effective systems, a number of technical challenges must be overcome. These include identification and characterization of semiconductors with appropriate band gaps; development of techniques for preparation and application of transparent catalytic coatings; evaluation of effects of pH, ionic strength, and solution composition on semiconductor energetics and stability, and on catalyst properties; and development of novel photovoltaic/photo-electrochemical system designs. NREL's approach to solving these challenges is to use the most efficient semiconductor materials available, consistent with the energy requirements for a water splitting system that is stable in an aqueous environment. To date, a photovoltaic/photo-electrochemical water splitting system with a solar-to-hydrogen efficiency of 12.4 per cent lower heating value (LHV) using concentrated light, has operated for over 20 h (Khaselev and Turner, 1998). HNEI is pursuing a low-cost amorphous silicon-based tandem cell design with appropriate stability and performance, and is developing protective coatings and effective catalysts. An outdoor test of the an Si cells resulted in a solar-to-hydrogen efficiency of 7.8 per cent LHV under natural sunlight.

10.6.4. Indirect hydrogen production technologies

These systems offer the opportunity to produce hydrogen from renewable resources in the mid-term (five to ten years). Using agricultural residues and wastes, or biomass specifically grown for energy uses, hydrogen can be produced using a variety of processes.

Biomass pyrolysis produces a bio-oil that, like petroleum, contains a wide spectrum of components. Unlike petroleum, bio-oil contains a significant number of highly reactive oxygenated components derived mainly from constitutive carbohydrates and lignin. These components can be transformed into hydrogen via catalytic steam reforming using Ni-based catalysts. By using high heat transfer rates and appropriate reactor configurations that facilitate contact with the catalyst, the formation of carbonaceous deposits (char) can be minimized. The resulting products from the thermal cracking of the bio-oils are steam reformed at temperatures ranging from 750–850°C. At these conditions, any char formed will also be gasified. At NREL and the Jet Propulsion Laboratory, research and modelling are underway to develop processing technologies that take advantage of the wide spectrum of components in the bio-oil, and address reactivity and reactor design issues (Miller and Bellan, 1997). Evaluation of co-product strategies indicates that high value chemicals, such as phenolic resins, can be economically produced in conjunction with hydrogen.

Biomass is typically 50 weight per cent (wt per cent) moisture (as received); biomass gasification and pyrolysis processes require drying of the feed to about 15 wt per cent moisture for efficient and sustained operation, in addition to requiring size reduction (particle size of ~1 cm). In supercritical gasification processes, feed drying is not required, although particle size reduction requirements are more severe. A slurry containing approximately 15 wt per cent biomass (required size reduction ~1 mm) is pumped at high pressure (>22 MPa, the critical pressure of water) into a reactor, where hydrothermolysis occurs, leading to extensive solubilization of the lignocellulosics at just above the supercritical conditions. If heat transfer rates to the slurry are sufficiently high, little char is formed, and the constituents of biomass are hydrolysed and solubilized in the supercritical medium. Increasing the temperature to ~700°C in the presence of catalysts results in the reforming of the hydrolysis products. Catalysts have been identified that are suitable for the steam reforming operation (Matsunaga *et al.* 1997). HNEI, Combustion Systems Inc., and General Atomics are investigating appropriate slurry compositions, reactor configurations, and operating parameters for supercritical water gasification of wet biomass.

10.6.5. Hydrogen storage, transport and delivery

The storage, transport and delivery of hydrogen are important elements in a hydrogen energy system. With keen interest in mobile applications of hydrogen systems, and as intermittent renewables penetration of the electric grid increases, storage becomes essential to a sustainable energy economy. Light weight and high energy density storage will enable the use of hydrogen as a transportation fuel. Efficient and cost effective stationary hydrogen storage will permit photovoltaic and wind to serve as base load power systems.

Compressed gas storage tanks

Currently, compressed gas is the only commercially available method for ambient-temperature hydrogen storage on a vehicle. Compressed hydrogen stored at 24.8 MPa in a conventional fibreglass-wrapped aluminum cylinder results in a volumetric storage density of 12 kg of hydrogen per m³ of storage volume and a gravimetric density of 2 wt per cent (grams of hydrogen per gram of system weight). Carbon fibre-wrapped polymer cylinders achieve higher densities (15 kg/m³ and 5 wt per cent), but are significantly below target values required for hydrogen to make major inroads in the transportation sector (62 kg/m³ and 6.5 wt per cent). Advanced lightweight pressure vessels have been designed and fabricated by Lawrence Livermore National Laboratory (Mitlitsky *et al.* 1998). These vessels use lightweight bladder liners that act as inflatable mandrels for composite overwrap and as permeation barriers for gas storage. These tank systems are expected to exceed 12 wt per cent hydrogen storage (at 33.8 MPa) when fully developed.

Carbon-based storage systems

Carbon-based hydrogen storage materials that can store significant amounts of hydrogen at room temperature are under investigation. Carbon nanostructures could provide the needed technological breakthrough that makes hydrogen powered vehicles practical. Two carbon nanostructures are of interest S single-walled nanotubes and graphite nanofibres. Single-walled carbon nanotubes, elongated pores with diameters of molecular dimensions (twelve-diamensional), adsorb hydrogen by capillary action at non-cryogenic temperatures. Single-walled nanotubes have recently been produced and tested at NREL in high yields using a number of production techniques, and have demonstrated hydrogen uptake at 5–10 wt per cent at room temperature (Dillon *et al.* 1997). Graphite nanofibres are a set of materials that are generated from the metal catalysed decomposition of hydrocarbon-containing mixtures. The structure of the nanofibres is controlled by the selection of catalytic species, reactant composition and temperature. The solid consists of an ordered stack of nanocrystals that are evenly spaced at 0.34–0.37 nanometers (depending on preparation conditions). These are bonded together by van der Waals forces to form a 'flexible wall' nanopore structure.

Metal hydride storage and delivery systems

Conventional high capacity metal hydrides require high temperatures (300–350°C) to liberate hydrogen, but sufficient heat is not generally available in fuel cell transportation applications. Low temperature hydrides, however, suffer from low gravimetric energy densities and require too much space on board or add significant weight to the vehicle. Sandia National Laboratories (SNL) and Energy Conversion Devices (ECD) are developing low-temperature metal hydride systems that can store 3–5 wt per cent hydrogen. Alloying techniques have been developed by ECD that result in high-capacity, multi-component alloys with excellent kinetics, albeit at high temperatures. Additional research is required to identify alloys with appropriate kinetics at low temperatures.

A new approach for the production, transmission and storage of hydrogen using a chemical hydride slurry as the hydrogen carrier and storage medium is under investigation by Thermo Power Corporation. The slurry protects the hydride from unanticipated contact with moisture and makes the hydride pumpable. At the point of storage and use, a chemical hydride/water reaction is used to produce high purity hydrogen. An essential feature of the process is recovery and reuse of spent hydride at a centralized processing plant. Research issues include the identification of safe, stable and pumpable slurries and the design of an appropriate high temperature reactor for regeneration of spent slurry.

10.6.6. End use technologies

Proton exchange membrane (PEM) fuel cells could provide low-cost, high-efficiency electric power, and be operated 'in reverse' as electrolysers to generate hydrogen. There has been a significant increase in industry activity for the development of PEM fuel cells for vehicular applications, with a number of active demonstration projects. Improvements in catalyst loading requirements, water management and temperature control have helped move these power units from mere curiosities to legitimate market successes. In order to increase the market penetration in both the transportation and utility sectors, additional improvements are required. Los Alamos National Laboratory is developing non-machined stainless steel hardware and membrane electrode assemblies with low catalyst loadings to achieve cost reductions and efficiency improvements (Cleghorn et al. 1997). The most important barriers to implementation of low-cost PEM fuel cells are susceptibility of the metal or alloy to corrosion, water management using metal screens as flow fields, and effective stack sealing. Operating the PEM fuel cell 'in reverse' as an electrolyser is possible, but optimum operating conditions for the power production mode and for the hydrogen production mode are significantly different. Design issues for the reversible fuel cell system include thermal management, humidification, and catalyst type and loading.

In an effort to promote near-term use of hydrogen as a transportation fuel, the Program is investigating the development of cost-effective, highly efficient, and ultra-low emission internal combustion engines (ICE) operating on pure hydrogen and hydrogen-blended fuels. Research at SNL is focused on the development of a hydrogen fueled ICE/generator set with an overall efficiency of >40 per cent while maintaining near zero NO_x emissions (van Blarigan, 1998).

10.6.7. Safety

Hydrogen leak detection is an essential element of safe systems. The development of low-cost fibre optic and thick film sensors by NREL and ORNL, respectively, will provide affordable and reliable options for hydrogen safety systems. NREL is using optical fibres with a thin film coating on the end that changes optic properties upon reversible reaction with hydrogen. Change in the reflected light signal is an indication of the presence of hydrogen. Sensitivity and selectivity are important research issues. ORNL is focused on the development of monolithic, resistive thick film sensors that are inherently robust, selective to hydrogen, and easy to manufacture. Research issues include developing appropriate techniques for active (versus traditional passive) thick film applications.

Recognizing the importance of safe use of hydrogen, the DOE, in conjunction with Natural Resources Canada, has compiled a comprehensive document of prevailing practices and applicable codes, standards, guidelines, and regulations for the safe use of hydrogen. The Sourcebook for Hydrogen Applications is intended to be a 'living document' that can be updated to reflect the current state of knowledge about, and experience with, safely using hydrogen in emerging applications. DOE also supports the development of codes and standards under the auspices of the International Standards Organization.

In conclusion, the DOE Hydrogen Program conducts R&D in the areas of production, storage, and utilization, for the purpose of making hydrogen a cost-effective energy carrier for utility, buildings, and transportation applications. Research is focused on the introduction of renewable-based options to produce hydrogen; development of hydrogen-based electricity storage and generation systems that enhance the use of distributed renewable-based utility systems; development of low-cost technologies that produce hydrogen directly from sunlight and water; and support of the introduction of safe and dependable hydrogen systems including the development of codes and standards for hydrogen technologies.

10.7. Japan: International cooperations/networks

Japan has been active in supporting the exploitation of bio-solar technology for H_2 production. The international project RITE (Research Innovative Technologies of the Earth), under the National Organization NEDO (New Energy Development Organization) was supported by MITI (Ministry of International Trade and Industry). A major initiative was the International program 'IEA Agreement of the Production and Utilization of Hydrogen' (Annex 10 'Photoproduction of hydrogen'; Annex 15 'Photobiological Hydrogen Production'). The main objectives were to investigate and to develop processes and equipment for the production of hydrogen by direct conversion of solar energy to molecular hydrogen using biological systems.

Many of the outcomes of this work, together with contributions from around the world, have been included in the recent book Biohydrogen II, edited by J. Miyake et al. (2000).

10.8. Future challenges

Bärbel Hüsing

The major rationale for bio-hydrogen R&D is the possible use of bio-hydrogen as an energy source and raw material as a component within a future solar energy system. This is, however, a very long-term option. Its realization is not likely before the year 2030–2040, and only if drastic reductions in CO_2 emissions must be reached (Hüsing and Reiß 1996). From today's knowledge, for thermodynamic and logistic reasons the technological option of bio-hydrogen production which seems to be the best for the above mentioned purpose is biophotolysis, i.e. splitting water into oxygen and hydrogen by absorption of sunlight. This process is catalysed by photosystems and hydrogenase.

Bio-hydrogen is part of the broader concept of developing environmentally friendly technologies, especially zero-emission technologies. The technological options of bio-hydrogen production that mainly contribute to this concept are photoproduction from biomass and production from biomass by anaerobic fermentation. Both processes depend on the supply of organic substrates and are therefore ideally suited for coupling with treatment of biomass, waste and waste water. H_2 production by these metabolic routes is catalysed by nitrogenase and/or hydrogenase, and it is much closer to practical application than bio-hydrogen production by biophotolysis.

However, hydrogenases should not only be considered in the context of large-scale hydrogen production. There is considerable progress in microsystems engineering and nanotechnology, and biotechnology is beginning to play a role in this field. It would be an interesting option for future hydrogenase research and development (R&D), to exploit the potential of hydrogenases also as an energy source in 'small-scale applications', e.g. in biomolecular devices in microsystems engineering or as power supply for microsurgery robots or artificial organs which operate inside the human body. First attempts in this direction are already underway (Sasaki und Karube 1999). Other application potentials are hydrogenases in enzymatic analytical test systems, in biosensors, and in electrochemical regeneration of redox enzymes used in chemical synthesis (Somers et al. 1997).

If these three goals in bio-hydrogen R&D are to be pursued, progress in the following fields of bio-hydrogen R&D is required.

10.8.1. Improving the performance of the production organism

Optimizing the performance of the bio-hydrogen production organism comprises two strategies which complement each other: first, isolation and selection of natural H_2-producing strains by conventional screening methods; and second, improvement of the H_2-producing organisms by genetic engineering.

In the past, screening for excellent natural H_2-producing organisms has yielded very promising strains, but was time consuming and labour intensive. New approaches and selection methods must be developed in order to speed up the screening process and to increase the likelihood of discovering excellent H_2 producers. This can be achieved by:

- screening the biodiversity of still untapped resources (e.g. extreme environments);
- applying automated, computer-controlled, miniaturized screening procedures;
- developing rapid identification tests based on information on the metabolic pathways involved in H_2 metabolism;
- developing rapid and miniaturized tests for the H_2-production ability of large numbers of organisms or mutants (ideally with positive selection); and
- screening for organisms with properties that are important for biotechnological exploitation, such as tolerance towards oxygen, pH, light, temperature, waste components, lack of adhesion to surfaces of culture vessels, and easy separation from the culture medium.

Once isolated, the natural H_2 producers can be optimized by conventional mutagenesis, and they should be studied so that we can understand those features that make them the best H_2 producers. This characterization would involve the analysis of metabolic fluxes (Stephanopoulos and Sinskey 1993; Schuster et al. 1999) and molecular genetics. It would result in new, previously unknown adaptations necessary for improved H_2 production, and could provide information on the most important mutations that are required to obtain excellent H_2 producers. Information obtained from these experiments should be used in genetic engineering approaches for optimizing H_2 producers. Moreover, excellent H_2 producers should be used in bioengineering approaches.

Parallel to the characterization of natural excellent H_2 producers is the rational optimization of H_2 producers by metabolic engineering, using genetic engineering approaches. This type of research can be performed in laboratory strains, which are relatively easy to handle. If certain approaches prove to be successful with respect to H_2 production they may also be transferred to the excellent natural H_2 producers. However, at present, there are hardly any model organisms or model hydrogenases that are well-characterized biochemically, chemically, spectroscopically, physiologically and by molecular genetics. If efforts were concentrated on a few model systems by combining different disciplines and approaches for a thorough investigation, synergies between the different methodological approaches could be exploited. This interdisciplinary approach could yield results that could not be obtained otherwise.

10.8.2. Improving the performance of hydrogenases in technical systems

At present, deepening our understanding for structure-function relationships in hydrogenases has mainly remained basic research. It is important for applications of hydrogenases in both 'large and small scale applications' to exploit this knowledge for the rational optimisation of hydrogenases with properties of biotechnological interest. This could be hydrogenases with e.g. enhanced stability and activity or biomimetic catalysts with these desired properties. Moreover, the 'irrational' design of hydrogenase properties could also be taken into consideration (Kuchner and Arnold 1997).

10.8.3. Bioprocess engineering, scaleup to a functional prototype of technical scale

The bioprocess engineering approach comprises the identification of factors influencing H_2 metabolism and growth, the optimization of culture conditions for H_2 producers, as well as bioreactor design, construction and operation. In the past, progress in this field has been hindered by the fact that research groups worked more or less isolated with their own system, so that comparisons and knowledge transfer between different systems were difficult to perform and therefore, possible synergies were not exploited. In order to provide data which allow the scaleup of bio-hydrogen production to a functional prototype of technical scale, future challenges of bio-hydrogen R&D are:

- to define standard culture conditions (e.g. light) and a standard report format for values (e.g. specific H_2-production rate, conversion efficiency, light conditions) to make comparisons between different strains and different R&D groups possible (especially for the evaluation of new screening isolates);
- to use the same strain wherever possible;
- to perform comparisons and 'benchmarking' of results wherever possible; and
- to systematically vary strains, culture conditions or bioreactor design while all other parameters are kept constant.

Up to now, there are hardly any photobioreactors which allow a cost-efficient production even of high-value products from phototrophic organisms. The development of such photobioreactors is an attractive goal. They could – as a 'spin-off' – prove useful for other production processes using phototrophic organisms. Another goal could be the design of a production process with minimal requirements of energy, water, cooling/heating, etc., minimal output of waste, and maximum product recovery. Technologies to achieve this goal (including e.g. membranes, electrodialysis) could also be applicable to other biotechnical production processes and would contribute to the implementation of the zero-emission concept.

10.8.4. Economic and ecological assessments parallel to experimental R&D

At the present state of bio-hydrogen R&D there are still many different options as to how a future bio-hydrogen production process might look; definite decisions about successful solutions cannot yet be made. Moreover, practical bio-hydrogen production will have to compete with alternative, often established technologies, and must therefore show clear comparative, specific advantages over these competing technologies. In addition, the need for bio-hydrogen is substantially influenced by priority-setting in environmental and energy policy.

Therefore, it is very important to periodically adjust bio-hydrogen R&D to new developments in the relevant frame conditions. For this purpose, relevant developments in bio-hydrogen R&D and related scientific fields, in environmental and energy policy, and in competing technologies should be monitored. Life-cycle assessments can contribute to defining the role of bio-hydrogen in relation to competing technologies, to identifying weak spots in technical processes and to improving process performance from the environmental perspective. Economic analysis can help to define the cost targets for bio-hydrogen in order to become economically competitive, and market studies can help in the identification of markets and niches for bio-hydrogen and its by-products (e.g. products from excess biomass). Information from such activities should support strategic decision-making within future bio-hydrogen R&D and its fine-tuning.

Finally, progress in the field of bio-hydrogen R&D will significantly depend on how well close interaction and feedback between these different fields of R&D can be achieved in order to obtain intensive knowledge transfer, make tacit knowledge available, and to exploit synergies.

10.9. Concluding remarks

K. Krishna Rao and Richard Cammack

It is clear that in the twenty-first century, much more effort will have to be directed at alternative energy sources, of which hydrogen is one of the most promising. This book has focused on the science that underpins the biological approach to the production of hydrogen energy.

Photosynthetic organisms have evolved for maximum efficiency at what they do, and are capable of light conversion efficiencies of over 10 per cent. This is comparable with the best solar cell systems for generating electricity. The practical pilot systems are only approaching 1 per cent. In order to bridge that gap, an enormous investment in the evolution of artificial organisms might be required. We should not forget that there has been an enormous investment of resources, time, effort and ingenuity into perfecting the way we currently harness fossil fuels as energy sources. The effort that will be needed to bring newer energy technologies into use is no less formidable.

Appendix 1

List of names of microorganisms

Name	Abbreviation	Notes
Allochromatium vinosum	A. vinosum	formerly Chromatium vinosum
Anabaena cylindrica	A. cylindrica	
Anabaena variabilis	A. variabilis	
Azotobacter vinelandii	A. vinelandii	
Azotobacter chroococcum	A. chroococcum	
Bradyrhizobium japonicum	B. japonicum	formerly Rhizobium japonicum
Bradyrhizobium leguminosarum	B. leguminosarum	formerly Rhizobium leguminosarum
Clostridium pasteurianum	C. pasteurianum	
Desulfomicrobium baculatum	D. baculatum	formerly Desulfovibrio desulfurican strain Norway 4
Desulfovibrio gigas	D. gigas	
Desulfovibrio indonensis	D. indonensis	
Desulfovibrio vulgaris (Miyazaki)		
Escherichia coli	E. coli	
Methanothermobacter gen. nov	Methanothermobacter gen. nov	formerly Methanobacterium thermoautotrophicum strain DH
Methanothermobacter marburgensis	M. marburgensis	formerly Methanobacterium thermoautotrophicum, Marburg stra
Methanococcus voltae	M. voltae	
Methanosarcina barkeri	M. barkeri	
Methanosarcina mazei	M. mazei	
Moorella thermoautotrophica	M. thermoautotrophica	formerly Clostridium thermoautotrophicum
Pyrococcus furiosus	P. furiosus	
Ralstonia eutropha	R. eutropha	formerly Acaligenes eutrophus
Rhodobacter capsulatus	R. capsulatus	
Rhodobacter sphaeroides	R. sphaeroides	formerly Rhodopseudomonas sphaeroides
Thiocapsa roseopersicina	T. roseopersicina	
Thermococcus litoralis	T. litoralis	

Appendix 2

Glossary

Activator   See Transcriptional activator.

Bioreactor   A reaction chamber for biological processes, mostly in biotechnology.

Cladistics   Method of classification employing genealogies alone in inferring phylogenetic relationships among organisms (see also Phylogeny).

Clade   Phylogenetic lineage of related taxa from a common ancestral taxon.

Cloning vector   Usually a plasmid or a viral genome adapted for the introduction (cloning) of foreign DNA and usually containing a selectable marker (e.g. antibiotic resistance gene). Foreign DNA is introduced at restriction enzyme cleavage sites.

cDNA   Copy DNA made from an RNA molecule.

Diazotroph   An organism that derives its nitrogen for growth by fixing atmospheric N2.

Domain   A region of a protein, often forming a self-contained folded unit.

Electron paramagnetic resonance (EPR) spectroscopy   The form of spectroscopy concerned with microwave-induced transitions between magnetic energy levels of unpaired electrons, i.e. those having a net spin. The spectrum is normally obtained by magnetic field scanning. Also known as electron spin resonance (ESR) spectroscopy or electron magnetic resonance (EMR) spectroscopy. The microwave frequency ν is measured in gigahertz (GHz) or megahertz (MHz). The following band designations are used: L(1.1 GHz), S(3.0 GHz), X(9.5 GHz), K(22.0 GHz) and Q(35.0 GHz). The static magnetic field at which the EPR spectrometer operates is measured by the magnetic flux density (B) and its recommended unit is the tesla (T). In the absence of nuclear hyperfine interactions, B and ν are related by: $h\nu = g\mu_B B$ where h is the Planck constant, μ_B is the Bohr magneton, and the g factor, g, is a parameter which is characteristic of the spin system.

Electron-nuclear double resonance (ENDOR) spectroscopy   A magnetic resonance spectroscopic technique for the determination of *hyperfine* interactions between electrons and nuclear spins. There are two principal techniques. In continuous-wave ENDOR the intensity of an *electron paramagnetic resonance* signal, partially saturated with microwave power, is measured as radio frequency is applied. In pulsed ENDOR the radio frequency is applied as pulses and the EPR signal is detected as a spin-echo. In each case an enhancement of the EPR signal is observed when the radiofrequency is in resonance with the coupled nuclei.

Exon (see also Intron)   That part of a eukaryotic gene which is both transcribed and expressed in the gene product.

Extended X-ray absorption fine structure (EXAFS)   A technique for observing the local structure around a metal centre, using X-rays from a synchrotron source. The atom of interest absorbs photons at a characteristic wavelength and the emitted electrons, undergoing constructive or destructive interference as they are scattered by the surrounding atoms, modulate the absorption spectrum. The modulation frequency corresponds directly to the distance of the surrounding atoms while the amplitude is related to the type and number of atoms. In particular, bond lengths and coordination numbers may be derived.

Fourier-transform infrared (FTIR) spectroscopy   Spectroscopy based on excitation of vibrational modes of chemical bonds in a molecule. The energy of the infrared radiation absorbed is expressed in inverse centimeters (cm^{-1}), which represents a frequency unit. For transition-metal complexes, the ligands $-C \equiv N$ and $-C = O$ have characteristic absorption bands at unusually high frequencies, so that they are easily distinguished from other bonds. The position of these bonds depends on the distribution of electron density between the metal and the ligand; an increase of charge density at the metal results in a shift of the bands to lower frequencies.

Gene Bank/Library   Collection of DNA fragments representing part or all of the genome of an organism held in a cloning vector in a suitable host. Used to isolate and replicate genes of interest independently of their original host. A *cDNA library* is constructed from RNA using reverse transcriptase and represents those genes which are expressed in that cell at a particular time but which may be only a small proportion of all the genes in the cells. A cDNA library reflects the abundance of mRNA molecules in the cell.

Genome   The total genetic information present in an organism. Viral genomes are unusual in that they can be diverse. Examples of viral genomes include single-stranded DNA, single- and double-stranded RNA molecules.

Hydron   The hydrogen ion, $^1H^+$, is generally referred to as the proton, which is the nucleus of hydrogen, ^1H. But since hydrogenase can also use deuterium ions (deuterons, $^2H^+$) and tritium ions (tritons, $^3H^+$) as substrates, the correct term is 'hydrons', which does not discriminate between the isotopes.

Hyperfine interaction, or hyperfine coupling (hfc)   The interaction between an electron in a paramagnetic center and nuclear spin. It can be observed as a splitting of lines in an EPR spectrum, or a pair of lines in an ENDOR spectrum. There are two components to the hyperfine interaction: through bonds (contact hyperfine interaction) and through space (dipolar interaction). The magnitude of the coupling can provide information about relative location of the nucleus relative to the paramagnetic centre.

In-frame mutation   A mutation, usually a deletion, which does not shift the reading frame downstream of the mutation.

Intron (see Exon)   That part of a eukaryotic gene which is transcribed but, as a result of editing through the spliceosome assembly, is not expressed in the final gene product.

Kilobase (Kb)   A thousand base pairs of DNA or a thousand bases of RNA. This is a significant number because the average gene is 1,000 base pairs long.

Knallgas   A mixture of hydrogen and oxygen that is too low in hydrogen to initiate an explosion. Hence knallgas reaction – the biological oxidation of low concentrations of hydrogen with oxygen.

Ligation   The joining of two DNA strands to each other to form a 5′ to 3′ phosphodiester linkage. The reaction is catalysed by the enzyme DNA-ligase which is important in DNA replication and repair and when purified it is used for joining DNA *in vitro*.

Lithotroph   An organism that derives its energy for growth from the conversion of inorganic substances.

Minimal hydrogenase   The part of a hydrogenase structure that is found in all hydrogenases of a particular type, and which is therefore proposed to be the essential part for its function. For example in the NiFe hydrogenases, this comprises the NiFe catalytic centre in the large subunit, and the proximal [4Fe-4S] cluster in the small subunit.

Mössbauer spectroscopy   The Mössbauer effect is resonance absorption of γ radiation of a precisely defined energy, by specific nuclei. It is the basis of a form of spectroscopy used for studying coordinated metal ions. The principal application in bioinorganic chemistry is ^{57}Fe. The source for the γ rays is ^{57}Co, and the frequency is shifted by the Doppler effect, moving it at defined velocities (in mm/s) relative to the sample. The parameters derived from the Mössbauer spectrum (isomer shift, quadrupole splitting, and the hyperfine coupling) provide information about the oxidation, spin and coordination state of the iron.

Motif   A sequence of amino acids in a protein, which is found in many different species, and which has a particular function such as binding a metal center.

Operon   A group of two or more adjacent genes which are transcribed from a single promoter on a single messenger RNA. Operons often contain functionally related genes. Messenger RNA which carries the information for two genes is known as polycistronic.

Open reading frame (ORF)   See Translation.

Overexpression   The production of abnormally high levels of foreign (usually) proteins or RNA molecules in a host cell usually by cloning the gene of interest into an overexpression vector.

Phototroph   An organism that derives its energy for growth from photosynthesis.

Phylogeny   Evolutionary history. The genealogical history of a group of organisms represented by its hypothesized ancestor–descendent relationships.

Plasmid   A relatively (cf. the chromosome) small (>20 Kb), usually circular, double-stranded DNA molecule found in prokaryotes capable of replicating independently of the chromosome. Plasmids carry genes which are usually not essential for the growth of the organism except under special conditions. Some plasmids carry genes for antibiotic resistance. See also Ti-plasmid. Some plasmids however can be very large, e.g. the plasmids in Rhizobium species.

Polarity   The phenomenon where a mutation in one gene exerts cis effects on downstream genes. Usually polar mutations lower expression of the downstream genes.

Polymerase chain reaction (PCR)   The process by which a specific sequence of DNA can be amplified (copied many times) *in vitro*. It requires a pair of primers and template DNA, thermostable DNA polymerase (e.g. *Taq* polymerase), deoxynucleotide triphosphates and a thermocycler. The process can amplify large amounts of a specific DNA sequence (an amplicon) given just a few molecules of template. A revolutionary technique in biology.

Primer A short length of single-stranded DNA or sometimes RNA (usually ~20 bases which can be synthesized) which is complementary to a known DNA sequence so that it can bind there and serve for the initiation of DNA replication.

Promoter Usually a specific region of DNA at which RNA-polymerase binds and initiates transcription.

Ribosome binding site (RBS) See Translation.

Redox potential The driving force for an oxidation–reduction reaction. Reduction can be considered either as the addition of hydrogen or electrons to a molecule (since $H^+ + e^- \rightleftharpoons H^{\cdot}$); oxidation is the opposite process. Any redox (reduction–oxidation) reaction can be divided into two half-reactions: one in which a chemical species, A, undergoes oxidation, and one in which another chemical species, B, undergoes reduction:

$$A_{ox} + B_{red} = A_{red} + B_{ox}.$$

The reducing equivalents transferred can be considered either as hydrogen atoms or electrons. The driving force for the reaction, E, is the reduction/oxidation (redox) potential, and can be measured by electrochemistry; it is often expressed in millivolts. The number of reducing equivalents transferred is n. The redox potential of a compound A depends on the concentrations of the oxidized and reduced species $[A_{ox}]$ and $[A_{red}]$ according to the Nernst equation:

$$E = E_m + RT/nF \, \ln([A_{ox}]/[A_{red}]).$$

The midpoint potential of a half-reaction E_m, is the value when the concentrations of oxidized and reduced species are equal, $[A_{ox}] = [A_{red}]$. In biological systems the standard redox potential of a compound is the reduction/oxidation potential measured under standard conditions, defined at $pH = 7.0$ versus the hydrogen electrode. On this scale, the potential of O_2/water is $+815$ mV, and the potential of water/H_2 is -414 mV. A characteristic of redox reactions involving hydrogen transfer is that the redox potential changes with pH. The oxidation of hydrogen $H_2 = 2H^+ + 2e^-$ is an $n = 2$ reaction, for which the potential is -414 mV at pH 7, changing by -59.2 mV per pH unit at 30°C.

Repressor See Transcriptional repressor.

Transformation Processes by which DNA is introduced into a host organism. Some organisms (cells) are naturally competent for DNA uptake. In most others the DNA has to be introduced into the cell by various artificial means including soaking cells in high levels of divalent cations (usually Ca^{2+}), treating with electric shock (electroporation) or by shooting DNA coated 'bullets' into cells (ballistic techniques or 'gene guns').

Transcription The process of copying a DNA sequence into an RNA molecule catalysed by the enzyme RNA polymerase.

Transcriptional activator A protein which activates (up regulates) transcription of a specific gene or group of genes. Transcriptional activators are DNA binding proteins which usually bind to specific sequences close to the promoter and enhance the binding of RNA polymerase and/or stimulate the rate of transcription initiation. Some activators bind at a distance from the promoter yet can interact with RNA polymerase to enhance transcription via a loop in the DNA.

Transcriptional repressor   A protein (usually) which binds at a specific sequence in DNA at, or close to, the promoter and blocks the binding of RNA polymerase and hence blocks transcription of the genes controlled by that promoter.

Translation   The complex process of interpreting the message in messenger RNA in the form of a polypeptide chain. The components of the system include: transfer RNAs, aminoacyl tRNA synthetases and the ribosome assembly. Messenger RNAs carry specific signals which allow ribosomes to recognize and bind to the message. These signals differ in eukaryotes and prokaryotes. In prokaryotes, the signal is usually a short purine-rich region (see ribosome binding site) approximately 10 bases upstream of the translation start site (usually, but not always, AUG). In eukaryotes the mRNA requires to be processed before it can be recognized by the ribosome. Translation terminates at a stop codon (UAA, UGA, UAG). The region between the initiation and termination of translation is usually known as an open reading frame (ORF).

Translational coupling   Two or more adjacent genes in an operon which overlap or are separated by a few base pairs so that, at the level of messenger RNA, ribosomes terminating translation of first gene reinitiate translation of the second, or subsequent gene. Translational coupling may allow stoichiometric amounts of gene products to be produced.

Transposon   A small mobile DNA element of no more than a few kbp which is capable of copying and inserting itself into a new location in a genome. When transposons insert into genes, they cause mutations (insertional inactivation). Tranposons may carry markers such as antibiotic resistance. Some have been engineered to carry reporter genes, e.g. the gene for β-galactosidase (*lacZ*) so that they can act as reporters of gene activity when inserted in the correct orientation.

X-ray absorption near edge structure (XANES)   The X-ray absorption spectrum, as for EXAFS, may also show detailed structure below the absorption edge. This arises from excitation of core electrons to high level vacant orbitals, and can be used to estimate the oxidation state of the metal ion.

Appendix 3

Websites for hydrogen research

http://www.dwv-info.de/indexe.htm German Hydrogen Network
http://www.eren.doe.gov/hydrogen/ Energy Efficiency and Renewable energy network
http://www.H$_2$eco.org/index.htm Hydrogen Energy Center
http://www.hydrogenus.com/nha/index.htm
http://www.iahe.org/ International Association for Hydrogen Energy
www.H$_2$forum.org Swedish Hydrogen Forum
www.H$_2$net.org.uk. UK Hydrogen Energy Network
www.hydrogen.no Norwegian Hydrogen Forum
www.iea.org/ International Energy Agency
www.inta.es Spanish Organization for Investigation of Technology
www.nedo.go.jp New Energy and Industrial Techology Development Organization (Japan)
www.novem.org Netherlands agency for energy and the environment
www.nrel.gov/ National Renewable Energy Laboratory (US)
www.nrel.gov/data/pix/searchpix.cgi pictures
www.stem.se Swedish National Energy Administration
www.ttcorp.com National Hydrogen Association

References

Adams, M. W. W. (1990) The metabolism of hydrogen by extremely thermophilic, sulfur-dependent Bacteria. *FEMS Microbiol. Rev.*, 75, 219–38.

Adams, M. W., Eccleston, E. and Howard, J. B. (1989) Iron–sulfur clusters of hydrogenase I and hydrogenase II of *Clostridium pasteurianum*. *Proc. Natl. Acad. Sci. USA*, 86, 4932–6.

Afting, C., Hochheimer, A. and Thauer, R. K. (1998) Function of H_2-forming methylene-tetrahydromethanopterin dehydrogenase from methanobacterium thermoautotrophicum in coenzyme F420 reduction with H_2. *Arch. Microbiol.*, 169, 206–10.

Afting, C., Kremmer, E., Brucker, C., Hochheimer, A. and Thauer, R. K. (2000) Regulation of the synthesis fo H_2-forming methylenetetrahydromethanopterin dehydrogenase (Hmd) and of Hmd2 and Hmd3 in *Methanothermobacter marburgensis. Arch. Microbiol.*, 174, 225–2.

Albracht, S. P. (1994) Nickel hydrogenases: In search of the active site. *Biochim. Biophys. Acta*, 1188, 167–204.

Albracht, S. P. J., Kröger, A., Van der Zwaan, J. W., Unden, G., Böcher, R., Mell, H. and Fontijn, R. D. (1986) Direct evidence for sulfur as a ligand to nickel in hydrogenase: An EPR study of the enzyme from *Wolinella succinogenes* enriched in O_2. *Biochim. Biophys. Acta*, 874, 116–27.

Albrecht, S. L., Maier, R. J., Hanus, R. J., Russell, S. A., Emerich, D. W. and Evans, H. J. (1979) Hydrogenase in *Rhizobium japonicum* increases nitrogen fixation by nodulated soybeans. *Science*, 203, 1255–7.

Alex, L. A., Reeve, J. N., Orme-Johnson, W. H. and Walsh, C. T. (1990) Cloning, sequence determination, and expression of the genes encoding the subunits of the nickel-containing 8-hydroxy-5-deazaflavin reducing hydrogenase from *Methanobacterium thermoautotrophicum* delta H. *Biochemistry*, 29, 7237–44.

Amara, P. V., Volbeda, A., Fontecilla-Camps, J. C. and Field, M. J. A. (1999) Hybrid density functional theory/molecular mechanics study of nickel-iron hydrogenase: Investigation of the active site redox states. *J. Am. Chem. Soc.*, 121, 4468–77.

Anderson, R. T., Chapelle, F. H. and Lovley, D. R. (1998) Evidence against hydrogen-based microbial ecosystems in basalt aquifers. *Science*, 28, 976–7.

Andrews, S. C. (1998) Iron storage in bacteria. *Adv. Microb. Physiol.*, 40, 281–351.

Andrews, S. C., Berks, B. C., McClay, J., Ambler, A., Quail, M. A., Golby, P. and Guest, J. R. (1997) A 12-cistron *Escherichia coli* operon (hyf) encoding a putative proton- translocating formate hydrogenlyase system. *Microbiology*, 143, 3633–47.

Appel, J. and Schulz, R. (1998) Hydrogen metabolism in organisms with oxygeneic photosynthesis: Hydrogenases as important regulatory devices for a proper redox poising. *J. Photobiochem. Photobiol.*, 47, 1–11.

Appel, J., Phunpruch., S., Steinmuller, K. and Schulz, R. (2000) The bidirectional hydrogenase of *Synechocystis* sp PCC 6803 works as an electron valve during photosynthesis. *Arch. Microbiol.*, 173, 333–8.

Appleby, A. J. (1999) The electrochemical engine for vehicles. *Sci. Am.*, 281, 58–63.

Arp, D. J. (1992) Hydrogen cycling in symbiotic bacteria. In G. Stacey, R. H. Burris and H. J. Evans (eds), *Biological Nitrogen Fixation*. New York: Chapman and Hall, pp. 432–60.

Atlung, T., Knudsen, K., Heerfordt, L. and Broendsted, L. (1997) Effects of sigma(S) and the transcriptional activator AppY on induction of the *Escherichia coli* hya and cbdAB-appA operons in response to carbon and phosphate starvation. *Journal of Bacteriology*, 179, 2141–6.

Axelsson, R., Oxelfelt, F. and Lindblad, P. (1999) Transcriptional regulation of Nostoc uptake hydrogenase. *FEMS Microbiol. Lett.*, 170, 77–81.

Bagley, K. A., Van Garderen, C. J., Chen, M., Duin, E. C., Albracht, S. P. and Woodruff, W. H. (1994) Infrared studies on the interaction of carbon monoxide with divalent nickel in hydrogenase from *Chromatium vinosum*. *Biochemistry*, 33, 9229–36.

Bagley, K. A., Duin, E. C., Roseboom, W., Albracht, S. P. J. and Woodruff, W. H. (1995) Infrared-detectable group senses changes in charge density on the nickel center in hydrogenase from *Chromatium vinosum*. *Biochemistry*, 34, 5527–35.

Bagyinka, C., Whitehead, J. P. and Maroney, M. J. (1993) An X-ray absorption spectroscopic study of nickel redox chemistry in hydrogenase. *J. Am. Chem. Soc.*, 115, 3576–85.

Baidya, N., Olmstead, M. M. and Mascharak, P. K. (1992) A mononuclear nickel(II) complex with [NiN$_3$S$_2$] chromophore that readily affords the Ni(I) and Ni(III) analogues: Probe into the redox behavior of the nickel site in [FeNi] hydrogenases. *J. Am. Chem. Soc.*, 114, 9666–8.

Ballantine, S. P. and Boxer, D. H. (1985) Nickel-containing hydrogenase isoenzymes from anaerobically grown *Escherichia coli* K-12. *J. Bacteriol.*, 163, 454–9.

Baron, S. F. and Ferry, J. G. (1989) Purification and properties of the membrane-associated coenzyme F420-reducing hydrogenase from *Methanobacterium-Formicicum*. *J. Bacteriol.*, 171, 3846–53.

Bartha, R. and Ordal, E. J. (1965) Nickel-dependent chemolithotrophic growth of two *Hydrogenomonas* strains. *J. Bacteriol.*, 89, 1015–19.

Bartoschek, S., Buurman, G., Thauer, R. K., Geierstanger, B. H., Weyrauch, J. P., Griesinger, C., Nilges, M., Hutter, M. C., Helms, V. (2001) Re-face stereospecificity of methylenetetrahydromethanopterin dehydrogenases and methylenetetrahydrofolate dehydrogenases is predetermined by intrinsic properties of the substrate. *Chem Bio Chem*, in press.

Báscones, E., Imperial, J., Ruiz-Argueso, T. and Palacios, J. M. (2000) Generation of new hydrogen-recycling rhizobiaceae strains by introduction of a novel hup minitransposon [In Process Citation]. *Appl. Environ. Microbiol.*, 66, 4292–9.

Bauschlicher, C. W. Jr, Langhoff, S. R. and Partridge, H. (1995) The application of Ab initio electronic structure calculations to molecules containing transition metal atoms. In D. R. Yarkony (ed.), *Modern Electronic Structure Theory. Part II*, Vol. 2. Singapore: World Scientific, pp. 1281–1374.

Becke, A. D. (1988) Density-functional exchange-energy approximation with correct asymptotic behavior. *Phys. Rev. A*, 38, 3098–3100.

Becke, A. D. (1998) A new inhomogeneity parameter in density-functional theory. *J. Chem. Phys.*, 109, 2092–8.

Belaich, J. P., Bruschi, M. and Garcia, J. L. (1990) *Microbiology and Biochemistry of Strict Anaerobes Involved in Interspecies Hydrogen Transfer*. New York: Plenum Press.

Bender, B. R., Kubas, G. J., Jones, L. H., Swanson, B. I., Eckert, J., Capps, K. B. and Hoff, C. D. (1997) Why does D-2 bind better than H-2? A theoretical and experimental study of the equilibrium isotope effect on H-2 binding in a M(eta(2)-H-2) complex. Normal coordinate analysis of W(CO)(3)(PCy3)(2)(eta(2)-H-2). *J. Am. Chem. Soc.*, 119, 9179–90.

Beneke, S., Bestgen, H. and Klein, A. (1995) Use of the *Escherichia coli* uidA gene as a reporter in *Methanococcus voltae* for the analysis of the regulatory function of the intergenic region between the operons encoding selenium-free hydrogenases. *Mol. Gen. Genet.*, 248, 225–8.

Benemann, J. R. and Weare, N. M. (1974) Hydrogen evolution by nitrogen-fixing *Anabaena cylindrica* cultures. *Science*, 184, 1917–19.

Benemann, J. R., Berenson, J. A., Kaplan, N. O. and Kamen, M. D. (1973) Hydrogen evolution by a chloroplast-ferredoxin-hydrogenase system. *Proc. Nat. Acad. Sci. USA*, 70, 2317–20.

Berger, D. J. (1999) Fuel cells and precious-metal catalysts. *Science*, 286, 49.

Berghofer, Y., Agha-Amiri, K. and Klein, A. (1994) Selenium is involved in the negative regulation of the expression of selenium-free [NiFe] hydrogenases in *Methanococcus voltae*. *Mol. Gen. Genet.*, 242, 369–73.

Bergman, B., Gallon, J. R., Rai, A. N. and Stal, L. J. (1997) N_2 fixation by non-heterocystous cyanobacteria. *FEMS Microbiol. Rev.*, 19, 139–85.

Berkessel, A. and Thauer, R. K. (1995) On the mechanism of catalysis by a metal-free hydrogenase from methanogenic archaea: Enzymic transformation of H_2 without a metal and its analogy to the chemistry of alkanes in superacidic solution. *Angew. Chem., Int. Ed. Engl.*, 34, 2247–50.

Berlier, Y., Fauque, G. D., LeGall, J., Choi, E. S., Peck, H. D. Jr and Lespinat, P. A. (1987) Inhibition studies of three classes of *Desulfovibrio* hydrogenase: Application to the further characterization of the multiple hydrogenases found in *Desulfovibrio vulgaris* Hildenborough. *Biochem. Biophys. Res. Commun.*, 146, 147–53.

Bernhard, M., Friedrich, B. and Siddiqui, R. A. (2000) *Ralstonia eutropha* TF93 is blocked in tat-mediated protein export. *J. Bacteriol.*, 182, 581–8.

Bishop, P. E. and Premakumar, R. (1992) Alternative nitrogen fixation systems. In G. Stacey, R. H. Burris and H. J. Evans (eds), *Biological Nitrogen Fixation*. New York, USA: Chapman and Hall, pp. 736–62.

Black, L. K. and Maier, R. J. (1995) IHF- and RpoN-dependent regulation of hydrogenase expression in *Bradyrhizobium japonicum*. *Mol. Microbiol.*, 16, 405–13.

Black, L. K., Fu, C. and Maier, R. J. (1994) Sequences and characterization of hupU and hupV genes of *Bradyrhizobium japonicum* encoding a possible nickel-sensing complex involved in hydrogenase expression. *J. Bacteriol.*, 176, 7102–6.

Blackburn, J. W. and Hafker, W. R. (1993) The impact of biochemistry, bioavailability and bioactivity on the selection of bioremediation techniques. *Trends Biotechnol.*, 11, 328–33.

Bohm, R., Sauter, M. and Böck, A. (1990) Nucleotide sequence and expression of an operon in *Escherichia coli* coding for formate hydrogenlyase components. *Mol. Microbiol.*, 4, 231–43.

Böhme, H. (1998) Regulation of nitrogen fixation in heterocyst forming bacteria. *Trends Plant Sci.*, 3, 346–51.

Boison, G., Schmitz, O., Mikheeva, L., Shestakov, S. and Bothe, H. (1996) Cloning, molecular analysis and insertional mutagenesis of the bidirectional hydrogenase genes from the cyanobacterium *Anacystis nidulans*. *FEBS Lett.*, 394, 153–8.

Boison, G., Schmitz, O., Schmitz, B. and Bothe, H. (1998) Unusual gene arrangement of the bidirectional hydrogenase and functional analysis of its diaphorase subunit HoxU in respiration of the unicellular cyanobacterium *Anacystis nidulans*. *Curr. Microbiol.*, 36, 253–8.

Bonomi, F., Pagani, S. and Kurtz, D. M. Jr (1985) Enzymic synthesis of the 4Fe-4S clusters of *Clostridium pasteurianum* ferredoxin. *Eur. J. Biochem.*, 148, 67–73.

Borodin, V. B., Tsygankov, A. A., Rao, K. K. and Hall, D. O. (2000) Hydrogen production by *Anabaena variabilis* PK84 under simulated outdoor conditions. *Biotechnol. Bioeng.*, 69, 478–85.

Bothe, H., Kentemich, T. and Heping, D. (1991) Recent aspects on the hydrogenase-nitrogenase relationship in cyanobacteria. In M. Polsinelli, R. Materassi and M. Vincenzini (eds), *Nitrogen Fixation*. Dordrecht: Kluwer Academic Publishers, pp. 367–75.

Bott, M. and Thauer, R. K. (1989) Proton translocation coupled to the oxidation of carbon monoxide to CO_2 and H_2 in *Methanosarcina barkeri*. *Eur. J. Biochem.*, 179, 469–72.

Brecht, M., Stein, M., Trofanchuk, O., Lendzian, F., Bittl, R., Higuchi, Y. and Lubitz, W. (1998) Catalytic center of the [NiFe] hydrogenase: A pulse ENDOR and ESEEM study. In D. Ziessow, W. Lubitz and F. Lendzian, (eds), *Magnetic Resonance and Related Phenomena*, TU Berlin, pp. 818–19.

Brito, B., Palacios, J. -M., Hidalgo, E., Imperial, J. and Ruiz-Argueso, T. (1994) Nickel availability to Pea (*Pisum sativum*) plants limits hydrogenase activity *of Rhizobium leguminosarum* viciae bacteroids, by affecting the processing of the hydrogenase structural subunits. *J. Bacteriol.*, 176, 5297–303.

Brito, B., Martinez, M., Fernandez, D., Rey, L., Cabrera, E., Palacios, J. M., Imperial, J. and Ruiz-Argueso, T. (1997) Hydrogenase genes from *Rhizobium leguminosarum* bv. viciae are controlled by the nitrogen fixation regulatory protein nifA. *Proc. Natl. Acad. Sci. USA*, 94, 6019–24.

Brito, B., Monza, J., Imperial, J., Ruiz-Argueso, T. and Palacios, J. M. (2000) Nickel availability and hupSL activation by heterologous regulators limit symbiotic expression of the *Rhizobium leguminosarum* bv. viciae hydrogenase system in Hup(−) rhizobia. *Appl. Environ. Microbiol.*, 66, 937–42.

Bryant, F. O. and Adams, M. W. (1989) Characterization of hydrogenase from the hyperthermophilic archaebacterium, *Pyrococcus furiosus*. *J. Biol. Chem.*, 264, 5070–9.

Buhrke, T. and Friedrich, B. (1998) hoxX (hypX) is a functional member of the *Alcaligenes eutrophus* hyp gene cluster. *Arch. Microbiol.*, 170, 460–3.

Bult, C. J., White, O., Olsen, G. J., Zhou, L., Fleischmann, R. D., Sutton, G. G., Blake, J. A., FitzGerald, L. M., Clayton, R. A., Gocayne, J. D., Kerlavage, A. R., Dougherty, B. A., Tomb, J. F., Adams, M. D., Reich, C. I., Overbeek, R., Kirkness, E. F., Weinstock, K. G., Merrick, J. M., Glodek, A., Scott, J. L., Geoghagen, N. S. M. and Venter, J. C. (1996) Complete genome sequence of the methanogenic archaeon, *Methanococcus jannaschii*. *Science*, 273, 1058–73.

Burris, R. H. (1991) Nitrogenases. *J. Biol. Chem.*, 266, 9339–42.

Butt, J. N., Filipiak, M. and Hagen, W. R. (1997) Direct electrochemistry of *Megasphaera elsdenii* iron hydrogenase. Definition of the enzymes catalytic operating potential and quantitation of the catalytic behaviour over a continuous potential range. *Eur. J. Biochem.*, 245, 116–22.

Buurman, G., Shima, S., Thauer, R. K. (2000) The metal-free hydrogenase from methanogenic archaea: evidence for a bound cofactor. *FEBS Lett.* 485, 200–4

Cammack, R. (1995) Redox enzymes. Splitting molecular hydrogen. *Nature*, 373, 556–7.

Cammack, R. and van Vliet, P. (1999) Catalysis by Nickel in Biological Systems. In J. Reedijk and E. Bouwman (eds), *Bioinorganic Catalysis*, 2nd edn. New York: Marcel Dekker, pp. 231–68.

Cammack, R., Patil, D., Aguirre, R. and Hatchikian, E. C. (1982) Redox properties of the ESR-detectable nickel in hydrogenase from *Desulfovibrio gigas*. *FEBS Lett.*, 142, 289–92.

Cammack, R., Patil, D. S. and Fernandez, V. M. (1985) Electron-spin-resonance/electron-para-magnetic-resonance spectroscopy of iron–sulfur enzymes. *Biochem. Soc. Trans.*, 13, 572–8.

Cammack, R. P. D., Hatchikian, E. C. and Fernandez, V. M. (1987) Nickel and iron–sulfur centres in *Desulfovibrio gigas* hydrogenase: ESR spectra, redox properties and interactions. *Biochim. Biophys. Acta*, 912, 98–109.

Cammack, R., Fernandez, V. M. and Hatchikian, E. C. (1994) Nickel–iron hydrogenase. *Meth. Enzymol.*, 243, 43–68.

Carrasco, C. D. and Golden, J. W. (1995) Two heterocyst-specific DNA rearrangements of nif operons in *Anabaena cylindrica and Nostoc* sp. strain Mac. *Microbiology*, 141, 2479–87.

Carrasco, C. D., Buettner, J. A. and Golden, J. W. (1995) Programmed DNA rearrangement of a cyanobacterial hupL gene in heterocysts. *Proc. Natl. Acad. Sci. USA*, 92, 791–5.

Chaddock, A. M., Mant, A., Karnauchov, I., Brink, S., Herrmann, R. G., Klosgen, R. B. and Robinson, C. (1995) A new type of signal peptide: Central role of a twin-arginine motif in transfer signals for the delta pH-dependent thylakoidal protein translocase. *Embo. J.*, 14, 2715–22.

Chapman, A., Cammack, R., Hatchikian, E. C., McCracken, J. and Peisach, J. (1988) A pulsed ESR study of redox-dependent hyperfine interactions for the nickel centre of *Desulfovibrio gigas* hydrogenase. *FEBS Lett.*, 242, 134–8.

Chivers, P. T. and Sauer, R. T. (1999) NikR is a ribbon-helix-helix DNA-binding protein. *Protein Sci.*, 8, 2494–2500.

Ciloslowski, J., Boche, G. (1997) Geometry-tunable Lewis acidity of amidinium cations and its relevance to redox reactions of the Thauer metal-free hydrogenase – a theoretical study. *Angew. Chem. Int. Ed. Engl.* 36, 107–9.

Claasen, P. A. M., vanLier, J. B., Lopez Contreras, A. M., vanNiel, E. W. J., Sijtsma, L., Stams, A. J. M., deVries, S. S. and Weusthuis, R. A. (1999) Utilization of biomass for the supply of energy carriers. *Appl. Microbiol. Biotechnol.* 52, 741–55.

Cleghorn, S., Ren, X., Springer, T., Wilson, M., Zawodzinski, C., Zawodzinski, T. and Gottesfeld, S. (1997) PEM fuel cells for transportation and stationary power generation applications. *Int. J. Hydrogen Ener.*, 23, 1137–44.

Colbeau, A., Elsen, S., Tomiyama, M., Zorin, N. A., Dimon, B. and Vignais, P. M. (1998) *Rhodobacter capsulatus* HypF is involved in regulation of hydrogenase synthesis through the HupUV proteins. *Eur. J. Biochem.*, 251, 65–71.

Colpas, G. J., Maroney, M. J., Bagyinka, C., Kumar, M., Willis, W. S., Suib, S. L., Mascharak, P. K. and Baidya, N. (1991) X-ray spectroscopic studies of nickel complexes, with application to the structure of nickel sites in hydrogenases. *Inorg. Chem.*, 30, 920–8.

Colpas, G. J., Day, R. O. and Maroney, M. J. (1992) Synthesis and structure of a Ni$_3$Fe cluster featuring single and double thiolato bridges. *Inorg. Chem.*, 31, 5053–55.

Coremans, J. M. C. C., Van der Zwaan, J. W. and Albracht, S. P. J. (1989) Redox behaviour of nickel in hydrogenase from *Methanobacterium thermoautotrophicum* (strain Marburg). Correlation between the nickel valence state and enzyme activity. *Biochim. Biophys. Acta*, 997, 256–67.

Coremans, J. M., van Garderen, C. J. and Albracht, S. P. (1992) On the redox equilibrium between H$_2$ and hydrogenase. *Biochim. Biophys. Acta*, 1119, 148–56.

Creighton, A. M., Hulford, A., Mant, A., Robinson, D. and Robinson, C. (1995) A monomeric, tightly folded stromal intermediate on the delta pH-dependent thylakoidal protein transport pathway. *J. Biol. Chem.*, 270, 1663–9.

Darensbourg, D. J., Reibenspies, J. H., Lai, C.-H., Lee, W.-Z. and Darensbourg, M. Y. (1997) Analysis of an organometallic iron site model for the heterodimetallic unit of [NiFe] hydrogenase. *J. Am. Chem. Soc.*, 119, 7903–4.

Darensbourg, M. Y., Lyon, E. J. and Smee, J. J. (2000) The bio-organometallic chemistry of active site iron in hydrogenases. *Coord. Chem. Rev.*, 206, 533–61.

Davidson, G., Choudhury, S. B., Gu, Z., Bose, K., Roseboom, W., Albracht, S. P. and Maroney, M. J. (2000) Structural examination of the nickel site in *chromatium vinosum* hydrogenase: Redox state oscillations and structural changes accompanying reductive activation and CO binding. *Biochemistry*, 39, 7468–79.

Davies, S. C., Evans, D. J., Hughes, D. L., Longhurst, S. and Sanders, J. R. (1999) Synthesis and structure of a thiolate-bridged nickel-iron complex: Towards a mimic of the active site of NiFe-hydrogenase. *Chem. Commun.*, 1999, 1935–6.

De Gioia, L., Fantucci, P., Guigliarelli, B. and Bertrand, P. (1999a) Ni-Fe hydrogenases: A density functional theory study of active site models. *Inorg. Chem.*, 38, 2658–62.

De Gioia, L., Fantucci, P., Guigliarelli, B. and Bertrand, P. (1999b) Ab initio invstigation of the structural and electronic differences between active site models of [NiFe] and [NiFeSe] hydrogenases. *Int. J. Quant. Chem.*, 73, 187–95.

De Lacey, A. L., Hatchikian, E. C., Volbeda, A., Frey, M., Fontecilla-Camps, J. C. and Fernandez, V. M. (1997) Infrared-spectroelectrochemical characterization of the [NiFe] hydrogenase of *Desulfovibrio gigas*. *J. Am. Chem. Soc.*, 119, 7181–9.

De Lacey, A. L., Stadler, C., Cavazza, C., Hatchikian, E. C. and Fernandez, V. M. (2000) FTIR characterization of the active site of the Fe-hydrogenase from *Desulfovibrio desulfuricans*. *J. Amer. Chem. Soc.*, 122, 11232–3.

de Pina, K., Desjardin, V., Mandrand-Berthelot, M. A., Giordano, G. and Wu, L. F. (1999) Isolation and characterization of the nikR gene encoding a nickel-responsive regulator in *Escherichia coli*. *J. Bacteriol.*, 181, 670–4.

de Pina, K., Navarro, C., McWalter, L., Boxer, D. H., Price, N. C., Kelly, S. M., Mandrand-Berthelot, M. A. and Wu, L. F. (1995) Purification and characterization of the periplasmic nickel-binding protein NikA of *Escherichia coli* K12. *Eur. J. Biochem.*, 227, 857–65.

Deckert, G., Warren, P. V., Gaasterland, T., Young, W. G., Lenox, A. L., Graham, D. E., Overbeek, R., Snead, M. A., Keller, M., Aujay, M., Huber, R., Feldman, R. A., Short, J. M., Olsen, G. J. and Swanson, R. V. (1998) The complete genome of the hyperthermophilic *bacterium Aquifex aeolicus*. *Nature*, 392, 353–8.

Delachapelle, S., Renaud, M. and Vignais, P. M. (1991) Etude de la production dhydrogène en bioréacteur par une bactérie photosynthétique *Rhodobacter capsulatus*. 1. Photobioréacteur et conditions optimales de production dhydrogène. *Revue des Sciences de lEau*, 4, 83–99.

Deppenmeier, U. (1995) Different structure and expression of the operons encoding the membrane-bound hydrogenases from *Methanosarcina mazei* Go1. *Arch. Microbiol.*, 164, 370–6.

Deppenmeier, U., Blaut, M., Schmidt, B. and Gottschalk, G. (1992) Purification and properties of a F420-nonreactive, membrane-bound hydrogenase from *Methanosarcina* strain Go1. *Arch. Microbiol.*, 157, 505–11.

Dillon, A. C., Jones, K. M., Bekkedahl, T. A., Kiang, C. H., Bethune, D. S. and Heben, M. J. (1997) Storage of hydrogen in single-walled carbon nanotubes. *Nature*, 386, 377–9.

Dischert, W., Vignais, P. M. and Colbeau, A. (1999) The synthesis of *Rhodobacter capsulatus* HupSL hydrogenase is regulated by the two-component HupT/HupR system. *Mol. Microbiol.*, 34, 995–1006.

Doan, P. E., Fan, C. and Hoffman, B. M. (1994) Pulsed proton-deuterium ^{1,2}H ENDOR and ^{2}H-^{2}H TRIPLE resonance of H-bonds and cysteinyl .beta.-CH$_2$ of the *D. gigas* hydrogenase [3Fe-4S] + cluster. *J. Am. Chem. Soc.*, 116, 1033–41.

Dole, F., Fournel, A., Magro, V., Hatchikian, E. C., Bertrand, P. and Guigliarelli, B. (1997) Nature and electronic structure of the Ni-X dinuclear center of desulfovibrio gigas hydrogenase. Implications for the enzymatic mechanism. *Biochemistry*, 36, 7847–54.

Drapal, N. and Böck, A. (1998) Interaction of the hydrogenase accessory protein HypC with HycE, the large subunit of *Escherichia coli* hydrogenase 3 during enzyme maturation. *Biochemistry*, 37, 2941–8.

Dross, F., Geisler, V., Lenger, R., Theis, F., Krafft, T., Fahrenholz, F., Kojro, E., Duchene, A., Tripier, D., Juvenal, K. *et al.* (1992) The quinone-reactive Ni/Fe-hydrogenase of *Wolinella succinogenes* (published erratum appears in *Eur. J. Biochem*. 1993 Jun 15;214(3), 949–50), *Eur. J. Biochem.*, 206, 93–102.

Ducruix, A. and Giégé, R. (1992) *Crystallization of Nucleic Acids and Proteins. A Practical Approach.* Oxford: IRL at Oxford University Press.

Durmowicz, M. C. and Maier, R. J. (1998) The FixK2 protein is involved in regulation of symbiotic hydrogenase expression *in Bradyrhizobium japonicum*. *J. Bacteriol.*, 180, 3253–6.

Earhart, C. F. (1996) Uptake and metabolism of iron and molybdenum. In F. C. Neidhardt *et al.* (eds), *Escherichia coli and Salmonella. Cellular and Molecular Biology*, 2nd edn. Washington, DC: ASM Press, pp. 1075–90.

Eberz, G., Eitinger, T. and Friedrich, B. (1989) Genetic determinants of a nickel-specific transport system are part of the plasmid-encoded hydrogenase gene cluster in *Alcaligenes eutrophus*. *J. Bacteriol.*, 171, 1340–5.

Eidsness, M. K., Scott, R. A., Prickril, B. C., DerVartanian, D. V., Legall, J., Moura, I., Moura, J. J. and Peck, H. D. Jr (1989) Evidence for selenocysteine coordination to the active site nickel in the [NiFeSe] hydrogenases from *Desulfovibrio baculatus*. *Proc. Natl. Acad. Sci. USA*, 86, 147–51.

Eitinger, T. and Friedrich, B. (1991) Cloning, nucleotide sequence, and heterologous expression of a high-affinity nickel transport gene from *Alcaligenes eutrophus*. *J. Biol. Chem.*, 266, 3222–7.

Eitinger, T. and Mandrand-Berthelot, M. A. (2000) Nickel transport systems in microorganisms. *Arch. Microbiol.*, 173, 1–9.

Elsen, S., Colbeau, A., Chabert, J. and Vignais, P. M. (1996) The hupTUV operon is involved in negative control of hydrogenase synthesis in *Rhodobacter capsulatus*. *J. Bacteriol.*, 178, 5174–81.

Elsen, S., Dischert, W., Colbeau, A., Bauer C. E. (2000) Expression of uptake hydrogenase and molybdenum nitrogenase in *Rhodobacter capsulatus* is coregulated by the RegB-RegA two-component regulatory system. *J. Bacteriol.*, 182, 2831–7.

Embley, T. M. and Martin, W. (1998) A hydrogen-producing mitochondrion. *Nature*, 396, 517–9.

Erbes, D. L., Burris, R. H. and Orme-Johnson, W. H. (1975) On the iron–sulfur cluster in hydrogenase from *Clostridium pasteurianum* W5. *Proc. Natl. Acad. Sci. USA*, 72, 4795–9.

Evans, H. J., Harker, A. R., Papen, H., Russell, S. A., Hanus, F. J. and Zuber, M. (1987) Physiology, biochemistry, and genetics of the uptake hydrogenase in *rhizobia*. *Annu. Rev. Microbiol.*, 41, 335–61.

Evans, H. J., Russell, S. A., Hanus, F. J. and Ruiz-Argüeso, T. (1988) The importance of hydrogen recycling in nitrogen fixation by legumes. In R. J. Summerfield (ed.), World Crops: Cool Season Food Legumes. Boston: Kluwer Academic Publisher, pp. 777–91.

Ewart, G. D. and Smith, G. D. (1989) Purification and properties of soluble hydrogenase from the cyanobacterium *Anabaena cylindrical*. *Arch. Biochem. Biophys.*, 268, 327–37.

Ewart, G. D., Reed, K. C. and Smith, G. D. (1990) Soluble hydrogenase of *Anabaena cylindrica*. Cloning and sequencing of a potential gene encoding the tritium exchange subunit. *Eur. J. Biochem.*, 187, 215–23.

Fan, C. L., Teixeira, M., Moura, J., Moura, I., Huynh, B. H., Legall, J., Peck, H. D. and Hoffman, B. M. (1991) Detection and characterization of exchangeable protons bound to the hydrogen-activation nickel site of *Desulfovibrio gigas hydrogenase* - a H-1 and H-2 Q-band ENDOR study. *J. Amer. Chem. Soc.*, 113, 1–24

Fan, C., Teixeira, M., Moura, J., Moura, I., Huynh, B. -H., LeGall, J., Peck, H. D. Jr and Hausinger, R. P. (1993) *Biochemistry of Nickel*. New York: Plenum Press.

Farmer, P. J., Solouki, T., Mills, D. K., Soma, T., Russel, D. H., Reibenspies, J. H. and Darensbourg, M. Y. (1992) Isotopic labeling investigation of the oxygenation of nickel-bound thiolates by molecular-oxygen. *J. Am. Chem. Soc.*, 114, 4601–5.

Fay, P. (1992) Oxygen relations of nitrogen fixation in cyanobacteria. *Microbiol. Rev.*, 56, 340–73.

Fernandez, V. M., Aguirre, R. and Hatchikian, E. C. (1984) Reductive activation and redox properties of hydrogenase from *Desulfovibrio gigas*. *Biochim. Biophys. Acta*, 790, 1–7.

Fernandez, V., Hatchikian, E. and Cammack, R. (1985) Properties and Reactivation of two different deactivated forms of *Desulfovibrio gigas hydrogenase*. *Biochim. Biophys. Acta*, 832, 69–79.

Fiebig, K. and Friedrich, B. (1989) Purification of the F420-reducing hydrogenase from *Methanosarcina barkeri* (strain Fusaro). *Eur. J. Biochem.*, 184, 79–88.

Flores, E. and Herrero, A. (1994) Molecular evolution and taxonomy of the cyanobacteria. In D. A. Bryan (ed.), *The Molecular Biology of Cyanobacteria*. Dordrecht, The Netherlands: Kluwer Academic Publishers, pp. 487–517.

Florin, L., Tsokoglou, A. and Happe, T. (2001) A novel type of Fe-hydrogenase in the green alga *Scenedesmus obliquus* is linked to the photosynthetic electron transport chain. *J. Biol. Chem.*, (in press).

Fodor, B., Rakhely, G., Kovacs, A. T. and Kovacs, K. L. (2001) Transposon mutagenesis in purple sulfur photosynthetic bacteria: Identification of hypF, encoding a protein capable of processing [NiFe] hydrogenases in alpha, beta, and gamma subdivisions of the proteobacteria. *Applied and Environmental Microbiology*, 67, 2476–83.

Fontecilla-Camps, J. C. (1996) The active site of Ni-Fe hydrogenases: Model chemistry and crystallographic results. *J. Biol. Inorg. Chem.*, 1, 91–8.

Fox, S., Wang, Y., Silver, A. and Millar, M. (1991) Viability of the $[Ni^{III}(SR)_4]$-unit in classical coordination compounds and the nickel sulfur center of hydrogenases. *J. Am. Chem. Soc.*, 112, 3218–20.

Fox, J. D., He, Y., Shelver, D., Roberts, G. P. and Ludden, P. W. (1996) Characterization of the region encoding the CO-induced hydrogenase of *Rhodospirillum rubrum*. *J. Bacteriol.*, 178, 6200–8.

Franco, R., Moura, I., LeGall, J., Peck, H. D. Jr, Huynh, B. H. and Moura, J. J. (1993) Characterization of *D. desulfuricans* (ATCC 27774) [NiFe] hydrogenase EPR and redox properties of the native and the dihydrogen reacted states. *Biochim. Biophys. Acta*, 1144, 302–8.

Franolic, J. D., Wang, W. Y. and Millar, M. (1992) Synthesis, structure and characterization of a mixed-valence [Ni(II)Ni(III)] thiolate dimer. *J. Amer. Chem. Soc.*, 114, 6587–8.

Friedrich, B., Heine, E., Finck, A. and Friedrich, C. G. (1981) Nickel requirement for active hydrogenase formation in *Alcaligenes eutrophus*. *J. Bacteriol.*, 145, 1144–9.

Friedrich, B., Bernhard, M., Dernedde, J., Eitinger, T., Lenz, O., Massanz, M. and Schwartz, E. (1996) Hydrogen oxidation by *Alcaligenes*. In M. E. Lindström and F. R. Tabita, (eds), *Microbial growth on C1 compounds*. Dordrecht, The Netherlands: Kluwer Academic. pp. 110–7.

Fritsche, E., Paschos, A., Beisel, H. G., Böck, A. and Huber, R. (1999) Crystal structure of the hydrogenase maturating endopeptidase HYBD from *Escherichia coli*. *J. Mol. Biol.*, 288, 989–98.

Fu, C. and Maier, R. J. (1993) A genetic region downstream of the hydrogenase structural genes of *Bradyrhizobium japonicum* that is required for hydrogenase processing. *J. Bacteriol.*, 175, 295–8.

Fu, W., Drozdzewski, P. M., Morgan, T. V., Mortenson, L. E., Juszczak, A., Adams, M. W. W., He, S. H., Peck, H. D. Jr, DerVartanian, D. V. *et al.* (1993) Resonance Raman studies of iron-only hydrogenases. *Biochemistry*, 32, 4813–9.

Gaffron, H. and Rubin, J. (1942) Fermentative and photosynthetic production of hydrogen in algae. *J. Gen. Physiol.*, 26, 219–40.

Gallon, J. R. (1992) Reconciling the incompatible: N_2 fixation and O_2. *New Phytol.*, 122, 571–609.

Garcin, E., Vernède, X., Hatchikian, E. C., Volbeda, A., Frey, M. and Fontecilla-Camps, J. C. (1999) The crystal structure of a reduced [NiFeSe] hydrogenase provides an image of the activated catalytic center. *Structure Fold. Des.*, 7, 557–66.

Geierstanger, B. H., Prasch, T., Griesinger, C., Hartmann, G. C., Buurman, G. and Thauer, R. K. (1998) Catalytic mechanism of the metal-free hydrogenase from Methanogenic Archaea: Reversed stereospecificity of the catalytic and non-catalytic reaction. *Angew. Chem. Int.*, 37, 3300–3.

Gessner, C., Trofanchuk, O., Kawagoe, K., Higuchi, Y., Yasuoka, N. and Lubitz, W. (1996) Single crystal EPR study of the Ni center of NiFe hydrogenase. *Chem. Phys. Lett.*, 256, 518–24.

Gessner, C., Stein, M., Albracht, S. P. and Lubitz, W. (1999) Orientation-selected ENDOR of the active center in *Chromatium vinosum* [NiFe] hydrogenase in the oxidized 'ready' state. *J. Biol. Inorg. Chem.*, 4, 379–89.

Ghirardi, M. L., Togasaki, R. K. and Seibert, M. (1997) Oxygen sensitivity of algal hydrogen production. *Appl. Biochem. Biotechnol.*, 63–5, 141–51.

Ghirardi, M. L., Zhang, L., Lee, J. W., Flynn, T., Seibert, M., Greenbaum, E. and Melis, A. (2000) Microalgae: A green source of renewable H(2). *Trends Biotechnol.*, 18, 506–11.

Godbout, N., Salahub, D. R., Andzelm, J. and Wimmer, E. (1992) Optimization of Gaussian-type basis-sets for local spin-density functional calculations. 1. Boron through neon, optimization technique and validation. *Can. J. Chem. Rev.*, 70, 560–71.

Golden, J. W., Robinson, S. J. and Haselkorn, R. (1985) Rearrangement of nitrogen fixation genes during heterocyst differentiation in the cyanobacterium *Anabaena*. *Nature*, 314, 419–23.

Golden, J. W., Carrasco, C. D., Mulligan, M. E., Schneider, G. J. and Haselkorn, R. (1988) Deletion of a 55-kilobase-pair DNA element from the chromosome during heterocyst differentiation of *Anabaena sp.* strain PCC 7120. *Journal of Bacteriology*, 170, 5034–41.

Golden, J. W., Mulligan, M. E. and Haselkorn, R. (1987) Different recombination site specificity of two developmentally regulated genome rearrangements. *Nature*, 327, 526–9.

Gollin, D. J., Mortenson, L. E. and Robson, R. L. (1992) Carboxyl-terminal processing may be essential for production of active NiFe hydrogenase in *Azotobacter vinelandii*. *FEBS Lett.*, 309, 371–5.

Gorwa, M.-F., Croux, C. and Soucaille, P. (1996) Molecular characterization and transcriptional analysis of the putative hydrogenase gene of *Clostridium acetobutylicum* ATCC 824. *J. Bacteriol.*, 178, 2668–75.

Graf, E.-G. and Thauer, R. K. (1981) Hydrogenase from *Methanobacterium thermoautotrophicum*, a nickel-containing enzyme. *FEBS Lett.*, 36, 165–9.

Grätzel, M. (1982) Artificial photosynthesis, light-driven electron transfer processes in organized molecular assemblies and colloidal semiconductors. *Pure. Appl. Chem*, 54, 2369–82.

Gu, Z., Dong, J., Allan, C. B., Choudhury, S. B., Franco, R., Moura, J. J. G., Moura, I., LeGall, J., Przybyla, A. E., Roseboom, W., Albracht, S. P. J., Axley, M. J., Scott, R. A. and Maroney, M. J. (1996) Structure of the Ni sites in hydrogenases by X-ray absorption spectroscopy. Species variation and the effects of redox poise. *J. Am. Chem. Soc.*, 118, 11155–65.

Gutierrez, D., Hernando, Y., Palacios, J.-M., Imperial, J. and Ruiz-Argueso, T. (1997) FnrN controls symbiotic nitrogen fixation and hydrogenase activities in *Rhizobium leguminosarum* biovar viciae UPM791. *J. Bacteriol.*, 179, 5264–70.

Hagen, W. R., van Berkel-Arts, A., Kruse-Wolters, K. M., Dunham, W. R. and Veeger, C. (1986) EPR of a novel high-spin component in activated hydrogenase from *Desulfovibrio vulgaris* (Hildenborough). *FEBS Lett.*, 201, 158–62.

Halboth, S. (1991) Molekulargenetische Untersuchung der Hydrogenasen aus *Methanococcus voltae*. Ph.D., University of Marburg.

Halboth, S. and Klein, A. (1992) *Methanococcus voltae* harbors four gene clusters potentially encoding two [NiFe] and two [NiFeSe] hydrogenases, each of the cofactor F420-reducing or F420-non-reducing types. *Mol. Gen. Genet.*, 233, 217–24.

Halcrow, M. A. and Christou, G. (1994) Biomimetic chemistry of nickel. *Chem. Rev.*, 94, 2421–81.

Halcrow, M. A. (1995) The Structure of the *D. gigas* [NiFe] hydrogenase and the nature of the hydrogenase Nickel complex. *Angew. Chem. Int. Ed.*, 11, 1193–6.

Hall, D. O., Cammack, R. and Rao, K. K. (1971) A role for ferredoxins in the origin of life and biological evolution. *Nature*, 233, 136–8.

Hall, D.O and Rao, K. K. (1999) *Photosynthesis*, 6th ed. Cambridge University Press, Cambridge.

Hansel, A. and Lindblad, P. (1998) Towards optimization of cyanobacteria as biotechnologically relevant producers of molecular hydrogen, a clean and renewable energy source. *Appl. Microbiol. Biotechnol.*, 50, 153–60.

Happe, R. P., Roseboom, W., Pierik, A. J., Albracht, S. P. and Bagley, K. A. (1997) Biological activation of hydrogen. *Nature*, 385, 126.

Happe, R. P., Roseboom, W. and Albracht, S. P. (1999) Pre-steady-state kinetics of the reactions of [NiFe]-hydrogenase from *Chromatium vinosum* with H_2 and CO. *Eur. J. Biochem.*, 259, 602–8.

Hartmann, G. C., Klein, A. R., Linder, M. and Thauer, R. K. (1996) Purification, properties and primary structure of H_2-forming N5, N10-methylenetetrahydromethanopterin dehydrogenase from *Methanococcus thermolithotrophicus*. *Arch. Microbiol.*, 165, 187–93.

Haselkorn, R. and Buikema, W. J. (1992) Nitrogen fixation in cyanobacteria. In G. Stacey, R. H. Burris and H. J. Evans (eds), *Biological Nitrogen Fixation*. London, Great Britain: Chapman and Hall, pp. 166–90.

Hatchikian, E. C., Magro, V., Forget, N. and Nicolet, Y. (1999) Carboxy-terminal processing of the large subunit of [Fe] hydrogenase from *Desulfovibrio desulfuricans* ATCC 7757. *J. Bacteriol.*, 181, 2947–52.

Hausinger, R. P. (1993). *Biochemistry of Nickel*. New York, Plenum Press.

He, S. H., Woo, S. B., DerVartanian, D. V., Le Gall, J. and Peck, H. D. Jr (1989a) Effects of acetylene of hydrogenases from the sulfate reducing and methanogenic bacteria. *Biochem Biophys. Res. Commun.*, 161, 127–33.

He, S. H., Teixeira, M., LeGall, J., Patil, D. S., Moura, I., Moura, J. J., DerVartanian, D. V., Huynh, B. H. and Peck, H. D. Jr (1989b) EPR studies with O_2-enriched (NiFeSe) hydrogenase of *Desulfovibrio baculatus*. Evidence for a selenium ligand to the active site nickel. *J. Biol. Chem.*, 264, 2678–82.

Hembre, R. T., McQueen, J. S. and Day, V. W. (1996) Coupling H_2 to electron transfer with a 17-electron heterobimetallic hydride: A 'Redox Switch' model for the H_2-activating center of hydrogenase. *J. Am. Chem. Soc.*, 118, 798–803.

Henderson, R. K., Bouwman, E., Spek, A. L. and Reedijk, J. (1997) A unique mononuclear nickel disulfonato complex obtained by oxidation of a mononuclear nickel dithiolate complex. *Inorg. Chem.*, 36, 4616–17.

Herskovitz, T., Averill, B. A., Holm, R. H., Ibers, J. A., Phillips, W. D. and Weiher, J. F. (1972) Structure and prop erties of a synthetic analogue of bacterial iron–sulfur proteins. *Proc. Natl. Acad. Sci. USA*, 69, 2437–41.

Higuchi, Y., Yasuoka, N., Kakudo, M., Katsube, Y., Yagi, T. and Inokuchi, H. (1987) Single crystals of hydrogenase from *Desulfovibrio vulgaris* Miyazaki F. *J. Biol. Chem.*, 262, 2823–25.

Higuchi, Y., Yagi, T. and Yasuoka, N. (1997) Unusual ligand structure in Ni-Fe active center and an additional Mg site in hydrogenase revealed by high resolution X-ray structure analysis. *Structure (London)*, 5, 1671–80.

Higuchi, Y. and Yagi, T. (1999a) Liberation of hydrogen sulfide during the catalytic action of *Desulfovibrio* hydrogenase under the atmosphere of hydrogen. *Biochem. Biophys. Res. Commun.*, 255, 295–9.

Higuchi, Y., Ogata, H., Miki, K., Yasuoka, N. and Yagi, T. (1999b) Removal of the bridging ligand atom at the Ni-Fe active site of [NiFe] hydrogenase upon reduction with H_2, as revealed by X-ray structure analysis at 1.4 A resolution. *Structure Fold. Des.*, 7, 549–56.

Higuchi, Y., Toujou, F., Tsukamoto, K. and Yagi, T. (2000) The presence of a SO molecule in [NiFe] hydrogenase from *Desulfovibrio vulgaris* Miyazaki as detected by mass spectrometry (In Process Citation). *J. Inorg. Biochem.*, 80, 205–11.

Hoch, J. A. and Silhavy, T. J. (1995) *Two Component Signal Transduction*. American Society for Microbiology.

Holliger, C. and Schumacher, W. (1994) Reductive dehalogenation as a respiratory process. *Antonie Van Leeuwenhoek*, 66, 239–46.

Holm, R. H. (1977) Synthetic approaches to the active sites of iron–sulfur proteins. *Acc. Chem. Res.*, 10, 427–34.

Hopper, S., Babst, M., Schlensog, V., Fischer, H. M., Hennecke, H. and Böck, A. (1994) Regulated expression *in vitro* of genes coding for formate hydrogenlyase components of *Escherichia coli. J. Biol. Chem.*, 269, 19597–604.

Hornhardt, S., Schneider, K., Friedrich, B., Vogt, B. and Schlegel, H. G. (1990) Identification of distinct NAD-linked hydrogenase protein species in mutants and nickel-deficient wild-type cells of *Alcaligenes eutrophus* H16. *Eur. J. Biochem.*, 189, 529–37.

Houchins, J. P. and Burris, R. H. (1981b) Comparative characterization of two distinct hydro-genases from *Anabaena* sp. strain 7120. *J. Bacteriol.*, 146, 215–21.

Hsu, H. F., Koch, S. A., Popescu, C. V. and Munck, E. (1997) Chemistry of iron thiolate complexes with CN- and CO. Models for the [Fe(CO)(CN)(2)] structural unit in Ni-Fe hydro-genase enzymes. *J. Am. Chem. Soc.*, 119, 8371–2.

Huang, T. C., Tu, J., Chow, T. J. and Chen, T. H. (1990) Circadian rhythm of the prokaryote *Synechococcus* sp. RF-1. *Plant Physiol.*, 92, 531–3.

Hüsing, B. and Reiß, T. (1996) Perspectives and limitation of biological hydrogen production. In T. N. Veziroglu, C.-J. Winter, J. P. Baselt and G. Kreysa (eds), *Hydrogen Energy Progress XI: Proceedings of the 11th World Hydrogen Energy Conference*, three vols. Stuttgart, Germany: DECHEMA Deutsche Gesellschaft für Chemisches Apparatewesen, Chemische Technik und Biotechnologie on Behalf of International Association for Hydrogen Energy. Frankfurt DECHEMA, pp. 355–360.

Huyett, J. E., Carepo, M., Pamplona, A., Franco, R., Moura, I., Moura, J. J. G. and Hoffman, B. M. (1997) ^{57}Fe Q-band pulsed ENDOR of the hetero-dinuclear site of nickel hydrogenase: Comparison of the NiA, NiB, and NiC states. *J. Am. Chem. Soc.*, 119, 9291–2.

Huynh, B. H., Patil, D. S., Moura, I., Teixeira, M., Moura, J. J. G., DerVartanian, D. V., Czechowski, M. H., Prickril, B. C., Peck, H. D. and LeGall, J. (1987) On the active sites of the [NiFe] hydrogenase from *Desulfovibrio gigas*. Mössbauer and redox-titration studies. *J. Biol. Chem.* 262, 795–800.

Ide, T., Baumer, S. and Deppenmeier, U. (1999) Energy conservation by the H$_2$: heterodisulfide oxidoreductase from *Methanosarcina mazei* Go1: identification of two proton-translocating segments. *J. Bacteriol.*, 181, 4076–80.

Jacobson, F. S., Daniels, L., Fox, J. A., Walsh, C. T. and Orme-Johnson, W. H. (1982) Purification and properties of an 8-hydroxy-5-deazaflavin-reducing hydrogenase from *Methanobacterium thermoautotrophicum. J. Biol. Chem.*, 257, 3385–8.

James, T. L., Cai, L., Muetterties, M. C. and Holm, R. H. (1996) Dihydrogen evolution by pro-tonation reactions of nickel(I). *Inorg. Chem.*, 35, 4148–61.

Kaasjager, V. E., Henderson, R. K., Bouwman, E., Lutz, M., Spek, A. L. and Reedijk, J. (1998) A structural model for [Fe]-only hydrogenases. *Angew. Chem., Int. Ed.*, 37, 1668–70.

Kaneko, T., Sato, S., Kotani, H., Tanaka, A., Asamizu, E., Nakamura, Y., Miyajima, N., Hirosawa, M., Sugiura, M., Sasamoto, S., Kimura, T., Hosouchi, T., Matsuno, A., Muraki, A., Nakazaki, N., Naruo, K., Okumura, S., Shimpo, S., Takeuchi, C., Wada, T., Watanabe, A., Yamada, M., Yasuda, M. and Tabata, S. (1996) Sequence analysis of the genome of the uni-cellular cyanobacterium *Synechocystis* sp. strain PCC6803. II. Sequence determination of the entire genome and assignment of potential protein-coding regions (supplement). *DNA Res.*, 3, 185–209.

Keasling, J. D., Benemann, J., Pramanik, J., Carrier, T. A., Jones, K. L. and vanDien, J. (1998) A toolkit for metabolic engineering of bacteria. Application to hydrogen production.

In O. R. Zaborsky, J. R. Benemann, T. Matsunaga, J. Miyake and A. SanPietro (eds), *BioHydrogen*. New York: Plenum Press, pp. 87–97.

Kengen, S. W. M., Stams, A. J. M. and deVos, W. M. (1996) Sugar metabolism of hyperthermophiles. *FEMS Microbiol. Rev.*, 18, 119–37.

Kentemich, T., Danneberg, G., Hundeshagen, B. and Bothe, H. (1988) Evidence for the occurrence of the alternative, vanadium-containing nitrogenase in the cyanobacterium *Anabaena variabilis*. *FEMS Microbiol. Lett.*, 51, 19–24.

Kentemich, T., Bahnweg, M., Mayer, F. and Bothe, H. (1989) Localization of the reversible hydrogenase in cyanobacteria. *Z. Naturforsch.*, 44c, 384–91.

Kentemich, T., Haverkamp, G. and Bothe, H. (1991) The expression of a third nitrogenase in the cyanobacterium *Anabaena variabilis*. *Z. Naturforsch.*, 46c, 217–22.

Kevan, L. and Bowman, M. K. (1990) *Modern pulsed and continuous-wave EPR spectroscopy*. Wiley, New York.

Khaselev, O. and Turner, J. A. (1998) A monolithic photovoltaic-photoelectrochemical device for hydrogen production via water splitting, *Science*, 280, 425.

Kleihues, L., Lenz, O., Bernhard, M., Buhrke, T. and Friedrich, B. (2000) The H_2 sensor of *Ralstonia eutropha* is a member of the subclass of regulatory [NiFe] hydrogenases. *J. Bacteriol.*, 182, 2716–24.

Knowles, R. (1982) Denitrification. *Microbiol. Rev.*, 46. 43–70.

Koch, W. and Hertwig, R. H. (1998) In P. V. R. Schleyer, N. L. Allinger, T. Clark, J. Gasteiger, P. A. Kollmann, P. A. Schaefer III and P. R. Schreiner (eds), *The Encyclopedia of Computational Chemistry*. Chichester: John Wiley.

Köckerling, M. and Henkel, G. (1993) Mononuclear nickel thiolate complexes containing nickel sites in different oxidation states – molecular definition of $[Ni(SC_6H_4O)_2]^{2-}$ and $[Ni(SC_6H_4O)2]^{-}$, *Chem. Ber.*, 126, 951–3.

Köckerling, M. and Henkel, G. (2000) Synthesis and structure of $[Ni_4(S_2C_7H_{10})_4]$, the first tetranuclear cyclic nickel complex with bifunctional thiolate ligands and of the mononuclear precursor compound $Na_2[Ni(S_2C_7H_{10})_2]$. 4MeOH. (PrOH)-Pr-i, *Inorg. Chem. Commun.*, 3, 117–9.

Kondo, T., Strayer, C. A., Kulkarni, R. D., Taylor, W., Ishiura, M., Golden, S. S. and Johnson, C. H. (1993) Circadian rhythms in prokaryotes: Luciferase as a reporter of circadian gene expression in cyanobacteria. *Proc. Natl. Acad. Sci. USA*, 90, 5672–6.

Kortlüke, C., Horstmann, K., Schwartz, E., Rohde, M., Binsack, R. and Friedrich, B. (1992) A gene complex coding for the membrane-bound hydrogenase of *Alcaligenes eutrophus* H16. *J. Bacteriol.*, 174, 6277–89.

Kovács, K. L. and Polyák, B. (1991) Hydrogenase reactions and utilization of hydrogen in biogas production and microbiological denitrification systems. *Proc. 4th IGT Symp. on Gas, Oil, and Environmental Biotechnology*, Colorado Springs, Chapter 5, pp. 1–16.

Kovács, K. L., Bodrossy, L., Bagyinka, Cs., Perei, K. and Polyák B. (1996) On the microbial contribution to practical solutions in bioremediation. *OECD Documents on Wider Application and Diffusion of Bioremediation Technologies*. Amsterdam, The Netherlands. pp. 381–390.

Krebs, B. and Henkel, G. (1991) Transition metal thiolates – from molecular fragments of sulfidic solids to models for active centers in biomolecules. *Angew. Chem. Int.*, 30, 769–88.

Krüger, H.-J. and Holm, R. H. (1989) Chemical and electrochemical reactivity of nickel(II,I) thiolate complexes – examples of ligand-based oxidation and metal-centered oxidative addition. *Inorg. Chem.*, 28, 1148–55.

Krüger, H. J., Huynh, B. H., Ljungdahl, P. O., Xavier, A. V., Der Vartanian, D. V., Moura, I., Peck, H. D. Jr, Teixeira, M., Moura, J. J. and LeGall, J. (1982) Evidence for nickel and a three-iron center in the hydrogenase of *Desulfovibrio desulfuricans*. *J. Biol. Chem.*, 257, 14620–3.

Krüger, T., Krebs, B. and Henkel, G. (1989) Nickel complexes containing sterically demanding thiolate ligands: [Ni8S(SC4H9)9]-, a mixed-valence nickel sulfide thiolate cluster, and [Ni4(SC3H7)8], a homoleptic nickel thiolate. *Angew. Chem. Int. Ed.*, 28, 61–2.

Krüger, H.-J. and Holm, R. H. (1990) Stabilization of trivalent nickel in tetragonal NiS_4N_2 and NiN_6 environments: Synhesis, structures, redox potentials and observations related to [Ni-Fe]-Hydrogenases. *J. Am. Chem. Soc.*, 112, 2955–63.

Krüger, H.-J., Peng, G. and Holm, R. H. (1991) Low-potential nickel(III,II) complexes – new systems based on tetradentàte amidate thiolate ligands and the influence of ligand structure on potentials in relation to the nickel site in [NiFe]-hydrogenases. *Inorg. Chem.*, 30, 734–42.

Kuchner, O. and Arnold, F. H. (1997) Directed evolution of enzyme catalysts. *Trends Biotechnol.*, 15, 523–30.

Kumar, M., Colpas, G., Day, R. and Maroney, M. (1989) Ligand oxidation in a nickel thiolate complex – a model for the deactivation of hydrogenase by O_2. *J. Am. Chem. Soc.*, 111, 8323–5.

Künkel, A., Vorholt, J. A., Thauer, R. K. and Hedderich, R. (1998) An *Escherichia coli* hydrogenase-3-type hydrogenase in methanogenic archaea. *Eur. J. Biochem.*, 252, 467–76.

Kurreck, H., Kirste, B. and Lubitz, W. (1988) *Electron Nuclear Double Resonance Spectroscopy of Radicals in Solution – Application to Organic and Biological Chemistry*. Weinheim, Germany: VCH.

Kurt, M., Dunn, I. J. and Bourne, J. R. (1987) Biological denitrification of drinking water using autotrophic organisms with H_2 in a fluidized- bed biofilm reactor. *Biotechnol. Bioeng.* 29, 493–501.

Lai, C.-H., Reibenspies, J. H. and Darensbourg, M. Y. (1996) Thiolate bridged nickel–iron complexes containing both iron(0) and iron(II) carbonyls. *Angew. Chem. Int. Ed. Engl.*, 35, 2390–3.

Lai, C. H., Lee, W. Z., Miller, M. L., Reibenspies, J. H., Darensbourg, D. J. and Darensbourg, M. Y. (1998) Responses of the Fe(CN)(2)(CO) unit to electronic changes as related to its role in [NiFe] hydrogenase. *J. Am. Chem. Soc.*, 120, 10103–14.

Leclerc, M., Colbeau, A., Cauvin, B. and Vignais, P. M. (1988) Cloning and sequencing of the genes encoding the large and the small subunits of the H_2 uptake hydrogenase (hup) of *Rhodobacter capsulatus* (published erratum appears in *Mol. Gen. Genet.*, 1989 215, 368). *Mol. Gen. Genet.*, 214, 97–107.

Lemon, B. J. and Peters, J. W. (1999) Binding of exogenously added carbon monoxide at the active site of the iron-only hydrogenase (CpI) from *Clostridium pasteurianum. Biochemistry*, 38, 12969–73.

Lenz, O. and Friedrich, B. (1998) A novel multicomponent regulatory system mediates H_2 sensing in *Alcaligenes eutrophus. Proc. Natl. Acad. Sci. USA*, 95, 12474–9.

Lenz, O., Strack, A., Tran-Betcke, A. and Friedrich, B. (1997) A hydrogen-sensing system in transcriptional regulation of hydrogenase gene expression in *Alcaligenes species. J. Bacteriol.*, 179, 1655–63.

Li, C., Peck, H. D. Jr, LeGall, J. and Przybyla, A. E. (1987) Cloning, characterization, and sequencing of the genes encoding the large and small subunits of the periplasmic [NiFe] hydrogenase of *Desulfovibrio gigas. Dna*, 6, 539–51.

Lindahl, P. A., Kojima, N., Hausinger, R. P., Fox, J. A., Teo, B. K., Walsh, C. T. and Orme-Johnson, W. H. (1984) Nickel and iron EXAFS of F-420-reducing hydrogenase from *Methanobacterium thermoautotrophicum. J. Am. Chem. Soc.*, 106, 3062–4.

Lippard, S. J. and Berg, J. M. (1994) *Principles of Bioinorganic Chemistry*. University Science Books, Sausalito, California, USA.

Lissolo, T., Pulvin, S. and Thomas, D. (1984) Reactivation of the hydrogenase from *Desulfovibrio gigas* by hydrogen. Influence of redox potential. *J. Biol. Chem.*, 259, 11725–9.

Lloyd, J. R., Yong, P. and Macaskie, L. E. (1998) Enzymatic recovery of elemental palladium by using sulfate-reducing bacteria. *Applied and Environmental Microbiology*, 64, 4607–9.

López, M., Carbonero, V., Cabrera, E. and Ruiz-Argüeso, T. (1983) Effects of host on the expression of the H₂-uptake hydrogenase of *Rhizobium* in legume nodules. *Plant Sci. Lett.*, 29, 191–9.

Luecke, H., Schobert, B., Richter, H. T., Cartailler, J. P. and Lanyi, J. K. (1999) Structural changes in bacteriorhodopsin during ion transport at 2 angstrom resolution. *Science*, 286, 255–61.

Lundegardh, H. (1940). *Ann. Agr. Coll. Sweden*, 8, 233.

Lusk, P. (1998) Methane recovery from animal manures. *The current opportunities casebook.* NREL, Golden, Colorado, pp. i–7.3.

Lutz, S., Jacobi, A., Schlensog, V., Bohm, R., Sawers, G. and Böck, A. (1991) Molecular characterization of an operon (hyp) necessary for the activity of the three hydrogenase isoenzymes in *Escherichia coli. Mol. Microbiol.*, 5, 123–35.

Lyon, E. J., Georgakaki, I. P., Reibenspies, J. H. and Darensbourg, M. Y. (1999) Carbon monoxide and cyanide ligands in a classical organometallic complex model for Fe-only hydrogenase. *Angew. Chem. Int. Ed. Engl.*, 38, 3178–80.

Ma, K., Schicho, R. N., Kelly, R. M. and Adams, M. W. W. (1993) Hydrogenase of the hyperthermophile Pyrococcus furiosus is an elemental sulfur reductase or sulfhydrogenase: Evidence for a sulfur-reducing hydrogenase ancestor. *Proc. Natl. Acad. Sci. USA*, 90, 5341–4.

McCarty, S., Chasteen, T., Marshall, M. *et al.* (1993) Phototrophic bacteria produce volatile, methylated sulfur and selenium compounds. *FEMS Lett.*, 112, 93–8.

Maier, T. and Böck, A. (1996) Generation of active [NiFe] hydrogenase *in vitro* from a nickel-free precursor form. *Biochemistry*, 35, 10089–93.

Maier, R. J. and Triplett, E. W. (1996) Toward more productive, efficient, and competitive nitrogen-fixing symbiotic bacteria. *CRC Crit. Rev. Plant Sci.*, 15, 191–234.

Maier, T., Jacobi, A., Sauter, M. and Böck, A. (1993) The product of the hypB gene, which is required for nickel incorporation into hydrogenases, is a novel guanine nucleotide-binding protein. *J. Bacteriol.*, 175, 630–5.

Maier, T., Lottspeich, F. and Böck, A. (1995) GTP hydrolysis by HypB is essential for nickel insertion into hydrogenases of *Escherichia coli. Eur. J. Biochem.*, 230, 133–8.

Maier, T., Binder, U. and Böck, A. (1996) Analysis of the hydA locus of *Escherichia coli*: Two genes (hydN and hypF) involved in formate and hydrogen metabolism. *Arch. Microbiol.*, 165, 333–41.

Marganian, C. A., Vazir, H., Baidya, N., Olmstead, M. M. and Mascharak, P. K. (1995) Toward functional models of the nickel sites in [FeNi] and [FeNiSe] hydrogenases: Syntheses, structures, and reactivities of nickel(II) complexes containing [NiN₃S₂] and [NiN₃Se₂] chromophores. *J. Am. Chem. Soc.*, 117, 1584–94.

Maroney, M. J. (1999) Structure/function relationships in nickel metallobiochemistry. *Curr. Opin. Chem. Biol.*, 3, 188–99.

Martins dos Santos, V. A. P., Vasilevska, T., Kajuk, B., Tramper, J., Wijffels, R. H. (1997) Production and characterization of double-layer beads for coimmobilization of microbial cells. *Biotechnology Annu. Rev.*, 3, 227–44.

Massanz, C. and Friedrich, B. (1999) Amino acid replacements at the H(2)-activating site of the NAD-reducing hydrogenase from *Alcaligenes eutrophus. Biochemistry*, 38, 14330–7.

Massanz, C., Fernandez, V. M. and Friedrich, B. (1997) C-terminal extension of the H₂-activating subunit, HoxH, directs maturation of the NAD-reducing hydrogenase in *Alcaligenes eutrophus. Eur. J. Biochem.*, 245, 441–8.

Massanz, C., Schmidt, S. and Friedrich, B. (1998) Subforms and *in vitro* reconstitution of the NAD-reducing hydrogenase of *Alcaligenes eutrophus. J. Bacteriol.*, 180, 1023–9.

Matias, P. M., Soares, C. M., Saraiva, L. M., Coelho, R., Morais, J., LeGall, J. and Carrondo, M. A. (2001) [NiFe] hydrogenase from *Desulfovibrio desulfuricans* ATCC 27774: gene sequencing, three-dimensional structure determination and refinement at 1.8 angstrom and modelling studies of its interaction with the tetrahaem cytochrome c(3). *J. Biol. Inorg. Chem.*, 6, 63–81.

Matsunaga, Y., Xu, X. and Antal, M. J. (1997) Gasification characteristics of an activated carbon in supercritical water, *Carbon*, 35, 819–24.

Matveyev, A. V., Rutgers, E., Söderbäck, E. and Bergman, B. (1994) A novel rearrangement involved in heterocyst differentiation of the cyanobacterium *Anabaena* sp. PCC7120. *FEMS Microbiol. Lett.*, 116, 201–8.

Medina, M., Hatchikian, E. C. and Cammack, R. (1996) Studies of light-induced nickel EPR signals in hydrogenase: Comparison of enzymes with and without selenium. *Biochim. Biophys. Acta*, 1275, 227–36.

Mege, R. M. and Bourdillon, C. (1985) Nickel controls the reversible anaerobic activation/inactivation of the *Desulfovibrio gigas* hydrogenase by the redox potential. *J. Biol. Chem.*, 260, 14701–6.

Melis, A., Zhang, L., Forestier, M., Ghirardi, M. L. and Seibert, M. (2000) Sustained photo-biological hydrogen gas production upon reversible inactivation of oxygen evolution in the green alga Chlamydomonas reinhardtii. *Plant Physiol.*, 122, 127–36.

Menon, A. L., Stults, L. W., Robson, R. L. and Mortenson, L. E. (1990a) Cloning, sequencing and characterization of the [NiFe] hydrogenase-encoding structural genes (hoxK and hoxG) from *Azotobacter vinelandii*. *Gene*, 96, 67–74.

Menon, N. K., Robbins, J., Peck, H. D. Jr, Chatelus, C. Y., Choi, E. S. and Przybyla, A. E. (1990b) Cloning and sequencing of a putative *Escherichia coli* [NiFe] hydrogenase-1 operon containing six open reading frames. *J. Bacteriol.*, 172, 1969–77.

Menon, N. K., Robbins, J., Wendt, J. C., Shanmugam, K. T. and Przybyla, A. E. (1991) Mutational analysis and characterization of the *Escherichia coli* hya operon, which encodes [NiFe] hydrogenase 1. *J. Bacteriol.*, 173, 4851–61.

Menon, N. K., Robbins, J., Der Vartanian, M., Patil, D., Peck, H. D. J., Menon, A. L., Robson, R. L. and Przybyla, A. E. (1993) Carboxy-terminal processing of the large subunit of [NiFe] hydrogenases. *FEBS Lett.*, 331, 91–5.

Menon, A. L. and Robson, R. L. (1994a) In vivo and in vitro nickel-dependent processing of the [NiFe] hydrogenase in *Azotobacter vinelandii*. *J. Bacteriol.*, 176, 291–5.

Menon, N. K., Chatelus, C. Y., Dervartanian, M., Wendt, J. C., Shanmugam, K. T., Peck, H. D. Jr and Przybyla, A. E. (1994b) Cloning, sequencing, and mutational analysis of the hyb operon encoding *Escherichia coli* hydrogenase 2. *J. Bacteriol.*, 176, 4416–23.

Meuer, J., Bartoschek, S., Koch, J., Kunkel A. and Hedderich, R. (1999) Purification and catalytic properties of Ech hydrogenase from *Methanosarcina barkeri*. *Eur. J. Biochem.*, 265, 325–35.

Meyer, J. and Gagnon, J. (1991) Primary structure of hydrogenase I from *Clostridium pasteurianum*. *Biochemistry*, 30, 9697–704.

Michel, R., Massanz, C., Kostka, S., Richter, M. and Fiebig, K. (1995) Biochemical characterization of the 8-hydroxy-5-deazaflavin-reactive hydrogenase from *Methanosarcina barkeri* Fusaro. *Eur. J. Biochem.*, 233, 727–35.

Miller, R. S. and Bellan, J. (1997) A generalized biomass pyrolysis model based on superimposed cellulose, hemicellulose and lignin kinetics, *Comb. Sci. and Techn.*, 126, 97–137.

Mills, D. K., Hsiao, Y. M., Farmer, P. J., Atnip, E. V., Reibenspies, J. H. and Darensbourg, M. Y. (1991) Applications of the N_2S_2 ligand, N,N-bis(mercaptoethyl)-1,5-diazacyclooctane, towards the formation of bimetallics and heterometallics – [(BME-DACO)Fe]$_2$ and [(BME-DACO)NiFeCl$_2$]$_2$. *J. Am. Chem. Soc.*, 113, 1421–3.

Mitchell, P. (1961). Coupling of phosphorylation to electron and hydrogen transfer by a chemiosmotic type of mechanism. *Nature*, 191, 144–48.

Mitlitsky, F., Myers, B. and Weisberg, A. H. (1998) Regenerative fuel cell systems. *Energy & Fuels*, 12, 56–71.

Miyake, J., Matsunaga, T. and San Pietro, A. (ed.) (2000) *Biohydrogen II*. Amsterdam: Elsevier.

Montet, Y., Amara, P., Volbeda, A., Vernede, X., Hatchikian, E. C., Field, M. J., Frey, M. and Fontecilla-Camps, J. C. (1997) Gas access to the active site of Ni-Fe hydrogenases probed by X-ray crystallography and molecular dynamics. *Nat. Struct. Biol.*, 4, 523–6.

Moore, M. D. and Kaplan, S. (1992) Identification of intrinsic high-level resistance to rare-earth oxides and oxyanions in members of the class Proteobacteria: Characterization of tellurite, selenite, and rhodium sesquioxide reduction in *Rhodobacter sphaeroides. J. Bacteriol.* 174, 1505–14.

Moura, J. J. G., Teixeira, M., Moura, I. and LeGall, J. (1988) (Ni,Fe)-hydrogenases from sulfate-reducing bacteria: Nickel catalytic and regulatory roles. In J. R. Lancaster (ed.), *Bioinorganic Chemistry of Nickel*. Deerfield Beach Florida: 91226, VCH Publishers.

Müller, A. and Henkel, G. (1995) $[Ni_2(SC_4H_9)_6]^{2-}$, a novel binuclear nickel-thiolate complex with NiS4 tetrahedra sharing edges and $[Ni(SC_6H_4SiMe_3)_4]^{2-}$, a structurally related mononuclear complex ion. *Z. Naturforsch.*, B50, 1464–8.

Müller, A. and Henkel, G. (1996) $[Ni_5S(Sbut)_5]^-$ and $[Ni_3S(Sbut)_3(CN)_3]^{2-}$, novel complexes with sulfur-capped, thiolate-bridged polygonal metal frames. *Chem. Commun.*, 1996, 1005–6.

Müller, A., Erkens, A., Schneider, K., Nolting, H. F., Sole, V. A. and Henkel, G. (1997a) NADH-induced changes of the nickel coordination within the active site of the soluble hydrogenase from *Alcaligenes eutrophus*: XAFS investigations on three states distinguishable by EPR spectroscopy. *Angew. Chem. Int. Ed.*, 36, 1747–50.

Müller, A., Kockerling, M. and Henkel, G. (1997b) X-ray absorption spectroscopic studies on the nickel center of the soluble Ni-Fe hydrogenase from *Alcaligenes eutrophus*: First evidence for ligand exchange at the active site within the catalytic cycle. In A. X. Trautwein (ed.), *Bioinorg. Chem.* Weinheim, Germany: Wiley-VCH Verlag GmbH pp. 234–43.

Muth, E., Morschel, E. and Klein, A. (1987) Purification and characterization of an 8-hydroxy-5-deazaflavin- reducing hydrogenase from the archaebacterium *Methanococcus voltae. Eur. J. Biochem.*, 169, 571–7.

Nakamura, M., Saeki, K., Takahashi, Y. (1999) Hyperproduction of recombinant ferredoxins in *Escherichia coli* by coexpression of the ORF1-ORF2-*iscS-iscU-iscA-hscB-hscA-fdx*-ORF3 gene cluster. *J. Biochem* 126, 10–18.

Navarro, C., Wu, L. F. and Mandrand-Berthelot, M. A. (1993) The nik operon of *Escherichia coli* encodes a periplasmic binding-protein-dependent transport system for nickel. *Mol. Microbiol.*, 9, 1181–91.

Neilands, J. B. (1981) Microbial iron compounds. *Annu. Rev. Biochem.*, 50, 715–31.

Nicholls, D. G. and Ferguson, S. J. (1992) *Bioenergetics 2*. London: Academic Press.

Nicolet, Y., Piras, C., Legrand, P., Hatchikian, C. E. and Fontecilla-Camps, J. C. (1999) *Desulfovibrio desulfuricans* iron hydrogenase: The structure shows unusual coordination to an active site Fe binuclear center. *Structure Fold. Des.*, 7, 13–23.

Nicole, Y., Lemon, B. J., Fontecilla-Camps, J. C. and Peters, J. W. (2000) A novel FeS cluster in Fe-only hydrogenases, *Trends Biochem Sci*, 25, 138–43.

Nikhandrov, V. V., Shlyk, M. A., Zorin, N. A., Gogotov, I. N. and Krasnovsky, A. A. (1988) Efficient photoinduced electron transfer from inorganic semiconduction TiO_2 to bacterial hydrogenase. *FEBS Lett.*, 234, 111–14.

Niu, S., Thomson, L. M., Hall, M. B., Brecht, M., Stein, M., Trofanchuk, O., Lendzian, F., Bittl, R., Higuchi, Y. and Lubitz, W. (1998). In D. Ziessow, F. Lendzian and W. Lubitz (eds), *Magnetic Resonance and Related Phenomena*, Vol. II. Berlin: TU Berlin, pp. 818–9.

Niu, S. T., Thomson, L. M. and Hall, M. B. (1999) Theoretical characterization of the reaction intermediates in a model of the nickel–iron hydrogenase of *Desulfovibrio gigas. J. Am. Chem. Soc.*, 121, 4000–7.

Nivière, V., Hatchikian, C., Cambillau, C. and Frey, M. (1987) Crystallization, preliminary X-ray study and crystal activity of the hydrogenase from *Desulfovibrio gigas. J. Mol. Biol.*, 195, 969–71.

Nivière, V., Wong, S. L. and Voordouw, G. (1992) Site-directed mutagenesis of the hydro-genase signal peptide consensus box prevents export of a beta-lactamase fusion protein. *J. Gen. Microbiol.*, **138**, 2173–83.

Odom, J. M. and Peck, H. D. (1984) Hydrogenase, electron-transfer proteins, and energy coupling in the sulfate reducing bacteria Desulfovibrio. *Ann. Rev. Microbiol.*, **38**, 551.

Ohtsuki, T., Shimosaka, M. and Okazaki, M. (1997) Expression of genes encoding membrane-bound hydrogenase in *Pseudomonas hydrogenovorans* under autotrophic condition is dependent on two different promoters. *Biosci. Biotechnol. Biochem.*, **61**, 1986–90.

Olson, J. W., Fu, C. and Maier, R. J. (1997) The HypB protein from *Bradyrhizobium japonicum* can store nickel and is required for the nickel-dependent transcriptional regulation of hydrogenase. *Mol. Microbiol.*, **24**, 119–28.

Osterloh, F., Saak, W., Haase, D. and Pohl, S. (1997) Synthesis, X-ray structure and electro-chemical characterization of a binuclear thiolate bridged Ni-Fe-nitrosyl complex, related to the active site of NiFe hydrogenase. *Chem. Commun. (Cambridge)*, 1997, 979–80.

Oxelfelt, F., Tamagnini, P. and Lindblad, P. (1998) Hydrogen uptake in *Nostoc* sp. strain PCC 73102. Cloning and characterization of a hupSL homolog. *Arch. Microbiol.*, **169**, 267–74.

Page, C. C., Moser, C. C., Chen, X. and Dutton, P. L. (1999) Natural engineering principles of electron tunnelling in biological oxidation–reduction. *Nature*, **402**, 47–52.

Page, R. D. M. (1996) TreeView: An application to display phylogenetic trees on personal com-puters. *Comput. Appl. Biol. Sci.*, **12**, 357–58.

Parr, R. G. and Young, W. (1989) *Density-Functional Theory of Atoms and Molecule*. New York: Oxford University Press.

Paschos, A, Class, R. S. and Bock, A. (2001) Carbamoylphosphate requirement for synthesis of the active center of [NiFe]-hydrogenases. *FEBS Lett.*, **488**, 9–12.

Patil, D. S., He, S. H., DerVartanian, D. V., Le Gall, J., Huynh, B. H. and Peck, H. D. Jr (1988) The relationship between activity and the axial g = 2.06 EPR signal induced by CO in the periplasmic (Fe) hydrogenase from *Desulfovibrio vulgaris*. *FEBS Lett.*, **228**, 85–8.

Pavlov, M., Siegbahn, P. E. M., Blomberg, M. R. A. and Crabtree, R. H. (1998) Mechanism of H-H activation by nickel–iron hydrogenase. *J. Am. Chem. Soc.*, **120**, 548–55.

Pavlov, M. B., Blomberg, M. R. A. and Siegbahn, P. E. M. (1999) New aspects of H_2 activa-tion by nickel–iron hydrogenase. *Int. J. Quant. Chem.*, **73**, 197–207.

Pedroni, P., Della Volpe, A., Galli, G., Mura, G. M., Pratesi, C. and Grandi, G. (1995) Characterization of the locus encoding the [Ni-Fe] sulfhydrogenase from the archaeon Pyrococcus furiosus: Evidence for a relationship to bacterial sulfite reductases. *Microbiology*, **141**, 449–58.

Perei, K., Polyák, B., Bagyinka, Cs., Bodrossy, L. and Kovács, K. L. (1995) Selected applica-tions of bioremediation in hazardous waste treatment. *NATO ASI Series, 2. Environment - Vol. 1. 'Clean-up of Former Soviet Military Installations'* (Eds. Herndon, R. C., Richter, P. I. Moerlins, J. E., Kuperberg, J. M., Biczo, I. L., Springer-Verlag, Berlin, Heidelberg. pp. 87–95.

Pershad, H. R., Duff, J. L., Heering, H. A., Duin, E. C., Albracht, S. P. and Armstrong, F. A. (1999) Catalytic electron transport in *Chromatium vinosum* [NiFe]-hydrogenase: Application of voltammetry in detecting redox-active centers and establishing that hydrogen oxidation is very fast even at potentials close to the reversible $H+/H_2$ value. *Biochemistry*, **38**, 8992–9.

Peschek, G. A. (1979a) Evidence for two functionally distinct hydrogenases in *Anacystis nidulans*. *Arch. Microbiol.*, **123**, 81–92.

Peschek, G. A. (1979b) Aerobic hydrogenase activity in *Anacystis nidulans*. The oxyhydrogen reaction. *Biochim. Biophys. Acta*, **548**, 203–15.

Peters, J. W., Lanzilotta, W. N., Lemon, B. J. and Seefeldt, L. C. (1998) X-ray crystal structure of the Fe-only hydrogenase (CpI) from *Clostridium pasteurianum* to 1.8 angstrom resolution (published erratum appears in *Science* 1999 Jan 1; 283(5398), 35], *Science*, **282**, 1853–8.

Pfeiffer, M., Klein, A., Steinert, P. and Schomburg, D. (1996) An all sulfur analog of the small-est subunit of F420-non-reducing hydrogenase from *Methanococcus voltae* – metal binding and structure. *BioFactors*, 5, 157–68.

Pfeiffer, M., Bingemann, R. and Klein, A. (1998) Fusion of two subunits does not impair the function of a [NiFeSe]-hydrogenase in the archaeon *Methanococcus voltae*. *Eur. J. Biochem.*, 256, 447–52.

Pierik, A. J., Hagen, W. R., Redeker, J. S., Wolbert, R. B., Boersma, M., Verhagen, M. F., Grande, H. J., Veeger, C., Mutsaers, P. H., Sands, R. H. *et al.* (1992) Redox properties of the iron–sulfur clusters in activated Fe-hydrogenase from *Desulfovibrio vulgaris* (Hildenborough). *Eur. J. Biochem.*, 209, 63–72.

Pierik, A. J., Hulstein, M., Hagen, W. R. and Albracht, S. P. (1998a) A low-spin iron with CN and CO as intrinsic ligands forms the core of the active site in [Fe]-hydrogenases. *Eur. J. Biochem.*, 258, 572–8.

Pierik, A. J., Schmelz, M., Lenz, O., Friedrich, B. and Albracht, S. P. (1998b) Characterization of the active site of a hydrogen sensor from *Alcaligenes eutrophus*. *FEBS Lett.*, 438, 231–5.

Pierik, A. J., Roseboom, W., Happe, R. P., Bagley, K. A. and Albracht, S. P. (1999) Carbon monoxide and cyanide as intrinsic ligands to iron in the active site of [NiFe]-hydrogenases. NiFe(CN)$_2$CO, Biology's way to activate H$_2$. *J. Biol. Chem.*, 274, 3331–7.

Postgate, J. (1987) *Nitrogen Fixation*, 2nd edn. *New Studies in Biology*, London, Great Britain: Edward Arnold.

Puustinen, A. and Wikstrom, M. (1999) Proton exit from the heme-copper oxidase of *Escherichia coli*. *Proc. Natl. Acad. Sci. USA*, 96, 35–7.

Rakhely, G., Zhou, Z. H., Adams, M. W. and Kovacs, K. L. (1999) Biochemical and molecu-lar characterization of the [NiFe] hydrogenase from the hyperthermophilic archaeon, *Thermococcus litoralis*. *Eur. J. Biochem.*, 266, 1158–65.

Rao, K. K. and Hall, D. O. (1996) Hydrogen production by cyanobacteria: Potential, problems and prospects. *J. Mar. Biotechnol.*, 4, 10–5.

Reeve, J. N. and Beckler, G. S. (1990) Conservation of primary structure in prokaryotic hydro-genases. *FEMS Microbiol. Rev.*, 7, 419–24.

Rey, L., Imperial, J., Palacios, J.-M. and Ruiz-Argueso, T. (1994) Purification of *Rhizobium leguminosarum* HypB, a nickel-binding protein required for hydrogenase synthesis. *J. Bacteriol.*, 176, 6066–73.

Rey, L., Fernandez, D., Brito, B., Hernando, Y., Palacios, J.-M., Imperial, J. and Ruiz-Argueso, T. (1996) The hydrogenase gene cluster of *Rhizobium leguminosarum* bv. viciae contains an additional gene (hypX), which encodes a protein with sequence similarity to the N10-formyl-tetrahydrofolate-dependent enzyme family and is required for nickel-dependent hydrogenase processing and activity. *Mol. Gen. Genet.*, 252, 237–48.

Richard, D. J., Sawers, G., Sargent, F., McWalter, L. and Boxer, D. H. (1999) Transcriptional regulation in response to oxygen and nitrate of the operons encoding the [NiFe] hydro-genases 1 and 2 of *Escherichia coli*. *Microbiology*, 145, 2903–12.

Rosentel, J. K., Healy, F., Maupin-Furlow, J. A., Lee, J. H. and Shanmugam, K. T. (1995) Molybdate and regulation of mod (molybdate transport), fdhF, and hyc (formate hydro-genlyase) operons in *Escherichia coli*. *J. Bacteriol.*, 177, 4857–64.

Rossmann, R., Sawers, G. and Böck, A. (1991) Mechanism of regulation of the formate-hydro-genlyase pathway by oxygen, nitrate, and pH: Definition of the formate regulon. *Mol. Microbiol.*, 5, 2807–14.

Rossmann, R., Maier, T., Lottspeich, F. and Böck, A. (1995) Characterization of a protease from *Escherichia coli* involved in hydrogenase maturation, *Eur. J. Biochem.*, 227, 545–50.

Rousset, M., Montet, Y., Guigliarelli, B., Forget, N., Asso, M., Bertrand, P., Fontecilla-Camps, J. C. and Hatchikian, E. C. (1998) [3Fe-4S] to [4Fe-4S] cluster conversion in desulfovibrio fructosovorans [NiFe] hydrogenase by site-directed mutagenesis. *Proc. Natl. Acad. Sci. USA*, 95, 11625–30.

Sanchez, J. M., Arijo, S., Munoz, M. A., Morinigo, M. A. and Borrego, J. J. (1994) Microbial colonization of different support materials used to enhance the methanogenic process. *Appl. Microbiol. Biotechnol.*, 41, 480–6.

Sankar, P. and Shanmugam, K. T. (1988a) Biochemical and genetic analysis of hydrogen metabolism in *Escherichia coli*: The hydB gene, *J. Bacteriol.*, 170, 5433–9.

Sankar, P. and Shanmugam, K. T. (1988b) Hydrogen metabolism in *Escherichia coli*: Biochemical and genetic evidence for a hydF gene. *J. Bacteriol.*, 170, 5446–51.

Sankar, P., Lee, J. H. and Shanmugam, K. T. (1985) Cloning of hydrogenase genes and fine structure analysis of an operon essential for H2 metabolism in *Escherichia coli*. *J. Bacteriol.*, 162, 353–60.

Santini, C. L., Ize, B., Chanal, A., Muller, M., Giordano, G. and Wu, L. F. (1998) A novel sec-independent periplasmic protein translocation pathway in *Escherichia coli*. *Embo J.*, 17, 101–12.

Sargent, F., Bogsch, E. G., Stanley, N. R., Wexler, M., Robinson, C., Berks, B. C. and Palmer, T. (1998) Overlapping functions of components of a bacterial Sec-independent protein export pathway. *Embo J.*, 17, 3640–50.

Sasaki, S. and Karube, I. (1999) The development of microfabricated biocatalytic fuel cells. *Trends Biotechnol.*, 17, 50–2.

Sasikala, K., Ramana, C. V., Rao, P. R. and Kovacs, K. L. (1993) Anoxygenic phototrophic bacteria: Physiology and advances in hydrogen technology. *Adv. Appl. Microbiol.*, 38, 211–95.

Sauter, M., Bohm, R. and Böck, A. (1992) Mutational analysis of the operon (hyc) determining hydrogenase 3 formation in *Escherichia coli*. *Mol. Microbiol.*, 6, 1523–32.

Sawers, R. G., Ballantine, S. P. and Boxer, D. H. (1985) Differential expression of hydrogenase isoenzymes in *Escherichia coli* K-12: Evidence for a third isoenzyme. *J. Bacteriol.*, 164, 1324–31.

Sawers, R. G., Jamieson, D. J., Higgins, C. F. and Boxer, D. H. (1986) Characterization and physiological roles of membrane-bound hydrogenase isoenzymes from *Salmonella typhimurium*. *J. Bacteriol.*, 168, 398–404.

Sayavedra-Soto, L. A., Powell, G. K., Evans, H. J. and Morris, R. O. (1988) Nucleotide sequence of the genetic loci encoding subunits of *Bradyrhizobium japonicum* uptake hydrogenase. *Proc. Natl. Acad. Sci. USA*, 85, 8395–9.

Schink, B. and Schlegel, H. G. (1979) The membrane-bound hydrogenase of *Alcaligenes eutrophus*. I. Solubilization, purification, and biochemical properties, *Biochim. Biophys. Acta*, 567, 315–24.

Schmitz, O., Boison, G., Hilscher, R., Hundeshagen, B., Zimmer, W., Lottspeich, F. and Bothe, H. (1995) Molecular biological analysis of a bidirectional hydrogenase from cyanobacteria. *Eur. J. Biochem.*, 233, 266–76.

Schmidt, M., Contakes, S. M. and Rauchfuss, T. B. (1999) First generation analogues of the binuclear site in the Fe-only hydrogenases: $Fe_2(\mu\text{-}SR)_2(CO)_4(CN)_2^{2-}$. *J. Am. Chem. Soc.*, 121, 9736–7.

Schneider, K. and Schlegel, H. G. (1976) Purification and properties of soluble hydrogenase from Alcaligenes eutrophus H 16. *Biochim. Biophys. Acta*, 452, 66–80.

Schneider, K., Cammack, R. and Schlegel, H. G. (1984) Content and localization of FMN, Fe-S clusters and nickel in the NAD-linked hydrogenase of *Nocardia opaca* 1b. *Eur. J. Biochem.*, 142, 75–84.

Schneider, K., Gollan, U., Drottboom, M. *et al.* Selsemaier-Voight, S. Plass, W. and Müller, A. (1997) Comparative biochemical characterization of the iron-only nitrogenase and the molybdenum nitrogenase from *Rhodobacter capsulatus*. *Eur. J. Biochem.*, 244, 789–800.

Schumacher, W. and Holliger, C. (1996) The proton/electron ratio of the menaquinone-dependent electron transport from dihydrogen to tetrachloroethene in 'Dehalobacter restrictus'. *J. Bacteriol.*, 178, 2328–33.

Schuster, S., Dandekar, T. and Fell, D. A. (1999) Detection of elementary flux modes in biochemical networks: A promising tool for pathway analysis and metabolic engineering. *Trends Biotechnol.*, 17, 53–60.

Schwartz, E., Gerischer, U. and Friedrich, B. (1998) Transcriptional regulation of *Alcaligenes eutrophus* hydrogenase genes. *J. Bacteriol.*, 180, 3197–204.

Scott, A. P. Golding, B. T., Random, L. (1998) Remarkable cleavage of molecular hydrogen without the use of metallic catalysts: a theoretical investigation. *New J. Chem.*, 22, 1171–3.

Scott, R. A. (1985) Measurement of metal-ligand distances by EXAFS. *Meth. Enzymol.*, 117, 414–59.

Scott, R. A., Wallin, S. A., Czechowski, M., DerVartanian, D. V., LeGall, J., Peck, H. D. Jr and Moura I. (1984) X-ray absorption spectroscopy of nickel in the hydrogenase from *Desulfovibrio gigas*. *J. Am. Chem. Soc.*, 106, 6864–5.

Seefeldt, L. C. and Arp, D. J. (1986) Purification to homogeneity of *Azotobacter vinelandii* hydrogenase: A nickel and iron containing alpha beta dimmer. *Biochimie*, 68, 25–34.

Sellmann, D., Becker, T. and Knoch, F. (1996) Protonation and alkylation of the thiolato donors in [Fe(CO)('N$_H$S$_4$')]: Effects on the structural, electronic, and redox properties of metal-sulfur complexes. *Chem. Eur. J.*, 2, 1092–8.

Sellmann, D., Geipel, F. and Moll, M. (2000) [Ni(NHPnPr$_3$)(S$_3$)], the first nickel thiolate complex modeling the nickel cysteinate site and reactivity of [NiFe] hydrogenase. *Angew. Chem. Int. Ed. Engl.*, 39, 561–3.

Sellstedt, A. and Lindblad, P. (1990) Activities, occurrence and localisation of hydrogenase in free-living and symbiotic Frankia. *Plant Physiol.*, 92, 809–15.

Serebryakova, L. T., Medina, M., Zorin, N. A., Gogotov, I. N. and Cammack, R. (1996) Reversible hydrogenase of *Anabaena variabilis* ATCC 29413: Catalytic properties and characterization of redox centers. *FEBS Lett.*, 383, 79–82.

Serebryakova, L. T., Sheremetieva, M. and Tsygankov, A. A. (1998) Reversible hydrogenase activity of *Gloeocapsa alpicola* in continuous culture. *FEMS Microbiol. Lett.*, 166, 89–94.

Settles, A. M., Yonetani, A., Baron, A., Bush, D. R., Cline, K. and Martienssen, R. (1997) Sec-independent protein translocation by the maize Hcf106 protein. *Science*, 278, 1467–70.

Setzke, E., Hedderich, R., Heiden, S. and Thauer, R. K. (1994) H$_2$:heterodisulfide oxidoreductase complex from *Methanobacterium thermoautotrophicum*. Composition and properties. *Eur. J. Biochem.*, 220, 139–48.

Seyferth, D., Henderson, R. S. and Song, L. C. (1982) Chemistry of m-Dithio-bis(tricarbonyliron), a mimic of inorganic disulfides. 1. Formation of Di-m-thiolato-bis(tricarbonyliron) Dianion. *Organometallics*, 1, 125–33.

Shoner, S. C., Olmstead, M. M. and Kovacs, J. A. (1994) Synthesis and structure of a water-soluble five-coordinate nickel alkanethiolate complex, *Inorg. Chem.*, 33, 7–8.

Siegbahn, P. E. M. (1996) In I. Prigogine and S. A. Rice (eds), *New Methods in Computational Quantum Mechanics*, Vol. XCIII. New York: John Wiley, pp. 333–88

Silver, A. and Millar, M. (1992) Synthesis and structure of a unique nickel-thiolate dimer, [(RS)Ni(μ-2-SR)$_3$Ni(SR)]$^{1-}$ – an example of face-sharing bitetrahedra. *J. Chem. Soc. Chem. Commun.*, 1992, 948–9.

Smith, B. E. (1999) Metallosulfur clusters in nitrogenase. *Adv. Inorg. Chem.*, 47, 159–218.

Somers, W. A. C., van Hartingsveldt, W., Stigter, E. C. A. and van der Lugt, J. P. (1997) Electrochemical regeneration of redox enzymes for continuous use in preparative processes. *Trends Biotechnol.*, 15, 495–500.

Sorgenfrei, O., Linder, D., Karas, M. and Klein, A. (1993a) A novel very small subunit of a selenium containing [nickel–iron] hydrogenase of *Methanococcus voltae* is postranslationally processed by cleavage at a defined position. *Eur. J. Biochem.*, 213, 1355–8.

Sorgenfrei, O., Klein, A. and Albracht, S. P. J. (1993b) Influence of illumination on the electronic interaction between O_2 and nickel in active F420-non-reducing hydrogenase from *Methanococcus voltae*. *FEBS Lett.*, 332, 291–7.

Sorgenfrei, O., Duin, E. C., Klein, A. and Albracht, S. P. J. (1996) Interactions of O_2 and 13CO with nickel in the active site of active F420-nonreducing hydrogenase from *Methanococcus voltae*. *J. Biol. Chem.*, 271, 23799–806.

Sorgenfrei, O., Muller, S., Pfeiffer, M., Sniezko, I. and Klein, A. (1997a) The [NiFe] hydrogenases of *Methanococcus voltae*. Genes, enzymes, and regulation. *Arch. Microbiol.*, 167, 189–95.

Sorgenfrei, O., Duin, E. C., Klein, A. and Albracht, S. P. J. (1997b) Changes in the electronic structure around Ni in oxidized and reduced selenium-containing hydrogenases from *Methanococcus voltae*. *Eur. J. Biochem.*, 247, 681–7.

Spiller, H., Bookjans, G. and Shanmugam, K. T. (1983) Regulation of hydrogenase activity in vegetative cells of *Anabaena variabilis*. *J. Bacteriol.*, 155, 129–37.

Stein, M. and Lubitz, W. (1998) Electronic structure of [NiFe] hydrogenase. In D. Ziessow, F. Lendzian and W. Lubitz (eds), *Magnetic Resonance and Related Phenomena*, Vol. II. Berlin: Technische Universitaet Berlin, pp. 820–1.

Stein, M., van Lenthe, E., Baerends, E. J. and Lubitz, W. (2001a) G- and A-tensor calculations in the zero-order approximation for relativistic effects of Ni-complexes as models for the active center of [NiFe]-hydrogenase. *J. Phys. Chem.*, A 105, 416–25.

Stein, M., van Lenthe, E., Baerends, E. J. and Lubitz, W. (2001b) Relativistic DFT calculations of the paramagnetic intermediates of [NiFe] hydrogenase. Implications for the enzymatic mechanism. *J. Am. Chem. Soc.* (in press).

Steinfeld, G. and Kersting, B. (2000) Characterisation of a triply thiolate-bridged Ni-Fe aminethiolate complex: Insights into the electronic structure of the active site of [NiFe] hydrogenase. *Chem. Commun.*, 2000, 205–6.

Stephanopoulos, G. and Sinskey, A. J. (1993) Metabolic engineering – methodologies and future prospects. *Trends Biotechnol.*, 11, 392–6.

Stephenson, M. and Stickland, L. H. (1931) XXVII Hydrogenase: a bacterial enzyme capable of activating molecular hydrogen. I. The properties of the enzyme. *Biochem. J. (London)*, 25, 205–14.

Stevens, T. O. and McKinley, J. P. (1995) Lithoautotrophic microbial ecosystems in deep basalt aquifers. *Science*, 270, 450–4.

Stock, D., Leslie, A. G. and Walker, J. E. (1999) Molecular architecture of the rotary motor in ATP synthase, *Science*, 286, 1700–5.

Surerus, K. K., Chen, M., van der Zwaan, J. W., Rusnak, F. M., Kolk, M., Duin, E. C., Albracht, S. P. J. and Muenck, E. (1994) Further characterization of the spin coupling observed in oxidized hydrogenase from *Chromatium vinosum*. A Mössbauer and multifrequency EPR study. *Biochemistry*, 33, 4980–93.

Takahashi, Y. and Nakamura, M. (1999) Functional assignment of the ORF2-iscS-iscU-iscA-hscB-hscA-fdx-ORF3 gene cluster involved in the assembly of Fe-S clusters in *Escherichia coli*. *J. Biochem. (Tokyo)*, 126, 917–26.

Teixeira, M., Moura, I., Xavier, A. V., Dervartanian, D. V., Legall, J., Peck, H. D. Jr, Huynh, B. H. and Moura, J. J. (1983) *Desulfovibrio Gigas* hydrogenase: Redox properties of the nickel and iron–sulfur centers. *Eur. J. Biochem.*, 130, 481–4.

Teixeira, M., Moura, I., Xavier, A. V., Moura, J. J., LeGall, J., DerVartanian, D. V., Peck, H. D., Jr and Huynh, B. H. (1989) Redox intermediates of *Desulfovibrio gigas* [NiFe] hydrogenase generated under hydrogen. Mossbauer and EPR characterization of the metal centers. *J. Biol. Chem.*, 264, 16435–50.

Teles, J. H., Borde, S., Berkessel, A. (1998) Hydrogenation without a metal catalyst – an ab initio study on the mechanism of the metal-free hydrogenase from *Methanobacterium thermoautotrophicum*. *J. Am. Chem. Soc.* 120, 1345–6.

Tersteegen, A. and Hedderich, R. (1999) *Methanobacterium thermoautotrophicum* encodes two multisubunit membrane-bound [NiFe] hydrogenases. Transcription of the operons and sequence analysis of the deduced proteins. *Eur. J. Biochem.* 264, 930–43.

Thauer, R. K. (1998) Biochemistry of methanogenesis: A tribute to marjory Stephenson. *Microbiology*, 144, 2377–406.

Thauer, R. K., Klein, A. R. and Hartmann, G. C. (1996) Reactions with molecular hydrogen in microorganisms: Evidence for a purely organic hydrogenation catalyst. *Chem. Rev.* (Washington, D. C.), 96, 3031–42.

Thauer, R. K., Klein, A. R., Hartmann, G. C. (1996) Reactions with molecular hydrogen in microorganisms. Evidence for a purely organic hydrogenation catalyst. *Chem. Rev.* 96, 3031–42.

Theodoratou, E., Paschos, A., Mintz, W. and Böck, A. (2000a) Analysis of the cleavage site specificity of the endopeptidase involved in the maturation of the large subunit of hydrogenase 3 from *Escherichia coli*. *Arch. Microbiol.*, 173, 110–16.

Theodoratou, E., Paschos, A., Magalon, A., Fritsche, E., Huber, R. and Böck, A. (2000b) Nickel serves as a substrate recognition motif for the endopeptidase involved in hydrogenase maturation. *Eur. J. Biochem.*, 267, 1995–9.

Thiel, T. (1993) Characterization of genes for an alternative nitrogenase in the cyanobacterium *Anabaena variabilis*. *J. Bacteriol.*, 175, 6276–86.

Thiemermann, S., Dernedde, J., Bernhard, M., Schroeder, W., Massanz, C. and Friedrich, B. (1996) Carboxyl-terminal processing of the cytoplasmic NAD-reducing hydrogenase of *Alcaligenes eutrophus* requires the hoxW gene product. *J. Bacteriol.*, 178, 2368–74.

Thompson, J. D., Higgins, D. G. and Gibson, T. J. (1994) CLUSTAL W: Improving the sensitivity of progressive multiple sequence alignment through sequence weighting, position-specific gap penalties and weight matrix choice. *Nucleic Acids Res.*, 22, 4673–80.

Toussaint, B., David, L., de Sury d'Aspremont, R. and vignais, P. M. (1994) The IHF proteins of *Rhodobacter capsulatus* and *Pseudomonas aeruginosa*. *Biochimie*, 76, 951–7.

Toussaint, B., de Sury d'Aspremont, R., Delic-Attree, I., Berchet, V., Elsen, S., Colbeau, A., Dischert, W., Lazzaroni, Y. and Vignais, P. M. (1997) The *Rhodobacter capsulatus* hupSLC promoter: Identification of cis-regulatory elements and of trans-activating factors involved in H₂ activation of hupSLC transcription. *Mol. Microbiol.*, 26, 927–37.

Tran-Betcke, A., Warnecke, U., Böcker, C., Zaborosch, C. and Friedrich, B. (1990) Cloning and nucleotide sequences of the genes for the subunits of NAD-reducing hydrogenase of *Alcaligenes eutrophus* H16. *J. Bacteriol.*, 172, 2920–9.

Trofanchuk, O., Stein, M., Gessner, C., Lendzian, F., Higuchi, Y. and Lubitz, W. (2000) Single crystal EPR studies of the oxidized active site of [NiFe] hydrogenase from *Desulfovibrio vulgaris* Miyazaki F. *J. Biol. Inorg. Chem.*, 5, 36–44.

Van Berkel-Arts, A., Dekker, M., Van Dijk, C., Grande, H., Hagen, W. R., Hilhorst, R., Kruse-Wolters, M., Laane, C. and Veeger, C. (1986) Application of hydrogenase in biotechnological conversions. *Biochimie* 68, 201–9.

Van Blarigan, P. (1998) Advanced hydrogen fueled internal combustion engines. *Energy & Fuels*, 12, 72–77.

van der Hoek, J. P. and Klapwijk, A. (1988a) The use of nitrate selective resin in the combined ion exchange/biological denitrification process for nitrate removal from ground water. *Water Supply*, 6, 57–62.

van der Hoek, J. P., Griffioen, A., Klapwijk, A. (1988b) Biological regeneration of nitrate loaded anion-exchange resins by denitrifying bacteria. *J. Chem. Technol. Biotechnol.* 43, 213–222.

Van der Zwaan, J. W., Albracht, S. P., Fontijn, R. D. and Slater, E. C. (1985) Monovalent nickel in hydrogenase from *Chromatium vinosum*. Light sensitivity and evidence for direct interaction with hydrogen. *FEBS Lett.*, 179, 271–7.

Van der Zwaan, J. W., Albracht, S. P., Fontijn, R. D. and Mul, P. (1987) On the anomalous temperature behaviour of the EPR signal of monovalent nickel in hydrogenase. *Eur. J. Biochem.*, 169, 377–84.

Van der Zwaan, J. W., Albracht, S. P. J., Fontijn, R. D. and Roelofs, Y. B. M. (1986) EPR evidence for direct interaction of carbon monoxide with nickel in hydrogenase from *Chromatium vinosum*. *Biochim. Biophys. Acta*, 872, 208–15.

Van der Zwaan, J. W., Coremans, J. M., Bouwens, E. C. and Albracht, S. P. (1990) Effect of $^{17}O_2$ and ^{13}CO on EPR spectra of nickel in hydrogenase from *Chromatium vinosum*. *Biochim. Biophys. Acta*, 1041, 101–10.

Van der Spek, T. M., Arendsen, A. F., Happe, R. P., Yun, S., Bagley, K. A., Stufkens, D. J., Hagen, W. R. and Albracht, S. P. J. (1996) Similarities in the architecture of the active sites of Ni-hydrogenases and Fe-hydrogenases detected by means of infrared spectroscopy. *Eur. J. Biochem.*, 237, 629–34.

van Elp, J., Peng, G., Zhou, Z. H., Adams, M. W. W., Baidya, N., Mascharak, P. K. and Cramer, S. P. (1995) Nickel L-edge X-ray absorption spectroscopy of *Pyrococcus furiosus* hydrogenase. *Inorg. Chem.*, 34, 2501–4.

Van Soom, C., Rumjanek, N., Vanderleyden, J. and Neves, M. C. P. (1993) Hydrogenase in *Bradyrhizobium japonicum*: Genetics, regulation and effect on plant growth. *World J. Microbiol. Biotechnol.*, 9, 615–24.

Van Soom, C., Lerouge, I., Vanderleyden, J., Ruiz-Argueso, T. and Palacios, J. M. (1999) Identification and characterization of hupT, a gene involved in negative regulation of hydrogen oxidation in *Bradyrhizobium japonicum*. *J. Bacteriol.*, 181, 5085–9.

Vaupel, M. and Thauer, R. K. (1998) Two F420-reducing hydrogenases in *Methanosarcina barkeri*. *Arch. Microbiol.*, 169, 201–5.

Verhagen, M. F., ORourke, T. and Adams, M. W. (1999) The hyperthermophilic bacterium, *Thermotoga maritime*, contains an unusually complex iron-hydrogenase: Amino acid sequence analyses versus biochemical characterization. *Biochim. Biophys. Acta*, 1412, 212–29.

Vignais, P. M., Dimon, B., Zorin, N. A., Colbeau, A. and Elsen, S. (1997) HupUV proteins of *Rhodobacter capsulatus* can bind H_2: Evidence from the H-D exchange reaction. *J. Bacteriol.*, 179, 290–2.

Vignais, P. M., Dimon, B., Zorin, N. A., Tomiyama, M. and Colbeau, A. (2000) Characterization of the hydrogen-deuterium exchange activities of the energy-transducing HupSL hydrogenase and H_2 signaling HupUV. *J. Bacteriol.*, 182, 5997–6004.

Volbeda, A., Charon, M.-H., Piras, C., Hatchikian, E. C., Frey, M. and Fontecilla-Camps, J. C. (1995) Crystal structure of the nickel-iron hydrogenase from *Desulfovibrio gigas*. *Nature (London)*, 373, 580–7.

Volbeda, A., Garcin, E., Piras, C., de Lacey, A. L., Fernandez, V. M., Hatchikian, E. C., Frey, M. and Fontecilla-Camps, J. C. (1996) Structure of the [NiFe] hydrogenase active site: Evidence for biologically uncommon Fe ligands. *J. Am. Chem. Soc.*, 118, 12989–96.

Voordouw, G. and Brenner, S. (1985) Nucleotide sequence of the gene encoding the hydrogenase from *Desulfovibrio vulgaris* (Hildenborough). *Eur. J. Biochem.*, 148, 515–20.

Voordouw, G., Hagen, W. R., Kruse-Wolters, K. M., van Berkel-Arts, A. and Veeger, C. (1987) Purification and characterization of *Desulfovibrio vulgaris* (Hildenborough) hydrogenase expressed in *Escherichia coli*. *Eur. J. Biochem.*, 162, 31–6.

Voordouw, G., Nivière, V., Ferris, F. G., Fedorak, P. M. and Westlake, D. W. (1990) Distribution of hydrogenase genes in *Desulfovibrio* spp. and their use in identification of species from oil fields environment. *Appl. Environ. Microbiol.*, 56, 3748–54.

Wang, H., Ralston, C. Y., Patil, D. S., Jones, R. M., Gu, W., Verhagen, M., Adams, M., Ge, P., Riordan, C., Marganian, C. A., Mascharak, P., Kovacs, J., Miller, C. G., Collins, T. J., Brooker, S., Croucher, P. D., Wang, K., Stiefel, E. I. and Cramer, S. P. (2000) Nickel L-edge soft X-ray spectroscopy of nickel–iron hydrogenases and model compounds: Evidence for high-spin nickel(II) in the active enzyme. *J. Am. Chem. Soc.*, 122, 10544–52.

Waugh, R. and Boxer, D. H. (1986) Pleiotropic hydrogenase mutants of *Escherichia coli* K12: Growth in the presence of nickel can restore hydrogenase activity. *Biochimie*, 68, 157–66.

Wawer, C., Jetten, M. S. M. and Muyzer, G. (1997) Genetic diversity and expression of the [NiFe] hydrogenase large-subunit gene of *Desulfovibrio* spp. in environmental samples. *Appl. Environ. Microbiol.*, 63, 4360–9.

Whitehead, J. P., Gurbiel, R. J., Bagyinka, C., Hoffman, B. M. and Maroney, M. J. (1993) The hydrogen binding site in hydrogenase: 35-GHz ENDOR and XAS studies of the nickel-C (reduced and active form) and the Ni-L photoproduct. *J. Am. Chem. Soc.*, 115, 5629–35.

Wolf, I., Buhrke, T., Dernedde, J., Pohlmann, A. and Friedrich, B. (1998) Duplication of hyp genes involved in maturation of [NiFe] hydrogenases in *Alcaligenes eutrophus* H16. *Arch. Microbiol.*, 170, 451–9.

Wolk, C. P. (1996) Heterocyst formation. *Annu. Rev. Genet.*, 30, 59–78.

Woo, G., Wasserfallen, A. and Wolfe, R. S. (1993) Methyl viologen hydrogenase II, a new member of the hydrogenase family from *Methanobacterium thermoautotrophicum*. DELTA. H. *J. Bacteriol.*, 175, 5970–7.

Woodward, J., Orr, M., Cordray, K. and Greenbaum, E. (2000) Enzymatic production of bio-hydrogen. *Nature*, 405, 1014–5.

Wu, L. F. and Mandrand-Berthelot, M. A. (1986) Genetic and physiological characterization of new *Escherichia coli* mutants impaired in hydrogenase activity. *Biochimie*, 68, 167–79.

Wu, L. F. and Mandrand, M. A. (1993) Microbial hydrogenases: Primary structure, classification, signatures and phylogeny. *FEMS Microbiol. Rev.*, 104, 243–69.

Zaborsky, O. R. (ed.) (1998) *Biohydrogen*, Plenum Press. New York and London.

Zheng, L., Cash, V. L., Flint, D. H. and Dean, D. R. (1998) Assembly of iron–sulfur clusters. Identification of an iscSUA-hscBA-fdx gene cluster from *Azotobacter vinelandii*. *J. Biol. Chem.*, 273, 13264–72.

Zirngibl, C., van Dongen, W., Schwörer, B., von Bünau, R., Richter, M., Klein, A., Thauer, R. K. (1992) H$_2$-forming methylenetetrahydromethanopterin dehydrogenase, a novel type of hydrogenase with iron-sulfur clusters in methanogenic archaea. *Eur. J. Biochem.* 208, 511–20.

Index